QUANTUM MECHANICS

QUANTUM MECHANICS

Based on the Principle of
Minimum Mean Deviation from
Statistical Equilibrium and Independence

SILVIU GUIASU

Nova Science Publishers, Inc.
Huntington, New York

Senior Editors: Susan Boriotti and Donna Dennis
Office Manager: Annette Hellinger
Graphics: Wanda Serrano
Information Editor: Tatiana Shohov
Book Production: Cathy DeGregory, Kay Seymour, Lynette Van Helden and Jennifer Vogt
Circulation: Ave Maria Gonzalez, Ron Hedges and Andre Tillman

Library of Congress Cataloging-in-Publication Data

Guiasu, Silviu
    Quantum Mechanics / Silviu Guiasu.
       p.cm.
    Includes bibliographical references and index.
    ISBN 1-56072-896-5.
      1. Quantum Theory. I. Title.

QC174.12 .G85 2001
530.12—dc21

                                  2001018734

Copyright © 2001 by Nova Science Publishers, Inc.
        227 Main Street, Suite 100
        Huntington, New York 11743
        Tele. 631-424-6682   Fax 631-425-5933
        E Mail Novascil@aol.com

Printed in the United States of America

Preface

The objective of the present monograph is to propose a new probabilistic model in quantum mechanics where the probability wave functions of the quantum systems are obtained as solutions of the variational principle of minimum mean deviation from statistical equilibrium and independence where the values of the variational parameters involved are determined looking for stationary points of the mean energy of the system.

Part I deals with the general model and its applications to the standard systems (one or two particles in a box, the harmonic oscillator, and the hydrogen atom) without using the corresponding Schrödinger equations, and to the ground state of the helium atom for which there are excellent trial functions proposed in the literature, based on different variational methods. It also deals with a simplified model of the lithium atom which shows the necessity of taking the shell structure of more complex atoms into account.

Part II deals with the linear approximation of the probability wave functions induced by the minimum mean deviation from statistical equilibrium which is applied to the study of the structure of more complex atoms (the lithium, beryllium, boron, carbon, nitrogen, and argon atoms) in the ground state. The new tool allows us to calculate different correlations between electrons, the degree of stability of systems of electrons, and the amount of interdependence among different subsystems of electrons inside the atom.

The effective implementation of the formulas obtained is based on the software package MATHEMATICA which does both numerical computations and symbolic mathematics as well. Detailed computer programs are given at the end of each of the two parts.

Part I is more theoretical and is meant mainly for physicists and those interested in mathematical physics. Part II is more applied and is meant mainly for those working in quantum chemistry. The equations are numbered independently in the two parts of the volume. Each part is selfcontained and the prerequisites are kept to a minimum.

"Can we actually 'know' the universe? My God, it's hard enough finding your way around in Chinatown." Woody Allen (1978, p.22)

"At the heart of physics is the process of building models of the world. Often these models have required the invention of new mathematical languages. The models physicists build can never be compared with the world, because as Einstein tells us, we do not even know what that would mean. Rather, the predictions derived from the models can be compared with observations of the world." Bruce Gregory (1990, p.196)

"The introduction of probability in the sense of quantum mechanics - that is, probability as an inherent feature of fundamental physical law - may well be the most drastic scientific change yet effected in the 20th century." Abraham Pais (1982, p.1193)

Contents

PART I.
THE GENERAL MODEL.

1. INTRODUCTION.

The two great revolutions in the twentieth century, relativity theory and quantum mechanics, required drastic revisions of the basic concepts of space, time, and causality. Among the major steps in building quantum mechanics we mention Max Planck's original quantum theory, Niels Bohr's planetary model for atom, Louis Victor de Broglie's wave mechanics, Werner Heisenberg's matrix mechanics and uncertainty principle, and Erwin Schrödinger's wave mechanics of the atom. With the failure of rigid models to explain the structure of atoms and the behavior of electrons, statistical methods and hence probability considerations appeared to offer the only avenue of advance. Quantum mechanics abolished clear-cut trajectories of Newton's classical mechanics and introduced a probabilistic fitfulness into nature. The intrusion of probability concepts in mathematical physics had proved successful in classical statistical mechanics. Nevertheless, the theory was based there on the assumption that the rigorous laws of mechanics controlled the motion of the gas molecules. The only reason for introducing probabilities in the study of gases was to obviate the mathematical difficulties which a direct application of the mechanical laws would have involved. But in quantum theory, the situation is entirely different. Cumulative evidence drawn from the quantum theory pointed to an inevitable uncertainty in describing the behaviour of the quantum systems and raised the possibility of the absence of rigorous laws for individual atomic processes. As A. D'Abro (1951, p.957) writes, we are thus prompted to adopt one of the following alternatives: (a) The uncertainties are due to incompleteness of quantum mechanics; (b) They express a fundamental vagueness in Nature. After 50 years, the controversy is still in place and matters of disputed interpretation remain unresolved, although many quantum theorists have accepted the second alternative. Whether they indeed reflect the fundamental behavior of quantum systems or are only a convenient description of what happens at the quantum level of reality, statistical inference and probability theory are the only tool we are using today to make predictions in quantum mechanics and their computations are not based on the assumption of rigorous laws controlling individual processes.

The objective of the present monograph is to propose a new probabilistic model in quantum mechanics where the probability wave functions of the quantum systems are obtained as solutions of the variational principle of minimum mean deviation from statistical equilibrium and independence. Part I deals with the general model and its applications to the standard systems (one or two particles in a box, the harmonic oscillator, and the hydrogen atom) without using the corresponding Schrödinger equations, and to the ground state of the helium atom for which there are excellent trial functions proposed in the literature and based on different variational methods. Part II deals with the linear approximation of the probability wave functions induced by the minimum mean deviation from statistical equilibrium, which is applied to the study of the structure of more complex atoms (the lithium, beryllium, boron, carbon, and argon atoms) in the ground state. The new tool allows us to calculate different correlations between electrons, the degree of stability of systems of electrons, and the amount of interdependence among different subsystems of electrons inside the atom. The effective implementation of the formulas obtained is based on the software package MATHEMATICA which does both numerical computations and symbolic mathematics as well. But let us go back to the initial statistical viewpoint in quantum mechanics.

Schrödinger equation plays a central role in the nonrelativistic quantum mechanics and some textbooks include it among the postulates of the theory. Introduced in 1926, its solution is the so-called ψ-function, initially thought of by Schrödinger to represent a real disturbance, a matter wave in space. He soon abandoned these matter waves and turned to an 'electromagnetic interpretation', in which $\psi^*\psi$ would be a measure of the density of electric charge, or rather a sort of 'weighting function'. He wrote (Schrödinger 1926, the fourth paper): "$\psi^*\psi$ is a sort of weight function in the configuration space of the system. The wave mechanical configuration of the system is a superposition of many, strictly speaking all, the kinematically possible point-mechanical configurations. Thereby every point-mechanical configuration contributes to the true wave mechanical configuration with a certain weight, which is given precisely by $\psi^*\psi$. If one likes paradoxes, one can say that the system is found simultaneously in all conceivable kinematic locations but not in all of them in 'equal strength'." It was Born (1926) who interpreted $\psi^*\psi$ as being a probability density. As mentioned by Moore (1989), for Schrödinger $\psi^*\psi$ pictures a continuous cloud of 'something' (presumably charge and mass), whereas Born assumes point particles (electrons) within the atom and interprets $\psi^*\psi$ as being the probability density of finding a particle at a certain location. In a letter to Einstein (June, 1946), Schrödinger wrote: "God knows I am no friend of the probability theory, I have hated it from the first moment when our dear friend Max Born gave it birth. For it could be seen how easy and simple it made everything, in principle, everything ironed out and the true problems concealed...And actually not a year passed before it became an official credo, and it still is." It seems today that in fact Schrödinger's objections rather referred to the interpretation of the concept of probability. A frequency interpretation of probability (if the result A of an experiment has the probability p, it means that if the experiment is repeated many times then the fraction of outcomes that give the result A approximately equals p) would suggest that Born's probability density $\psi^*\psi$ rather refers to a statistical ensemble of systems. To Schrödinger, adopting such a viewpoint "we cut ourselves off from ever applying rational probability considerations to a single event." On the other hand, a purely subjective interpretation of probability, apparently more acceptable to Schrödinger, would suggest that Born's probabilistic interpretation of the wave function ψ does not relate to a system but to our knowledge about a system. Summarizing, the interpretation of the solution ψ of the Schrödinger equation is somewhat mysterious, but $|\psi|^2 = \psi^*\psi$ has a clear meaning: it is the probability density of the particle position at given time. Details about Max Born's statistical interpretation of quantum mechanics may be found in Pais (1982). In 1954 Born was awarded the Nobel Prize "for his fundamental research, especially for his statistical interpretation of the wave function."

As the ultimate use of the Schrödinger equation is to provide us with probabilistic models for the behavior of quantum systems, a natural question comes up: Is it possible to build up such probabilistic models without having to write and solve the corresponding Schrödinger equation? The question is even more justified if we take into account that the Schrödinger equation may be solved exactly only for a very limited number of quantum systems, namely the free particle in a box, the harmonic oscillator, and the hydrogen atom.

There have been attempts to deduce the Schrödinger equation from several variational principles or from prior suppositions instead of taking it as a postulate of quantum mechanics. Here are some of them:

(a) As shown in Flügge (1974, vol.1, pp.2-5), the Schrödinger equation may be obtained from the Euler equations corresponding to the minimization of the energy integral.

(b) Bohm (1984) suggested: "Instead of starting from Born's probability distribution $|\psi|^2$ as an absolute and final and unexplainable property of matter, we have [to show] how his property could come out of random motions originating in a subquantum mechanical level." Several papers (Nelson, 1985; Baublitz, 1988) have attempted to derive the Schrödinger equation from classical mechanics and an assumed Markov diffusion stochastic process induced by random fluctuations of a submicroscopic medium. The difficulty in accepting such an approach resides in the fact that in quantum mechanics it is not possible to assume that the sample space on which all random variables of the stochastic process are supposed to be defined remains unique and invariable.

(c) Frieden (1989, 1990, 1991) aimed at building up a probabilistic model based on the statistical estimation theory from which the Schrödinger equation could be derived as a consequence. Dealing with the position of a particle on the real line, for instance, the Schrödinger equation is obtained in this approach by minimizing a linear combination of the Fisher (1922) information, measuring the degree of ruggedness of a probability distribution, and the mean kinetic energy of the particle.

(d) Guiasu (1992) used the minimum $\tilde{\chi}^2$-deviation from the maximum entropy probability distribution subject to given mean values and the quantization rules of the old quantum theory in order to derive the same conclusions as those induced by the corresponding Schrödinger equation for a free particle in a box, the harmonic oscillator, and the hydrogen atom. Other results, in the same direction of thought, may be found in Slater (1994) and Preda, Bulacu, and Bulacu (1999).

The objective of this monograph is to combine the approaches (a) and (d) but without using the Euler equations from (a) and the old quantization rules from (d). The new approach obtained this way is simple and straightforward. The probabilistic model obtained gets the same conclusions as those induced by solving the Schrödinger equation for a free particle in a box, two interacting particles in a box, the harmonic oscillator, or the hydrogen atom and, moreover, it may be extended to the study of quantum systems for which the corresponding Schrödinger equation cannot be solved exactly, as the helium and lithium atoms, for instance. In a personal communication made in 1985 to Moore (1989, p.416), H.W. Peng mentioned that Schrödinger, in spite of his much discussed reservations about the probabilistic interpretation of the undulatory mechanics, once had told him that "Quantum mechanics was born in statistics and it will end in statistics." The objective of this monograph is to show that Schrödinger was right when he made such a statement.

Our program may be summarized in one sentence: we are looking for a probabilistic model of stationary energy obtained by minimizing the mean deviation from statistical equilibrium and independence subject to given generalized correlations induced by random internal or/and external fluctuations. Let us explain each step of it:

1) *Statistical equilibrium.* We start from statistical equilibrium described by a maximum entropy probability distribution subject to constraints induced by given mean values of some random variables. This is a problem well studied in the literature. About 125 years ago, Boltzmann (1872) introduced the famous **H**-function in his research on the behavior of the molecules of a gas. After 76 years, Shannon (1948) showed that the discrete analogue of Boltzmann's **H**-function, called the probabilistic entropy, may be accepted as to be a measure of the amount of uncertainty contained by any probability distribution. If the probability distribution is given, then its entropy is a number showing the amount of uncertainty contained by it. If, for instance, the set of possible outcomes is given and we know nothing more, then the entropy is maximum for the uniform distribution. The inverse problem is much more important: if the probability distribution is unknown and the only information available consists of one or several mean values of one or several random variables, then from the infinite set of feasible probability distributions compatible with such given constraints, we choose the one that maximizes entropy. The solution of this variational problem is the most uncertain, or unbiased (i.e., ignoring no possibility) probability distribution compatible with the given constraints represented by mean values. It is the closest probability distribution to the uniform distribution when the given mean values are known because, as we have said before, the uniform distribution maximizes entropy when there are no constraints imposed. This variational problem is known as the maximum entropy principle (MEP) and

it was explicitly formulated by Jaynes (1957) and implicitly by von Neumann (1932) and Kullback (1959). Eventually, it was surprising to see that well-known probability distributions could be rediscovered as solutions of MEP in a natural way. Thus, on the positive real axis $[0, +\infty)$, the solution of MEP subject to the mean value μ is the exponential probability distribution $\mathbf{E}(\mu)$. On the real axis $(-\infty, +\infty)$, the solution of MEP subject to the mean μ and the variance σ^2 is the normal distribution $\mathbf{N}(\mu, \sigma^2)$. On an arbitrary interval $[a, b]$ of the real axis, if we have no constraints attached, then the probability distribution of maximum entropy is the uniform distribution $\mathbf{U}(a, b)$. As shown in Guiasu (1990a), practically all the main probability distributions may be obtained as solutions of constrained variational problems similar to MEP.

2) *Systems of orthonormal functions.* Once the probability distribution that describes statistical equilibrium is obtained, and u is its density, we choose a sequence of orthogonal functions with the weight u as a system of generalized coordinates. Such systems of orthogonal functions are the Laguerre polynomials, for the exponential distribution $\mathbf{E}(\mu)$, the Hermite polynomials, for the normal distribution $\mathbf{N}(\mu, \sigma^2)$, the generalized Laguerre polynomials for the Gamma distribution $\mathbf{G}(\alpha, \beta + 1)$, and either the trigonometric system or the Legendre (spherical) polynomials, for the uniform distribution $\mathbf{U}(a, b)$, for instance.

3) *One-dimensional probability wave functions.* As long as the statistical equilibrium is not perturbed, the mean values of the generalized coordinates remain equal to zero. If, however, random perturbations induced by internal or external interactions alter the statistical equilibrium, then some of these mean values cease to be equal to zero. We minimize the mean deviation from statistical equilibrium subject to correlations between generalized coordinates induced by the random perturbations induced by internal or/and external interactions. The mean deviation is measured using Pearson's $\tilde{\chi}^2$-indicator, which is a weighted Euclidean distance. By minimizing $\tilde{\chi}^2$ subject to given mean values, we force the small initial probabilities to remain basically the same, focusing on changes induced by constraints on the probabilities of the most probable outcomes. In general, a wave is just any disturbance to a field. In our context, we introduce a probability wave function to be the minimum deviation from statistical equilibrium due to random internal or/and external fluctuations. Once the probability wave function χ is obtained, the normed square of its absolute value is used as a proba-

bility density function in the set of possible configurations. Bohm (1984) argued in favour of paying attention to random fluctuations in general and at the quantum level in particular: "It is not relevant where such fluctuations come from; all that is important is to assume that they exist and to see their effects." Certainly, at the quantum level the statistical equilibrium of a system may be easily affected by interactions with other quantum systems or by the measurement process performed by an external macroscopic observer. Minimizing the $\tilde{\chi}^2$-indicator, a slight generalization of the well-known least square method of Legendre (1805) and Gauss (1809), we want to determine the probability wave of minimum deviation from equilibrium subject to the mean effects of random fluctuations generated by perturbations or interactions, focusing on changes induced by the given constraints on the most probable outcomes. As mentioned by Bühler (1981): "The method of least squares was one of Gauss's most efficient tools in his research ... Least squares were Gauss's indispensable theoretical tool in experimental research; increasingly, he came to see it as the most important witness to the connection between mathematics and nature." The weighted least squares generalization gives more flexibility and makes a smooth connection between the Euclidean distance and the mean deviation.

Before ending this subsection, let us mention that, according to Stigler (1981), Adrien Marie Legendre published the least square method in 1805 and Carl Friedrich Gauss published the method in 1809, claiming, however, that he had been using the method since 1795. Chi-square, as a generalization of the Euclidian distance for measuring how different a probability distribution is from another probability distribution taken as point of reference, was introduced by Karl Pearson (1900). Sir Ronald Fisher, one of the founders of modern statistics, wrote (Fisher (1925)) that chi-square "is the great contribution to statistical methods by which the unsurpassed energy of Professor Pearson's work will be remembered." Strong arguments in favour of using the least chi-square method in statistical inference may be found in Berkson (1980).

4) *Multi-dimensional probability waves.* For incompatible entities (like position and momentum, for instance) it is senseless to talk about a joint probability distribution. For independent compatible entities (like the components of the position of a particle in a three-dimensional Euclidean space, for instance), the joint probability density is simply the product of the probability densities (i.e. the marginals) of the corresponding entities. For depen-

9

dent compatible entities (like the positions of several interacting particles, for instance) the joint probability distribution is not uniquely determined by the marginals and the partial information available about their dependence (like covariances, or correlations, for instance). We construct here the multi-dimensional probability wave obtained by minimizing the mean $\tilde{\chi}^2$- deviation from the direct independent product of the marginals subject to the given mixed moments (or generalized covariances). The solution of such a variational problem assumes nothing about the dependence among the components beyond the given mixed moments.

5) *Stationary values of the mean energy.* Once the one-dimensional or multi-dimensional probability wave is obtained by solving the above mentioned variational problem, we determine the stationary points of the corresponding mean energy. In this way we determine the stationary probability waves whose normed square may be used as a probability density for predicting the mean values of some observables of interest.

6) *Applications.* The formalism is applied to the study of a free particle in a box, of two independent particles in a box, of the harmonic oscillator, of the hydrogen atom, and of the ground state of the helium and lithium atoms.

The subtle relationship between statistical equilibrium and random fluctuations has been a topic of major interest for centuries. In 1773, Laplace proved that the mean distances of the planets from the Sun are invariable to within certain slight periodic variations. Newton himself believed that divine intervention might be necessary from time to time to put the solar system back in order and prevent it from destruction or dissolution. As mentioned by McMullin (1978), Newton's interest in alchemy and his chemical experimentation made him notice the turbulence of natural process, leading to his conviction about the presence in Nature of perpetually working active principles that, somehow mysteriously, reestablish and maintain global stability. Later, the attempt to reduce thermodynamics to mechanics led to the introduction of statistical ideas (Boltzmann (1872), Gibbs (1902)) which changed the absolute character of thermodynamics. As a direct consequence, a generalized thermodynamics was applied to phenomena involving fluctuation phenomena, such as Brownian motion, that led to the final establishment of the existence of atoms. Since then, statistical equilibrium and random fluctuations perturbing statistical equilibrium have remained topics of uninterrupted interest and scrutiny. The main conclusion of this study is that whereas the maximum entropy principle, often abbreviated as maxent principle, is a widely accepted method for describing the state of statistical equilibrium, the minimization of the mean deviation from statistical equilibrium, or the mindevstateq principle, proves to be a suitable mathematical tool for describing the natural evolution of a system in coping with the perturbing effects of random fluctuations induced by inner and outer interdependences and interactions. When statistical equilibrium is described by using the maximum entropy principle, then mindevstateq becomes mindevmaxent principle. Whereas maxent principle describes the maximum freedom under given mean constraints, mindevmaxent principle describes the strive for maximum freedom under changed constraints due to random fluctuations with known mixed moments or generalized correlations. Maxent principle is a tool for getting a stationary, static model of statistical equilibrium. Mindevmaxent principle is a tool for getting a dynamic model of evolution towards reobtaining a new statistical equilibrium meant to replace an initial statistical equilibrium perturbed by random fluctuations.

2. THE MATHEMATICAL MODEL

Let D be a domain in the k-dimensional Euclidean space $\mathbf{R}^k$. We denote by

$$< f >= \int_D f(x)dx, \quad < f^*g >= \int_D f^*(x)g(x)dx,$$

$$\|f\| =< f^* f >^{1/2},$$

provided that the above integrals exist, where f^* means the complex conjugate of f.

1) **Statistical equilibrium.** Long ago, Boltzmann (1872) defined the expression

$$\mathbf{E} = \int_0^{+\infty} f(x,t) \left\{ \log \left[\frac{f(x,t)}{\sqrt{x}} \right] - 1 \right\} dx$$

and showed that it possessed the properties of a negative entropy: thus $-\mathbf{E}$ always increased monotonically with time t, if the distribution function $f(x,t)$ deviated from the Maxwellian distribution $\sqrt{x} \exp\{-x/(kT)\}$, with

x denoting the kinetic energy, T the absolute temperature, and k a positive constant; it stayed constant if the latter distribution was assumed. This result, i.e.,

$$-\frac{d\mathbf{E}}{dt} \geq 0,$$

was later called the 'H-theorem'.

Let u be an arbitrary probability density function defined on $D \subseteq \mathbf{R}^1$. The amount of uncertainty contained by u is measured by the (absolute) entropy $\mathbf{H}(u) = - < u \log u >$, introduced by Shannon (1948) by analogy with Boltzmann's function from statistical mechanics. When u is given, then $\mathbf{H}(u)$ is a number. In many applications, however, u is not known and the only information available is provided by mean values of some random variables. In an everchanging world, the mean values seem to be the only kind of data accessible to us on which we can rely in building up a relatively stable model of reality. Generally, however, there are several (in fact, infinitely many) probability density functions u compatible with the given mean values. According to the principle of maximum entropy, used implicitly by von Neumann (1932) in quantum mechanics and by Kullback (1959) in statistical inference, and formulated as a general principle by Jaynes (1957, 1979), we choose the probability density function u which maximizes $\mathbf{H}(u)$ subject to the constraints induced by the given mean values. Such a solution u is the most unbiased probability density function (i.e. the most uncertain, ignoring no possibility) compatible with the given constraints. The principle of maximum entropy is an objective criterion for constructing a subjective probabilistic model when some mean values of some random variables are known. A probability density function u, solution of the principle of maximum entropy, describes the statistical equilibrium corresponding to the given mean values used as constraints of this nonlinear variational problem. The following results are known in literature and may be obtained without difficulty by applying the classic Lagrange multipliers method from calculus of variation, as shown in Appendix 2.

Proposition 1: *If $D = [a, b]$, then the solution of the nonlinear program*

$$\max_{u} \mathbf{H}(u)$$

subject to the constraint

$$< u >= 1$$

is the uniform distribution $\mathbf{U}(a,b)$ *with parameters* a *and* b, *whose density is*

$$u(x) = \frac{1}{b-a}, \quad x \in [a,b]. \tag{1}$$

Proposition 2: *If* $D = [0,+\infty)$, *then the solution of the nonlinear program*

$$\max_u \mathbf{H}(u)$$

subject to the constraints

$$< u >= 1,$$

$$< xu(x) >= \mu,$$

is the exponential distribution $\mathbf{E}(\mu)$ *whose density is*

$$u(x) = \frac{1}{\mu} e^{-x/\mu}, \qquad x \in [0,+\infty). \tag{2}$$

Proposition 3: *If* $D = (-\infty,+\infty)$, *then the solution of the nonlinear program*

$$\max_u \mathbf{H}(u)$$

subject to the constraints

$$< u >= 1,$$

$$< x\,u(x) >= \mu,$$

$$< (x-\mu)^2\,u(x) >= \sigma^2,$$

is the normal distribution $\mathbf{N}(\mu,\sigma^2)$ *with parameters* μ *and* σ^2, *whose density is*

$$u(x) = \frac{1}{\sigma\sqrt{2\pi}} e^{-(x-\mu)^2/(2\sigma^2)}, \qquad x \in (-\infty,+\infty). \tag{3}$$

Let u be a probability density function and v a nonnegative function, both defined on $D \subseteq \mathbf{R}^1$, such that u is absolutely continuous with respect to v, which means that $u(x) = 0$ if $v(x) = 0$. The relative entropy of u with

13

respect to the reference measure of density v (also called the Kullback-Leibler (1951) indicator, or the divergence of u with respect to v) is defined by

$$\mathbf{H}(u \mid v) = < u \log \frac{u}{v} > = - < u \log v > - \mathbf{H}(u),$$

provided that the integrals exist. We have $\mathbf{H}(u \mid v) \geq 0$ with equality if and only if $u = v$, u-almost everywhere. Obviously, if the reference function v is constant on D, i.e. $v(x) = c$, then the relative entropy is just the negative absolute entropy up to an additive constant, namely,

$$\mathbf{H}(u \mid v) = - \log c - \mathbf{H}(u).$$

In particular, if $D = [a, b]$ and v is the probability density of the uniform distribution on $[a, b]$, then u is absolutely continuous with respect to v and

$$\mathbf{H}(u \mid v) = \log(b - a) - \mathbf{H}(u),$$

which shows that $\mathbf{H}(u)$ measures how much the probability density u differs from the uniform distribution with respect to the logarithmic mean. According to the principle of minimum relative entropy, we determine the probability density u which is the closest one to the reference measure of density v subject to given mean values of some random variables. Thus, the principle of maximum (absolute) entropy may be viewed as a special case of the principle of minimum relative entropy when the reference measure is just the uniform distribution. The next proposition (Guiasu, 1990a) may be proved using the standard Lagrange multipliers method. Mathematical details may be found in Appendix 2.

Proposition 4: *If $D = [0, +\infty)$, and $v(x) = x^\beta, (\beta > -1)$, then the solution of the nonlinear program*

$$\min_{u} \mathbf{H}(u \mid v)$$

subject to the constraints

$$< u > = 1,$$

$$< x\, u(x) > = \mu,$$

14

is the Gamma distribution $\mathbf{G}(\alpha, \beta + 1)$, *with parameters* α *and* $\beta + 1$, *whose density is*

$$u(x) = \frac{1}{\alpha^{\beta+1}\Gamma(\beta+1)} x^\beta e^{-x/\alpha}, \qquad x \in [0, +\infty), \tag{4}$$

where $\alpha = \mu/(\beta + 1)$ *and the Gamma function is*

$$\Gamma(\beta + 1) = \int_0^{+\infty} x^\beta e^{-x} dx.$$

Before ending this subsection, let us mention that details about the maximum entropy principle may be found in Jaynes (1957, 1979), Guiasu (1977), Guiasu and Shenitzer (1985), Csiszár (1991). There are applications of MEP and related variational problems in quantum chemistry. Thus, Bernstein and Levine (1972), Ben-Shaul, Levine, and Bernstein (1972), Levine and Bernstein (1973), Alhassid and Levine (1977), applied the techniques of entropy maximization and what was called 'surprisal analysis' to the prediction of product state distributions in chemical reactions, and discussed the relationship between the maximum entropy approach to collision phenomena and the dynamical description using equation of motion. In Sears, Parr, and Dinur (1980), standard concepts of information theory (Shannon's entropy, Fisher's information, Kullback-Leibler indicator, and Jaynes' MEP) were extended to many dimensions in order to establish connections between the many body quantum-mechanical kinetic energy functional and information measures. In Sears and Gadre (1981), the information theoretic technique of entropy maximization was applied to Compton profile data. The paper Morrison and Parr (1991) studied the entropy of an electronic system. Finally, the papers Ghosh, Berkowitz, and Parr (1984) and Nagy and Parr (1996) used information theory to derive a macroscopic local thermodynamic language for describing the behavior of the microscopic electron fluid.

2) **Systems of orthonormal functions**. Let u be a probability density function on the set $D \subseteq \mathbf{R}^1$, and $\{U_n; n = 0, 1, \ldots\}, U_0 \equiv 1$, a sequence of orthonormal functions with the weight u, i.e.,

$$< U_n U_\ell \, u > = \begin{cases} 1, & \text{if } \ell = n, \\ 0, & \text{if } \ell \neq n. \end{cases}$$

15

which implies that $\|U_n\| = 1$, for all n, $(n = 0, 1, \ldots)$, with $< U_0\, u >= 1$, and $< U_n\, u >= 0$ for all n, $(n = 1, 2, \ldots)$. We call $\{U_n;\ n = 0, 1, \ldots\}$ a system of generalized coordinates associated to the probability density function u.

The four probability density functions describing statistical equilibrium listed above have well-known corresponding systems of orthonormal functions:

(a) A system of orthonormal polynomials with the weight (1) is:

$$U_n(x) = (2n + 1)^{1/2} P_n \left(\frac{2}{b - a}x - \frac{a + b}{b - a} \right), (n = 0, 1, \ldots), \tag{5}$$

where $P_n(x)$ is the Legendre (spherical) polynomial of degree n. The first ones are:

$$P_0(x) = 1, \quad P_1(x) = x, \quad P_2(x) = \frac{1}{2}(3x^2 - 1),$$

$$P_3(x) = \frac{1}{2}(5x^3 - 3x), \quad P_4(x) = \frac{1}{8}(35x^4 - 30x^2 + 3),$$

$$P_5(x) = \frac{1}{8}(63x^5 - 70x^3 + 15x).$$

A system of orthonormal functions with the weight $u(x) = 1/a$ on $D = [0, a]$ is the trigonometric system:

$$U_0 \equiv 1, \ U_n(x) = \sqrt{2} \sin \frac{n\pi}{a}x, \quad (n = 1, 2, \ldots). \tag{6}$$

A system of orthonormal functions with the weight $u(x) = 1/a$ on $D = [-a/2, a/2]$ is the trigonometric system:

$$U_0(x) = 1, \ U_n(x) = \begin{cases} \sqrt{2} \sin \frac{n\pi}{a}x, & \text{if } n \text{ is even}, \\ \\ \sqrt{2} \cos \frac{n\pi}{a}x, & \text{if } n \text{ is odd}, \end{cases}$$

for $(n = 1, 2, \ldots)$.

(b) A system of orthonormal polynomials with the weight (2) is:

$$U_n(x) = L_n \left(\frac{x}{\mu} \right), \qquad (n = 0, 1, \ldots), \tag{7}$$

16

where $L_n(x)$ is the Laguerre polynomial of degree n. The first ones are:

$$L_0(x) = 1, \quad L_1(x) = -x + 1,$$

$$L_2(x) = \frac{1}{2}(x^2 - 4x + 2),$$

$$L_3(x) = \frac{1}{6}(-x^3 + 9x^2 - 18x + 6),$$

$$L_4(x) = \frac{1}{24}(x^4 - 16x^3 + 72x^2 - 96x + 24),$$

$$L_5(x) = \frac{1}{120}(-x^5 + 25x^4 - 200x^3 + 600x^2 - 600x + 120).$$

Computer tip: Legendre polynomials may be generated and plotted easily using the symbolic software MATHEMATICA. Here is a MATHEMATICA session for generating the above mentioned Legendre polynomials and their plot shown in Figure 1:

```
math
LegendreP[0,x]
LegendreP[1,x]
LegendreP[2,x]
LegendreP[3,x]
LegendreP[4,x]
LegendreP[5,x]
f=Plot[{LegendreP[1,x],LegendreP[2,x],LegendreP[3,x],
LegendreP[4,x],LegendreP[5,x]},{x,−1.5,1.5}];
Display["john",f]
Quit
hardcopy john
```

(c) A system of orthonormal polynomials with the weight (3) is:

$$U_n(x) = (n!)^{-1/2} He_n\left(\frac{x - \mu}{\sigma}\right), \tag{8}$$

with

$$He_n(x) = 2^{-n/2} H_n\left(\frac{x}{\sqrt{2}}\right),$$

17

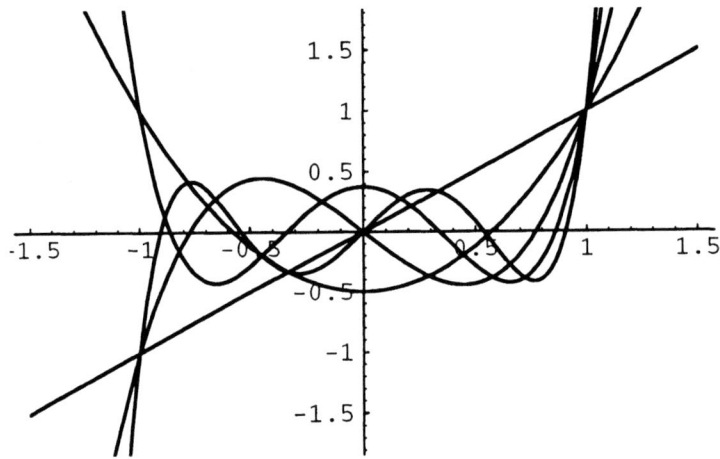

Figure 1: The first five Legendre polynomials.

where $(n = 0, 1, \ldots)$ and $H_n(x)$ is the Hermite polynomial of degree n. The first ones are:

$$H_0(x) = 1, \quad H_1(x) = 2x,$$
$$H_2(x) = 4x^2 - 2, \quad H_3(x) = 8x^3 - 12x,$$
$$H_4(x) = 16x^4 - 48x^2 + 12,$$
$$H_5(x) = 32x^5 - 160x^3 + 120x.$$

Computer tip: Hermite polynomials may be generated and plotted easily using the symbolic software MATHEMATICA. Here is a MATHEMATICA session for generating the above mentioned Hermite polynomials and their plot shown in Figure 2:

```
math
HermiteH[0,x]
HermiteH[1,x]
HermiteH[2,x]
HermiteH[3,x]
HermiteH[4,x]
```

18

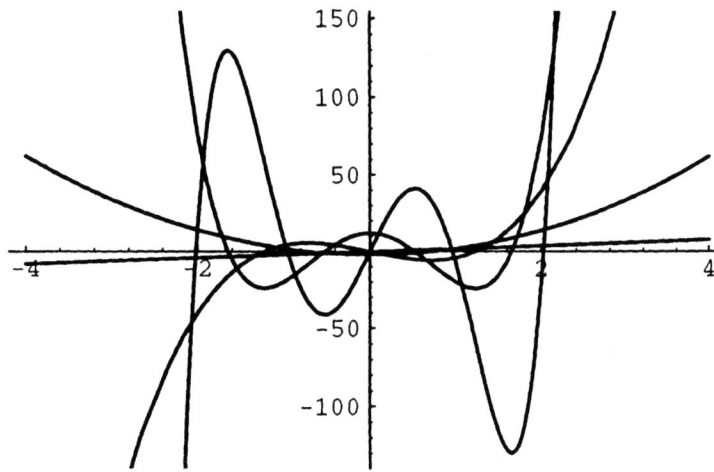

Figure 2: The first five Hermite polynomials.

```
HermiteH[5,x]
g=Plot[{HermiteH[1,x],HermiteH[2,x],HermiteH[3,x],
HermiteH[4,x],HermiteH[5,x]},{x,−4,4}];
Display["mary",g]
Quit
hardcopy mary
```

(d) A system of orthonormal polynomials with the weight (4) is

$$U_n(x) = \left(\frac{n!\Gamma(\beta + 1)}{\Gamma(\beta + n + 1)} \right)^{1/2} L_n^{(\beta)} \left(\frac{x}{\alpha} \right), \quad (n = 0, 1, \ldots), \qquad (9)$$

where $L_n^{(\beta)}(x)$ is the generalized Laguerre polynomial of degree n and order β. We have:

$$L_n^{(k)}(x) = \sum_{\ell=0}^{n} (-1)^\ell \binom{n + k}{n - \ell} \frac{1}{\ell!} x^\ell, \quad L_n(x) = L_n^{(0)}(x).$$

19

Some such polynomials are:

$$L_n^{(k)}(0) = \binom{n+k}{n},$$

$$L_0^{(k)}(x) = 1,$$

$$L_1^{(k)}(x) = 1 + k - x,$$

$$L_2^{(1)}(x) = \frac{1}{2}(6 - 6x + x^2),$$

$$L_3^{(1)}(x) = \frac{1}{6}(24 - 36x + 12x^2 - x^3),$$

$$L_4^{(1)}(x) = \frac{1}{24}(120 - 240x + 120x^2 - 20x^3 + x^4),$$

$$L_5^{(1)}(x) = \frac{1}{120}(720 - 1800x + 1200x^2 - 300x^3 + 30x^4 - x^5),$$

$$L_2^{(2)}(x) = \frac{1}{2}(12 - 8x + x^2),$$

$$L_3^{(2)}(x) = \frac{1}{6}(60 - 60x + 15x^2 - x^3),$$

$$L_4^{(2)}(x) = \frac{1}{24}(360 - 480x + 180x^2 - 24x^3 + x^4),$$

$$L_5^{(2)}(x) = \frac{1}{120}(2520 - 4200x + 2100x^2 - 420x^3 + 35x^4 - x^5).$$

Computer tip: Laguerre polynomials may be generated and plotted easily using the symbolic software MATHEMATICA. Here is a MATHEMATICA session for generating the Laguerre polynomials $\{L_n^{(0)}(x); n = 1, 2, 3, 4, 5\}$ and their plot shown in Figure 3. For the general Laguerre polynomial $L_n^{(\beta)}(x)$ the corresponding symbol is LaguerreL[n,β,x].

```
math
LaguerreL[1,0,x]
LaguerreL[2,0,x]
LaguerreL[3,0,x]
LaguerreL[4,0,x]
LaguerreL[5,0,x]
h=Plot[{LaguerreL[1,0,x],LaguerreL[2,0,x],LaguerreL[3,0,x],
```

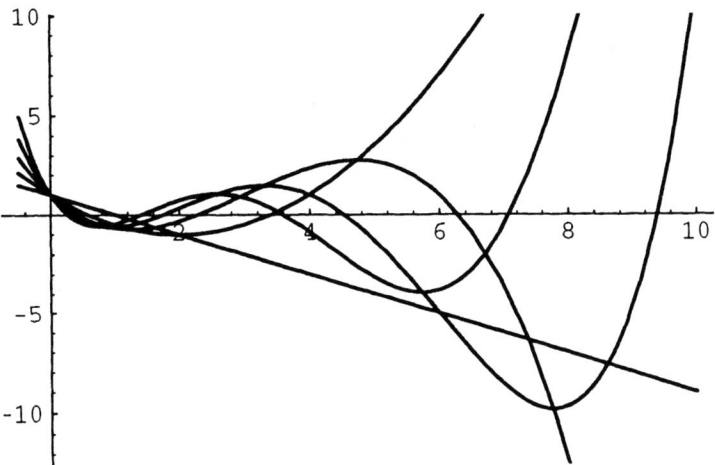

Figure 3: Five Laguerre polynomials.

LaguerreL[4,0,x],LaguerreL[5,0,x]},{x,−0.5,10}];
Display["terri",h]
Quit
hardcopy terri

The generalized Laguerre polynomials satisfy the following equalities:

$$L_n^{(\beta-1)}(x) = L_n^{(\beta)}(x) - L_{n-1}^{(\beta)}(x), \tag{10}$$

$$x\frac{d^2}{dx^2}L_n^{(\beta)}(x) + (\beta + 1 - x)\frac{d}{dx}L_n^{(\beta)}(x) + nL_n^{(\beta)}(x) = 0. \tag{11}$$

Details about the orthonormal polynomials mentioned above may be found in Abramowitz and Stegun (1972) or in Gradshteyn and Ryzhik (1980); they may be easily generated using the computer package MATHEMATICA (Wolfram (1991)), as shown above.

21

3) **One-dimensional probability waves**. Let u be a probability density on $D \subseteq \mathbf{R}^1$ and let

$$\{U_n; n = 0, 1, \ldots\}, \quad U_0 \equiv 1,$$

be a sequence of orthonormal polynomials with the weight u. As mentioned before, as long as nothing alters the statistical equilibrium described by the probability density function u, we have

$$< U_n\, u >= 0, \quad (n = 1, 2, \ldots), \quad < U_0\, u >=< u >= 1,$$

$$< U_n^2 u >= 1, \quad (n = 0, 1, \ldots), \quad < U_n U_k\, u >= 0, \quad (n \neq k).$$

Very often, however, random fluctuations alter such a statistical equilibrium and the probability density function u has to be replaced by another probability density function. If the sequence $\{U_n; n = 0, 1, \ldots\}$ is complete with respect to the weight u, then another probability density function f on D may be written as

$$f = u\left(1 + \sum_{n=1}^{+\infty} c_n U_n\right), \tag{12}$$

where $c_n =< U_n\, f >$ is the U_n-moment of f or the mean fluctuation in the direction U_n, and the sum is taken with respect to the values of $n, (n = 1, 2, \ldots)$. Convergence is assumed in the sense of topology induced by the norm $\| \cdot \|$, i.e.,

$$\left\| f - u\left(1 + \sum_{n=1}^{m} c_n U_n\right) \right\|$$

tends to zero when m tends to infinity, where

$$\|f\| =< f^* f >^{1/2}= \left(\int_D f^*(x) f(x)\, dx\right)^{1/2}.$$

The weighted deviation of f from u is the *probability wave function*

$$\chi = \frac{f - u}{\sqrt{u}} = \sqrt{u} \sum_{n=1}^{+\infty} c_n U_n. \tag{13}$$

The probability density function generated by the probability wave function of the minimum weighted deviation from u induced by the mean fluctuations $\{c_n; n = 1, 2, , \ldots\}$ is

$$\left(\sum_{n=1}^{+\infty} c_n^2\right)^{-1} \chi^2.$$

22

The elementary probability wave functions are:

$$\psi_0 = \sqrt{u}, \qquad \chi_n = \sqrt{u}\, U_n, \quad (n = 1, 2, \ldots),$$

where ψ_0 is the ground probability wave function induced by the probability density function u, and χ_n is the elementary wave function of level n. Therefore, the probability wave χ is a linear combination of the elementary probability wave functions, namely,

$$\chi = \sum_{n=1}^{+\infty} c_n \chi_n.$$

The jump from level ℓ to level n is defined by $\psi_{\ell,n} = \chi_n - \chi_\ell$. The elementary jumps are

$$\psi_n = \psi_{n-1,n} = \chi_n - \chi_{n-1}, \quad \text{where} \quad \chi_0 = \psi_0.$$

Using the Lagrange multipliers method, whose details are given in the second part of Appendix 4, we obtain:

Proposition 5: *The solution of the quadratic program:*

$$\min_f \tilde{\chi}^2 = <\chi^2> = <\left(\frac{f}{u} - 1\right)^2 u>$$

subject to the constraints

$$< U_n f > = c_n, \qquad (n = 1, 2, \ldots, N) \qquad (14)$$

is

$$f = u\left(1 + \sum_{n=1}^{N} c_n U_n\right). \qquad (15)$$

The least $\tilde{\chi}^2$ technique is a weighted variant of the least squares method of Legendre and Gauss. Minimizing $\tilde{\chi}^2$ subject to given mean values, we force the small initial probabilities to remain basically the same, focusing on changes induced by the constraints on the most probable outcomes.

Practically, c_n is estimated by calculating a confidence interval for a mean value. If $\{x_1, \ldots, x_M\}$ is a random sample of size M from the population to whom both u and f refer, we calculate the sample mean

$$\overline{U}_n^{(M)} = \frac{1}{M}[U_n(x_1) + \ldots + U_n(x_M)]$$

An $100(1 - \alpha)\%$ confidence interval for c_n is

$$\left(\overline{U}_n^{(M)} - t_{\alpha/2,M-1} \frac{s_n^{(M)}}{\sqrt{M}}, \quad \overline{U}_n^{(M)} + t_{\alpha/2,M-1} \frac{s_n^{(M)}}{\sqrt{M}} \right),$$

where $t_{\alpha/2,M-1}$ is the critical point of the t-distribution with $M - 1$ degrees of freedom, corresponding to the significance level $\alpha/2$, and $s_n^{(M)}$ is the sample standard deviation

$$s_n^{(M)} = \frac{1}{M-1} \sum_{k=1}^{M} \left[U_n(x_k) - \overline{U}_n^{(M)} \right]^2.$$

Introducing (15) into (13) we obtain the one-dimensional probability wave

$$\chi = \sqrt{u} \sum_{n=1}^{N} c_n U_n \tag{16}$$

generated by the deviation from u due to the generalized moments (14).

Introducing $U_n(x)$ given by (5),(7), or (8) into (12) (or into (15), as an approximation of (12)) we obtain the closest probability density function (or an approximation of it when using (15)) to u given by (1),(2), or (3), respectively, when the generalized moments c_n are given, where closeness is measured using Pearson's $\tilde{\chi}^2$ indicator, expressed by the squared amplitude of the probability wave (13). As long as the statistical equilibrium described by the probability density function u remains unchanged, no probability wave is generated. When the occurrence of random fluctuations is detected by estimating the generalized moments (14), then the probability wave (16) shows the deviation from the statistical equilibrium described by u. An external observer can generally observe only a change in an equilibrium state.

4) **Multi-dimensional probability wave.** Let u and v be two probability density functions on $D_1 \subseteq \mathbf{R}^1$, and $D_2 \subseteq \mathbf{R}^1$, respectively, and let

$$\{U_n; n = 0, 1, \ldots\}, \quad \{V_\ell; \ell = 0, 1, \ldots\}, \quad (U_0 \equiv 1, V_0 \equiv 1)$$

be two complete systems of orthonormal functions on D_1 and D_2 with the weights u and v, respectively. If there is independence between marginals, then the joint probability density on $D_1 \times D_2$ is simply the direct product

24

uv. But what does it happen when there is interdependence between the two components?

Consider the system of functions $\{U_n V_\ell; n, \ell = 0, 1, \ldots\}$ on $D_1 \times D_2$ with the weight uv. A joint probability density function f on $D_1 \times D_2$ has the form

$$f = u\,v\,(1+ \sum_{\substack{n=0 \\ (n,\ell)\neq(0,0)}}^{+\infty} \sum_{\ell=0}^{+\infty} c_{n\ell} U_n V_\ell). \tag{17}$$

Convergence is assumed in the sense of topology induced by the norm $\| \cdot \|$, i.e.,

$$\|f - uv(1+ \sum_{\substack{n=0 \\ (n,\ell)\neq(0,0)}}^{m} \sum_{\ell=0}^{s} c_{n\ell} U_n V_\ell)\|$$

tends to zero when m and s tend to infinity, where

$$\|f\| =< f^* f >^{1/2}= \left(\int_{D_1 \times D_2} f^*(x,y) f(x,y)\, dx\, dy \right)^{1/2}.$$

Such a joint probability density function is the closest one, in the $\tilde{\chi}^2$ sense, to the direct independent product uv subject to the generalized mixed moments (or correlations)

$$c_{n\ell} =< U_n V_\ell f >,$$

as shown by the next proposition whose proof is a standard application of the calculus of variation. Details may be found in the second part of Appendix 4.

Proposition 6: *Let*

$$\chi = \frac{f - uv}{\sqrt{uv}} = \left(\frac{f}{uv} - 1 \right)\sqrt{uv} \tag{18}$$

be the weighted mean deviation of f from the independent direct product uv. The solution of the quadratic program

$$\min_{f} \tilde{\chi}^2 =< \chi^2 >$$

subject to the constraints

$$< U_n V_\ell\, f >= c_{n\ell}, \tag{19}$$

$$(n = 0, 1, \ldots, N; \ell = 0, 1, \ldots, L; (n, \ell) \neq (0, 0)),$$

is the density

$$f = u\,v\,(1 + \sum_{\substack{n=0 \\ (n,\ell)\neq(0,0)}}^{N} \sum_{\ell=0}^{L} c_{n\ell}\,U_n V_\ell). \qquad (20)$$

Introducing (20) into (18) we obtain the two-dimensional probability wave

$$\chi = \sqrt{uv}\, \sum_{\substack{n=0 \\ (n,\ell)\neq(0,0)}}^{N} \sum_{\ell=0}^{L} c_{n\ell}\,U_n V_\ell \qquad (21)$$

generated by the deviation from independence of u and v due to the generalized correlations (19). The elementary probability wave functions are:

$$\psi_{00} = \sqrt{uv}, \quad \chi_{n\ell} = \sqrt{uv}\,U_n V_\ell,$$

$$(n = 0, 1, \ldots; \ell = 0, 1, \ldots; (n, \ell) \neq (0, 0)),$$

where ψ_{00} is the ground probability wave function and $\chi_{n\ell}$ is the elementary wave function of level (n, ℓ). The jump from level (i, j) to level (k, ℓ) is defined by: $\psi_{ij,kl} = \chi_{k\ell} - \chi_{ij}$. Thus, the probability wave function is a linear combination of the elementary probability wave functions, namely,

$$\chi = \sum_{\substack{n=0 \\ (n,\ell)\neq(0,0)}}^{+\infty} \sum_{\ell=0}^{+\infty} c_{n\ell}\,\chi_{n\ell}.$$

The generalization to more than two components is straightforward. Thus, if u, v, and w are three probability density functions on $D_1 \subseteq \mathbf{R}^1$, $D_2 \subseteq \mathbf{R}^1$, and $D_3 \subseteq \mathbf{R}^1$, respectively, and

$$\{U_n;\, n = 0, 1, \ldots\}, \quad \{V_\ell;\, \ell = 0, 1, \ldots\}, \quad \{W_k;\, k = 0, 1, \ldots\},$$

$$(U_0 \equiv 1, \quad V_0 \equiv 1, \quad W_0 \equiv 1),$$

are three complete systems of orthonormal functions on D_1, D_2, and D_3 with the weights u, v, and w, respectively, then the joint probability density function f that is the closest one, in the $\tilde{\chi}^2$ sense, to the direct independent product uvw subject to the generalized mixed moments (or correlations)

$$< U_n V_\ell W_k\, f > = c_{n\ell k},$$

$(n = 0, 1, \ldots, N; \ell = 0, 1, \ldots, L; k = 0, 1, \ldots, K; (n, \ell, k) \neq (0, 0, 0)),$

is given by the expression

$$f = uvw(1 + \sum_{\substack{n=0 \\ (n,\ell,k) \neq (0,0,0)}}^{N} \sum_{\ell=0}^{L} \sum_{k=0}^{K} c_{n\ell k} U_n V_\ell W_k).$$

Such a density f minimizes the chi-square deviation

$$\tilde{\chi}^2 = < \frac{(f - uvw)^2}{uvw} >$$

from the independent direct product uvw of the marginals, subject to the given generalized moments (correlations)

$$< U_n V_\ell W_k f > = c_{n\ell k},$$

$(n = 0, 1, \ldots, N; \ell = 0, 1, \ldots, L; k = 0, 1, \ldots, K; (n, \ell, k) \neq (0, 0, 0)).$

5) **Stationary values of the mean energy.** If χ is a probability wave function and $\hat{A}$ is an operator, then the mean value of $\hat{A}$ is defined by

$$< \hat{A} > = \frac{< \chi^* \hat{A} \chi >}{< \chi^* \chi >}.$$

The formalism discussed so far may be applied for building up a probabilistic model for describing the behavior of an arbitrary system. In order to apply it to a quantum system we take into account that the Hamiltonian operator $\hat{H}$ is the starting point of all applications of quantum mechanics. When the one-dimensional (or multi-dimensional) wave function χ is given, like (16) (or (21)) for instance, we are looking for the stationary points of the mean energy of the quantum system defined by

$$< \hat{E} > = \frac{< \chi^* \hat{H} \chi >}{< \chi^* \chi >}. \tag{22}$$

For the probability wave functions (16) and (21), the corresponding mean energy becomes

$$< \hat{E} > (c_1, c_2, \ldots, c_N) \text{ and } < \hat{E} > (c_{01}, c_{10}, c_{11}, \ldots, c_{NL}),$$

respectively. We determine the stationary points of the mean energy by analyzing the system of equations

$$\frac{\partial < \hat{E} >}{\partial c_n} = 0, \quad (n = 1, \ldots, N),$$

or the system of equations

$$\frac{\partial < \hat{E} >}{\partial c_{n\ell}} = 0,$$

$$(n = 0, 1, \ldots, N; \ell = 0, 1, \ldots, L; (n, \ell) \neq (0, 0)),$$

respectively. Obviously, if the functions $\sqrt{u}\, U_n$ are eigenfunctions of the operator $\hat{H}$, i.e.,

$$\hat{H}(\sqrt{u}\, U_n) = E_n(\sqrt{u}\, U_n),$$

then the mean value of energy is

$$< \hat{E} >= \sum_{n=1}^{N} E_n\, p_n,$$

where the probability of the energy value E_n is

$$p_n = \frac{c_n^2}{\sum_{\ell=1}^{N} c_\ell^2}.$$

Similarly, in the two-dimensional case, if the functions $\sqrt{uv}\, U_n V_\ell$ are eigenfunctions of the operator $\hat{H}$, i.e.,

$$\hat{H}(\sqrt{uv}\, U_n V_\ell) = E_{n\ell}\sqrt{uv}\, U_n V_\ell,$$

then the mean value of energy is

$$< \hat{E} >= \sum_{\substack{n=0 \\ (n,\ell)\neq(0,0)}}^{N} \sum_{\ell=0}^{L} E_{n\ell}\, p_{n\ell},$$

where the probability of the energy value $E_{n\ell}$ is

$$p_{n\ell} = c_{n\ell}^2 / \left(\sum_{\substack{n'=0 \\ (n',\ell')\neq(0,0)}}^{N} \sum_{\ell'=0}^{L} c_{n'\ell'}^2 \right).$$

Therefore, another approach is to see whether the functions $\sqrt{u}\,U_n$ in the one-dimensional case, or $\sqrt{uv}\,U_n V_\ell$ in the multi-dimensional case, are eigenfunctions of the Hamiltonian $\hat{H}$ of the system and find the corresponding eigenvalues as possible values of the mean energy of the respective quantum system.

3. APPLICATION 1: A FREE PARTICLE IN A ONE-DIMENSIONAL BOX

The quantum system consists of a free particle in a box $D = [0, a]$. Having no constraints except the box itself, the statistical equilibrium is given by the uniform probability density function $u(x) = 1/a$ in D and let us take

$$U_0 \equiv 1, \; U_n(x) = \sqrt{2}\sin\frac{n\pi}{a}x, \quad (n = 1, 2, \ldots)$$

as a system of orthonormal functions with the weight u. The probability wave function is approximated by

$$\chi(x) = \sqrt{u(x)}\sum_{n=1}^{N} c_n\,U_n(x) = \left(\frac{2}{a}\right)^{1/2}\sum_{n=1}^{N} c_n \sin\frac{n\pi}{a}x,$$

with at least one coefficient c_n different from zero.

Computer tip: Functions depending on the generalized coordinates U_n may be plotted using the symbolic software MATHEMATICA. Here is a MATHEMATICA session for getting the plot of $\frac{1}{2}U_2^2$, for $a = 3$, represented in Figure 4:

```
math
m[x_]:=(Sin[2*Pi*x/3])^2
a=Plot[m[x],{x,0,3}];
Display["john",a]
Quit
hardcopy john
```

As the potential energy is equal to zero, the Hamiltonian operator is

$$\hat{H} = -\frac{\hbar^2}{2m}\frac{d^2}{dx^2}$$

29

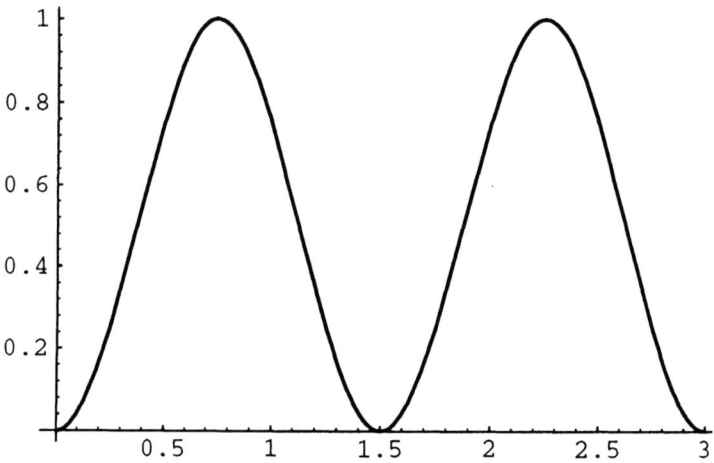

Figure 4: The plot of $\frac{1}{2}U_2^2$ for $a = 3$.

where m is the mass of the particle and

$$\hbar = \frac{h}{2\pi},$$

h being the Planck constant. We have

$$\hat{H}\chi(x) = \frac{\hbar^2}{2m}\left(\frac{2}{a}\right)^{1/2}\left(\frac{\pi}{a}\right)^2 \sum_{n=1}^{N} c_n n^2 \sin\frac{n\pi}{a}x.$$

Standard integration of trigonometric functions, namely

$$\int \cos \alpha x \, dx = \frac{1}{\alpha}\sin \alpha x,$$

$$\int \sin \alpha x \sin \beta x \, dx =$$

$$= \frac{\sin(\alpha - \beta)x}{2(\alpha - \beta)} - \frac{\sin(\alpha + \beta)x}{2(\alpha + \beta)}, \quad (\alpha^2 \neq \beta^2),$$

$$\int \sin^2 \alpha x \, dx = \int \frac{1}{2}(1 - \cos 2\alpha x)dx = \frac{x}{2} - \frac{\sin 2\alpha x}{4\alpha},$$

30

gives:

$$< \chi \hat{H}\chi >= \int_0^a \chi(x)\, \hat{H}\chi(x)\, dx = \frac{\hbar^2}{2m}\left(\frac{\pi}{a}\right)^2 \sum_{n=1}^N c_n^2 n^2$$

$$< \chi^2 >= \int_0^a \chi^2(x)\, dx = \sum_{n=1}^N c_n^2.$$

Introducing these two results into

$$< \hat{E} >= \frac{< \chi \hat{H}\chi >}{< \chi^2 >}$$

we get

$$< \hat{E} > \sum_{n=1}^N c_n^2 = \frac{\hbar^2}{2m}\left(\frac{\pi}{a}\right)^2 \sum_{n=1}^N c_n^2 n^2. \tag{23}$$

Taking the partial derivatives with respect to c_n, $(n = 1,\ldots,N)$, and looking for a stationary point of the mean energy, i.e.,

$$\frac{\partial < \hat{E} >}{\partial c_n} = 0, \quad (n = 1,\ldots,N),$$

we get

$$2c_n\left[\frac{\hbar^2}{2m}\left(\frac{\pi}{a}\right)^2 n^2 - < \hat{E} >\right] = 0,$$

which implies that either $c_n = 0$ or

$$< \hat{E} >= \frac{\hbar^2}{2m}\left(\frac{\pi}{a}\right)^2 n^2 = \frac{h^2 n^2}{8ma^2}.$$

Therefore, the possible values of the energy are

$$E_n = \frac{h^2 n^2}{8ma^2}, \quad (n = 1,2,\ldots,N). \tag{24}$$

Introducing these values into (23) we get an equivalent expression for the mean energy, namely,

$$< \hat{E} >= \sum_{n=1}^N E_n \frac{c_n^2}{\sum_{\ell=1}^N c_\ell^2}, \tag{25}$$

31

which shows that

$$\frac{c_n^2}{\sum_{\ell=1}^{N} c_\ell^2}$$

may be interpreted as being the probability that the value of the energy of the system is E_n. Obviously, if only one coefficient c_n is different from zero, then $< \hat{E} >= E_n$, with certainty.

Remark 1: The above conclusions may be obtained without looking for the stationary points of the energy by simply noticing that $\sqrt{u(x)}\, U_n(x)$ is an eigenfunction of $\hat{H}$ and the corresponding eigenvalue is just E_n. Indeed,

$$\hat{H}(\sqrt{u(x)}U_n(x)) = \left(\frac{2}{a}\right)^{1/2} \hat{H}(\sin\frac{n\pi}{a}x) =$$

$$= -\left(\frac{2}{a}\right)^{1/2} \frac{\hbar^2}{2m} \frac{d^2}{dx^2}\left(\sin\frac{n\pi}{a}x\right) =$$

$$= \frac{\hbar^2\pi^2 n^2}{2ma^2}\left(\frac{2}{a}\right)^{1/2} \sin\frac{n\pi}{a}x = E_n\sqrt{u(x)}U_n(x).$$

Then, from (22) we get (25), because

$$< \hat{E} >= \frac{<(\sqrt{u}\sum_{\ell=1}^{N} c_\ell U_\ell)(\sqrt{u}\sum_{n=1}^{N} c_n E_n U_n)>}{<(\sqrt{u}\sum_{\ell=1}^{N} c_\ell U_\ell)(\sqrt{u}\sum_{n=1}^{N} c_n U_n)>} =$$

$$= \sum_{n=1}^{N} E_n \frac{c_n^2}{\sum_{\ell=1}^{N} c_\ell^2}.$$

Remark 2: The above results have been obtained without solving the corresponding Schrödinger equation. It is somehow unexpected to see that the eigenfunctions of the Hamiltonian of this quantum system are just generalized coordinates of a maximum entropy probability distribution. But even more unexpected is to subsequently see that this happens with other quantum systems as well, which eventually will allow us to apply the same approach even to quantum systems for which the corresponding Schrödinger equation cannot be solved exactly.

4. APPLICATION 2: A FREE PARTICLE IN A THREE-DIMENSIONAL BOX

The quantum system consists of a free particle in a three-dimensional box $D = [0, a] \times [0, b] \times [0, c]$. Having no constraints except the box itself, the statistical equilibrium is given by the independent product of the uniform marginals and, therefore, the joint probability density function is

$$u(x)v(y)w(z) = \frac{1}{a}\frac{1}{b}\frac{1}{c}, \quad (0 \leq x \leq a, 0 \leq y \leq b, 0 \leq z \leq c). \tag{26}$$

We take again the trigonometric system (6) as the generalized coordinates whose weights are the uniform marginals on $[0, a], [0, b]$, and $[0, c]$, respectively. The corresponding probability three-dimensional wave function induced by the minimum $\tilde{\chi}^2$ deviation from statistical equilibrium described by (26) is

$$\chi(x, y, z) = \tag{27}$$

$$= \sqrt{u(x)v(y)w(z)} \ \sum_{n=0}^{N} \sum_{\ell=0}^{L} \sum_{k=0}^{K} \ c_{n\ell k} U_n(x) V_\ell(y) W_k(z) =$$
$$\scriptstyle (n, \ell, k) \neq (0,0,0)$$

$$= \left(\frac{8}{abc}\right)^{1/2} \ \sum_{n=0}^{N} \sum_{\ell=0}^{L} \sum_{k=0}^{K} \ c_{n\ell k} \sin\frac{n\pi}{a}x \sin\frac{\ell\pi}{b}y \sin\frac{k\pi}{c}z,$$
$$\scriptstyle (n, \ell, k) \neq (0,0,0)$$

with at least one coefficient $c_{n\ell k}$ different from zero.

Computer tip: Functions depending on the generalized coordinates U_n, V_ℓ, and W_k, involving one or two spatial variables, may be plotted using the symbolic software MATHEMATICA. Here is a MATHEMATICA session for getting the three-dimensional plot, the contour plot, and the density plot of

$$\frac{1}{8}U_1^2(x)V_2^2(y)W_3^2(0.5), \quad x \in [0, a], y \in [0, b],$$

with $a = b = c = 3$, represented in Figures 5-7:
math
n[x_,y_,z_]:=(Sin[Pi⋆x/3]⋆
Sin[2⋆Pi⋆y/3]⋆

33

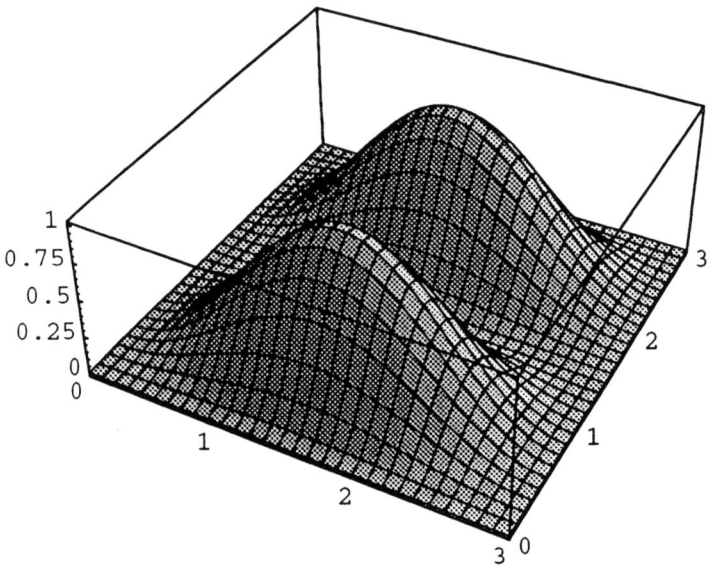

Figure 5: The plot of $\frac{1}{8}U_1^2(x)V_2^2(y)W_3^2(0.5)$.

```
Sin[3*Pi*z/3])^2
b=Plot3D[n[x,y,0.5],{x,0,3},{y,0,3},PlotPoints− >30];
c=ContourPlot[n[x,y,0.5],{x,0,3},{y,0,3}];
d=DensityPlot[n[x,y,0.5],{x,0,3},{y,0,3},Mesh− >False];
Display["john1",b]
Display["john2",c]
Display["john3",d]
Quit
hardcopy john1
hardcopy john2
hardcopy john3
```

As the potential energy is equal to zero in D, the Hamiltonian operator

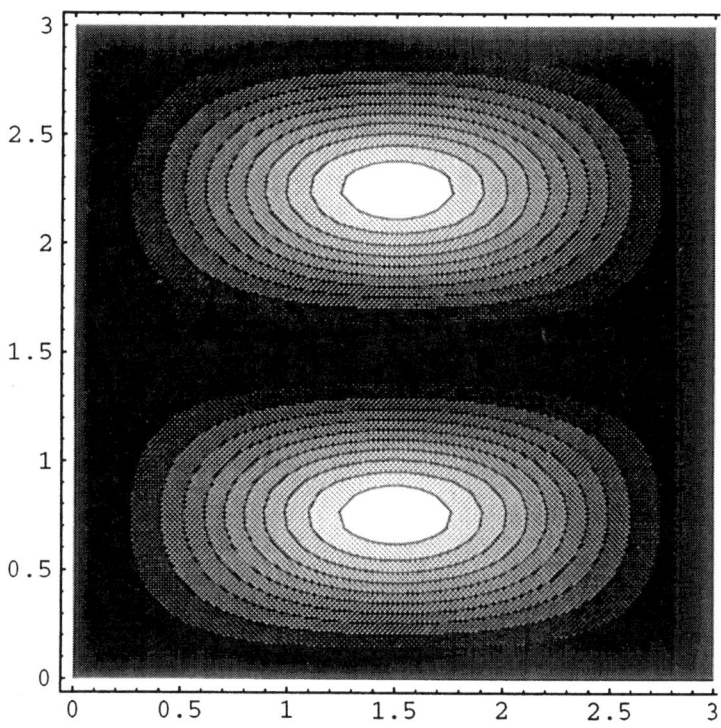

Figure 6: The contour plot of $\frac{1}{8}U_1^2(x)V_2^2(y)W_3^2(0.5)$.

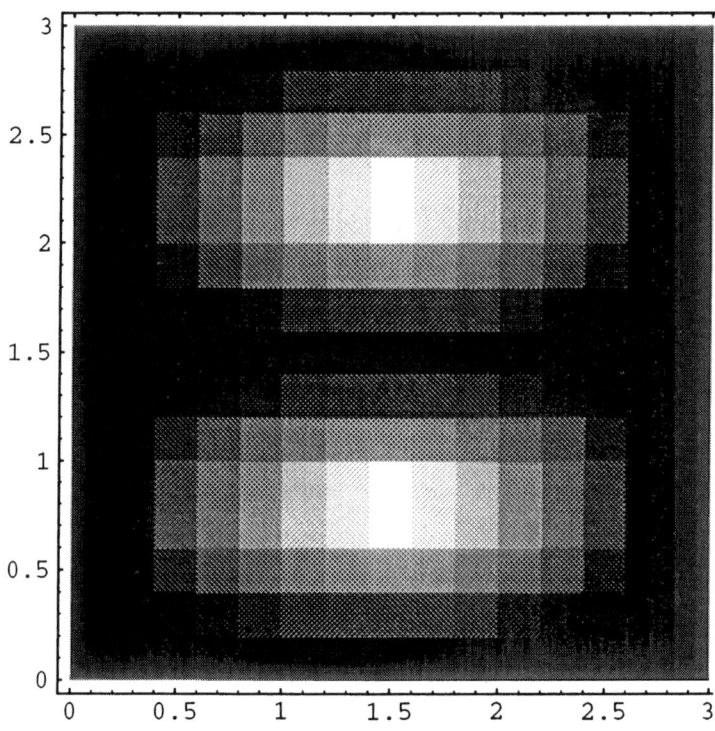

Figure 7: The density plot of $\frac{1}{8}U_1^2(x)V_2^2(y)W_3^2(0.5)$.

is

$$\hat{H} = -\frac{\hbar^2}{2m}\nabla^2$$

where ∇^2 is the Laplacian operator. We have

$$\hat{H}\chi(x,y,z) =$$

$$= C \ \sum_{n=0}^{N}\sum_{\ell=0}^{L}\sum_{k=0}^{K} \ c_{n\ell k} \left[\left(\frac{n}{a}\right)^2 + \left(\frac{\ell}{b}\right)^2 + \left(\frac{k}{c}\right)^2\right] \times$$

$$(n,\ell,k)\neq(0,0,0)$$

$$\times \sin\frac{n\pi}{a}x \sin\frac{\ell\pi}{b}y \sin\frac{k\pi}{c}z,$$

where

$$C = \frac{\hbar^2\pi^2}{2m}\left(\frac{8}{abc}\right)^{1/2}.$$

Standard integration of trigonometric functions, namely

$$\int \sin\alpha x \sin\beta x\, dx =$$

$$= \frac{\sin(\alpha-\beta)x}{2(\alpha-\beta)} - \frac{\sin(\alpha+\beta)x}{2(\alpha+\beta)}, \quad (\alpha^2 \neq \beta^2),$$

$$\int \sin^2\alpha x\, dx = \frac{x}{2} - \frac{\sin 2\alpha x}{4\alpha},$$

gives:

$$< \chi\hat{H}\chi > = \frac{\hbar^2\pi^2}{2m} \ \sum_{n=0}^{N}\sum_{\ell=0}^{L}\sum_{k=0}^{K} \ c_{n\ell k}^2 \left[\left(\frac{n}{a}\right)^2 + \left(\frac{\ell}{b}\right)^2 + \left(\frac{k}{c}\right)^2\right],$$

$$(n,\ell,k)\neq(0,0,0)$$

$$< \chi\chi > = \ \sum_{n=0}^{N}\sum_{\ell=0}^{L}\sum_{k=0}^{K} \ c_{n\ell k}^2$$

$$(n,\ell,k)\neq(0,0,0)$$

Introducing these two results into

$$< \hat{E} > = \frac{< \chi\hat{H}\chi >}{< \chi^2 >} \tag{28}$$

37

we get

$$\sum_{n=0}^{N} \sum_{\ell=0}^{L} \sum_{k=0}^{K} \left\{ \frac{h^2}{8m} \left[\left(\frac{n}{a}\right)^2 + \left(\frac{\ell}{b}\right)^2 + \left(\frac{k}{c}\right)^2 \right] - <\hat{E}> \right\} c_{n\ell k}^2 = 0.$$
$$(n,\ell,k) \neq (0,0,0)$$

Taking the partial derivatives with respect to $c_{n\ell k}$, and looking for a stationary point of the mean energy, i.e.,

$$\frac{\partial <\hat{E}>}{\partial c_{n\ell k}} = 0,$$

for the values

$$n = 0, 1, \ldots, N; \quad \ell = 0, 1, \ldots, L; \quad k = 0, 1, \ldots, K;$$

$$(n, \ell, k) \neq (0, 0, 0),$$

we get either $c_{n\ell k} = 0$ or

$$<\hat{E}> = E_{n\ell k} = \frac{h^2}{8m} \left[\left(\frac{n}{a}\right)^2 + \left(\frac{\ell}{b}\right)^2 + \left(\frac{k}{c}\right)^2 \right],$$

which are the possible values of the energy of the system. Introducing these values into (28) we get an equivalent expression for the mean energy, namely,

$$<\hat{E}> = \sum_{n=0}^{N} \sum_{\ell=0}^{L} \sum_{k=0}^{K} E_{n\ell k}\, p_{n\ell k}, \tag{29}$$
$$(n,\ell,k) \neq (0,0,0)$$

which shows that

$$p_{n\ell k} = c_{n\ell k}^2 / \left(\sum_{n'=0}^{N} \sum_{\ell'=0}^{L} \sum_{k'=0}^{K} c_{n'\ell'k'}^2 \right)$$
$$(n',\ell',k') \neq (0,0,0)$$

may be interpreted as being the probability that the value of the energy of the system is $E_{n\ell k}$. Obviously, if only one coefficient $c_{n\ell k}$ is different from zero, then $<\hat{E}> = E_{n\ell k}$, with certainty.

Remark: The above conclusions may be obtained without looking for the stationary points of the energy by simply noticing that

$$-\frac{\hbar^2}{2m} \nabla^2 \left[\left(\frac{8}{abc}\right)^{1/2} \sin\frac{n\pi}{a}x \sin\frac{\ell\pi}{b}y \sin\frac{k\pi}{c}z \right] =$$

$$= \frac{h^2}{8m} \left[\left(\frac{n}{a}\right)^2 + \left(\frac{\ell}{b}\right)^2 + \left(\frac{k}{c}\right)^2 \right] \left(\frac{8}{abc}\right)^{1/2} \sin\frac{n\pi}{a}x \sin\frac{\ell\pi}{b}y \sin\frac{k\pi}{c}z,$$

which shows that $\sqrt{uvw}\, U_n V_\ell W_k$ is an eigenfunction of $\hat{H}$ and the corresponding eigenvalue is just $E_{n\ell k}$. Then, from (28) we get (29).

5. APPLICATION 3: TWO NONINTERACTING FREE PARTICLES IN A ONE-DIMENSIONAL BOX

The quantum system consists of two noninteracting free particles in a one-dimensional box $[0, a]$. The domain is $D = [0, a] \times [0, a]$. The two particles being independent and having no constraints except the box itself, the statistical equilibrium is given by the independent product of the uniform marginals and, therefore, the joint probability density function is

$$u(x)v(y) = \frac{1}{a}\frac{1}{a}, \quad (0 \le x \le a, 0 \le y \le a). \tag{30}$$

We take again the trigonometric system (6) as the generalized coordinates whose weights are the uniform marginals on $[0, a]$. The corresponding probability two-dimensional wave function induced by the minimum $\tilde{\chi}^2$ deviation from statistical equilibrium described by (30) is

$$\chi(x, y) = \sqrt{u(x)v(y)} \sum_{\substack{n=0 \\ (n,\ell)\neq(0,0)}}^{N} \sum_{\ell=0}^{L} c_{n\ell} U_n(x) V_\ell(y) = \tag{31}$$

$$= \frac{2}{a} \sum_{\substack{n=0 \\ (n,\ell)\neq(0,0)}}^{N} \sum_{\ell=0}^{L} c_{n\ell} \sin\frac{n\pi}{a}x \sin\frac{\ell\pi}{a}y,$$

with at least one coefficient $c_{n\ell}$ different from zero.

Computer tip: Functions depending on the generalized coordinates U_n, and V_ℓ may be plotted using the symbolic software MATHEMATICA. Here is a MATHEMATICA session for getting the three- dimensional plot, the contour plot, and the density plot of

$$\frac{1}{2}U_2^2(x)V_2^2(y), \quad x \in [0,a], y \in [0,a],$$

with $a = 3$, represented in Figures 8-10:

```
math
o[x_,y_]:=(Sin[2*Pi*x/3]*
Sin[2*Pi*y/3])^2
r=Plot3D[o[x,y],{x,0,3},{y,0,3},PlotPoints- >30];
s=ContourPlot[o[x,y],{x,0,3},{y,0,3}];
t=DensityPlot[o[x,y],{x,0,3},{y,0,3},Mesh- >False];
Display["mary1",r]
Display["mary2",s]
Display["mary3",t]
Quit
hardcopy mary1
hardcopy mary2
hardcopy mary3
```

As the potential energy is equal to zero in D, the Hamiltonian operator is

$$\hat{H} = -\frac{\hbar^2}{2m_1}\frac{\partial^2}{\partial x^2} - \frac{\hbar^2}{2m_2}\frac{\partial^2}{\partial y^2}$$

where m_1 and m_2 are the mass of the two particles. We have

$$\hat{H}\chi(x,y) = \frac{\hbar^2\pi^2}{a^3} \sum_{n=0}^{N}\sum_{\ell=0}^{L} \underset{(n,\ell)\neq(0,0)}{c_{n\ell}} \left(\frac{n^2}{m_1} + \frac{\ell^2}{m_2}\right) \sin\frac{n\pi}{a}x \sin\frac{\ell\pi}{a}y.$$

Standard integration of trigonometric functions, namely,

$$\int \sin \alpha x \sin \beta x \, dx =$$

40

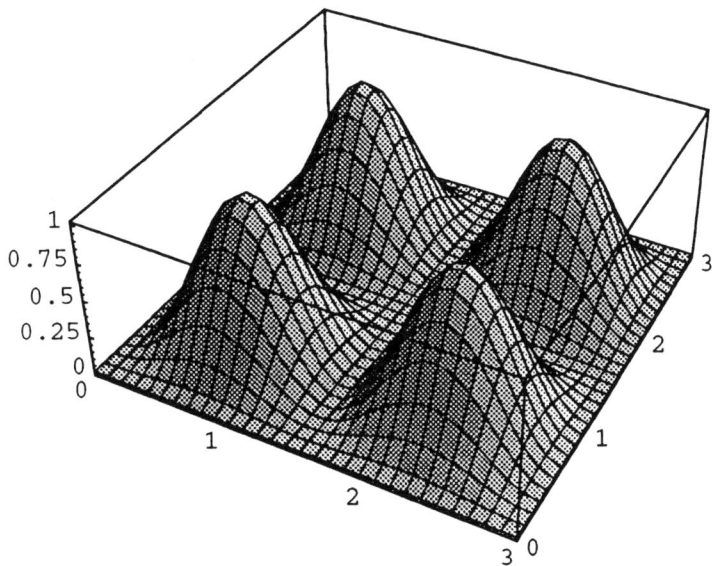

Figure 8: The plot of $\frac{1}{2}U_2^2(x)V_2^2(y)$.

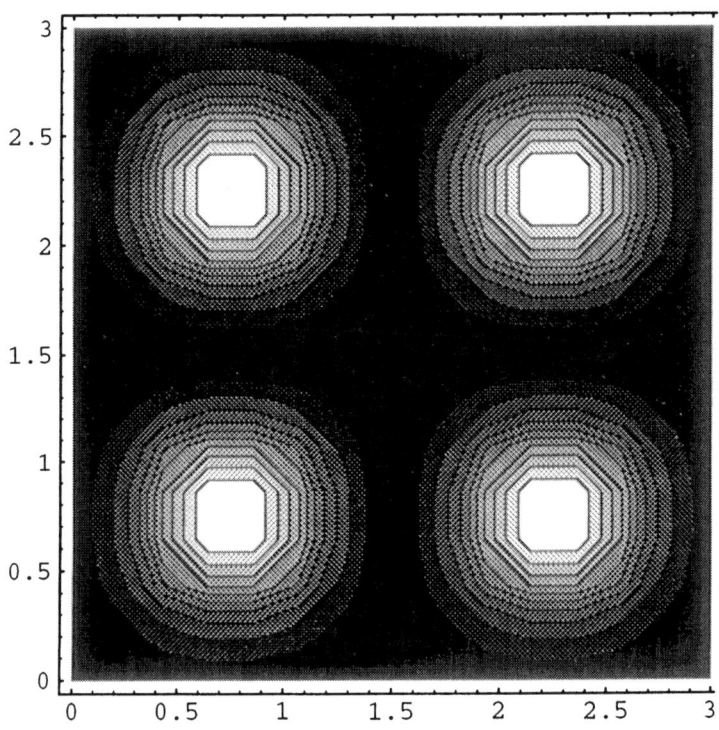

Figure 9: The contour plot of $\frac{1}{2}U_2^2(x)V_2^2(y)$.

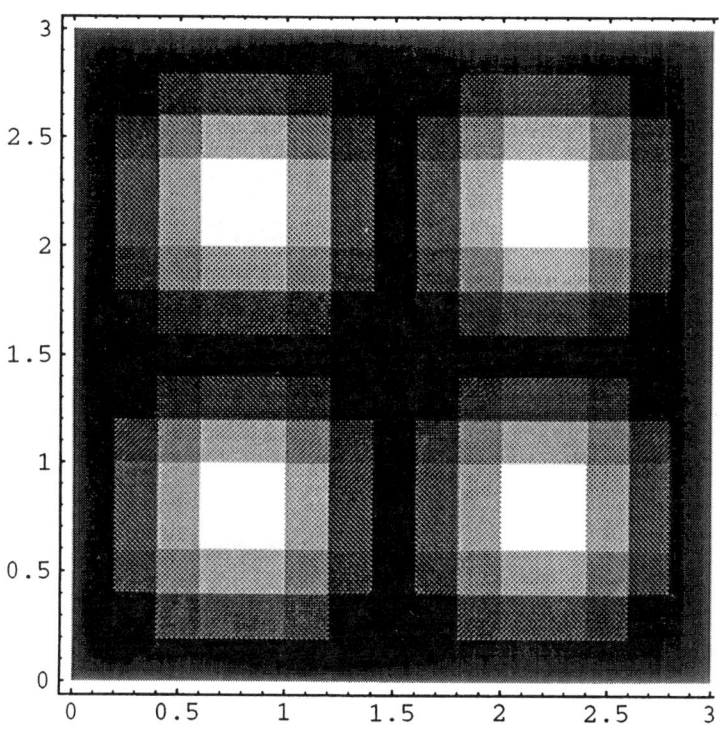

Figure 10: The density plot of $\frac{1}{2}U_2^2(x)V_2^2(y)$.

$$= \frac{\sin(\alpha - \beta)x}{2(\alpha - \beta)} - \frac{\sin(\alpha + \beta)x}{2(\alpha + \beta)}, \quad (\alpha^2 \neq \beta^2),$$

$$\int \sin^2 \alpha x \, dx = \frac{x}{2} - \frac{\sin 2\alpha x}{4\alpha},$$

gives:

$$< \chi \hat{H} \chi > = \left(\frac{\hbar \pi}{a} \right)^2 \frac{1}{2} \sum_{n=0}^{N} \sum_{\ell=0}^{L} c_{n\ell}^2 \left(\frac{n^2}{m_1} + \frac{\ell^2}{m_2} \right),$$
$$(n,\ell) \neq (0,0)$$

$$< \chi^2 > = \sum_{n=0}^{N} \sum_{\ell=0}^{L} c_{n\ell}^2.$$
$$(n,\ell) \neq (0,0)$$

Introducing these two results into

$$< \hat{E} > = \frac{< \chi \hat{H} \chi >}{< \chi^2 >}, \tag{32}$$

we get

$$\sum_{n=0}^{N} \sum_{\ell=0}^{L} \left[\frac{1}{2} \left(\frac{\hbar \pi}{a} \right)^2 \left(\frac{n^2}{m_1} + \frac{\ell^2}{m_2} \right) - < \hat{E} > \right] c_{n\ell}^2 = 0.$$
$$(n,\ell) \neq (0,0)$$

Taking the partial derivatives with respect to $c_{n\ell}$, and looking for a stationary point of the mean energy, i.e.,

$$\frac{\partial < \hat{E} >}{\partial c_{n\ell}} = 0,$$

$$(n = 0, 1, \ldots, N; \quad \ell = 0, 1, \ldots, L; \quad (n, \ell) \neq (0, 0)),$$

we get either $c_{n\ell} = 0$, or

$$< \hat{E} > = E_{n\ell} = \frac{1}{2} \left(\frac{\hbar \pi}{a} \right)^2 \left(\frac{n^2}{m_1} + \frac{\ell^2}{m_2} \right).$$

44

which are the possible values of the energy of the system. Introducing these values into (32) we get an equivalent expression for the mean energy, namely,

$$< \hat{E} >= \sum_{n=0}^{N} \sum_{\ell=0}^{L} \; E_{n\ell} \, p_{n\ell}, \qquad (33)$$
$$\text{\small$(n,\ell)\neq(0,0)$}$$

which shows that

$$p_{n\ell} = c_{n\ell}^2 / \; (\sum_{n'=0}^{N} \sum_{\ell'=0}^{L} \; c_{n'\ell'}^2)$$
$$\text{\small$(n',\ell')\neq(0,0)$}$$

may be interpreted as being the probability that the value of the energy of the system is $E_{n\ell}$. Obviously, if only one coefficient $c_{n\ell}$ is different from zero, then $< \hat{E} >= E_{n\ell}$, with certainty.

Remark: The above conclusions may be obtained without looking for the stationary points of the energy by simply noticing that

$$\left[-\frac{\hbar^2}{2m_1} \frac{\partial^2}{\partial x^2} - \frac{\hbar^2}{2m_2} \frac{\partial^2}{\partial y^2} \right] \frac{2}{a} \sin \frac{n\pi}{a} x \sin \frac{\ell\pi}{a} y =$$

$$= \frac{1}{2} \left(\frac{\hbar\pi}{a} \right)^2 \left(\frac{n^2}{m_1} + \frac{\ell^2}{m_2} \right) \frac{2}{a} \sin \frac{n\pi}{a} x \sin \frac{\ell\pi}{a} y,$$

which shows that $\sqrt{uv} \, U_n V_\ell$ is an eigenfunction of $\hat{H}$ and the corresponding eigenvalue is just $E_{n\ell}$. Then, from (32) we get (33).

6. APPLICATION 4: THE HARMONIC OSCILLATOR

This application deals with a quantum system which randomly oscillates around the origin on the real axis with the variance σ^2. The domain is $D = (-\infty, +\infty)$. As shown by Proposition 3, the statistical equilibrium on the real axis corresponding to the mean value $\mu = 0$ and variance σ^2 is described by the normal probability distribution $\mathbf{N}(0, \sigma^2)$ whose density is

$$u(x) = \frac{1}{\sigma\sqrt{2\pi}} e^{-x^2/2\sigma^2}, \quad (-\infty < x < +\infty).$$

We take (8) with $\mu = 0$ as the system of generalized coordinates associated to u, namely,

$$U_n(x) = (2^n n!)^{-1/2} H_n \left(\frac{x}{\sigma\sqrt{2}} \right), \quad (n = 0, 1, \ldots).$$

The probability wave function is approximated by

$$\chi(x) = \sqrt{u(x)} \sum_{n=1}^{N} c_n U_n(x) = \tag{34}$$

$$= (2\pi\sigma^2)^{-1/4} \sum_{n=1}^{N} (2^n n!)^{-1/2} c_n e^{-x^2/4\sigma^2} H_n \left(\frac{x}{\sigma\sqrt{2}} \right),$$

with at least one coefficient c_n different from zero.

Computer tip: The elementary wave functions $\sqrt{u}U_n$ may be plotted using the symbolic software MATHEMATICA. Here is a MATHEMATICA session for getting the plot of

$$\left[\sqrt{u}U_3(x) \right]^2 =$$

$$= (2\pi)^{-1/2} (2^3 3!)^{-1} e^{-x^2/2} H_3^2 \left(\frac{x}{\sqrt{2}} \right),$$

for $\sigma = 1$, represented in Figure 11:
math
ff[x_]:=(((2*Pi)^(−1/4))*
(((2^3)*6)^(−1/2))*
Exp[−(x^2)/4]*
HermiteH[3,x/(2^(1/2))])^2
a1=Plot[ff[x],{x,−6,6}];
Display["lily",a1]
Quit
hardcopy lily

The Hamiltonian operator of the harmonic oscillator is

$$\hat{H} = -\frac{\hbar^2}{2m} \frac{d^2}{dx^2} + \frac{1}{2} kx^2, \tag{35}$$

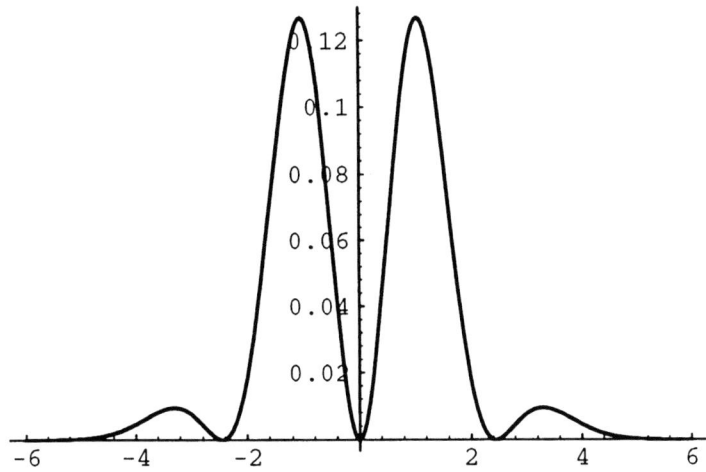

Figure 11: The plot of $u(x)U_3^2(x)$ for $\sigma = 1$.

where m is the reduced mass and k is the force constant. Hermite polynomials satisfy the differential equation (Abramowitz and Stegun (1972), p.781)

$$\frac{d^2}{dx^2}\left[e^{-x^2/2}H_n(x)\right] + (2n+1-x^2)e^{-x^2/2}H_n(x) = 0,$$

which implies

$$2\sigma^2 \frac{d^2}{dx^2}\left[e^{-x^2/(4\sigma^2)}H_n\left(\frac{x}{\sigma\sqrt{2}}\right)\right] + \tag{36}$$

$$+ \left(2n+1-\frac{x^2}{2\sigma^2}\right)e^{-x^2/(4\sigma^2)}H_n\left(\frac{x}{\sigma\sqrt{2}}\right) = 0.$$

Applying the operator (35) to the probability wave function (34) and taking (36) into account, we get

$$\hat{H}\chi(x) = -\frac{\hbar^2}{2m}\frac{d^2}{dx^2}\chi(x) + \frac{1}{2}kx^2\chi(x) = \sum_{n=1}^{N}(2^n n!)^{-1/2}c_n \times \tag{37}$$

$$\times \left[\frac{\hbar^2}{2\sigma^2 m}\left(n+\frac{1}{2}\right) - \left(\frac{\hbar^2}{8\sigma^4 m}-\frac{k}{2}\right)x^2\right](2\pi\sigma^2)^{-1/4}e^{-x^2/(4\sigma^2)}H_n\left(\frac{x}{\sigma\sqrt{2}}\right).$$

47

The Hermite polynomials satisfy the following recurrence formula (Abramowitz and Stegun (1972), p.782):

$$\xi H_n(\xi) = n H_{n-1}(\xi) + \frac{1}{2} H_{n+1}(\xi),$$

from which we get

$$\xi^2 H_n(\xi) = n\xi H_{n-1}(\xi) + \frac{1}{2}\xi H_{n+1}(\xi) = \tag{38}$$

$$= n(n-1)H_{n-2}(\xi) + \left(n + \frac{1}{2}\right)H_n(\xi) + \frac{1}{4}H_{n+2}(\xi).$$

Denote by

$$\chi_n(x) = (2^n n!)^{-1/2}(2\pi\sigma^2)^{-1/4} e^{-x^2/4\sigma^2} H_n\left(\frac{x}{\sigma\sqrt{2}}\right).$$

Obviously, $< \chi_\ell \chi_n >$ is equal to 1 if $\ell = n$ and to 0 if $\ell \neq n$. Then, (34) and (37) may be written as

$$\chi(x) = \sum_{n=1}^{N} c_n \chi_n(x), \qquad < \chi^2 > = \sum_{n=1}^{N} c_n^2, \tag{39}$$

$$\hat{H}\chi(x) = \tag{40}$$

$$= \sum_{n=1}^{N} c_n \left[\frac{\hbar^2}{2\sigma^2 m}\left(n + \frac{1}{2}\right) - \left(\frac{\hbar^2}{8\sigma^4 m} - \frac{k}{2}\right)x^2\right]\chi_n(x).$$

Applying the recurrence relation (38), we get

$$x^2 \chi_n(x) =$$

$$= (2^n n!)^{-1/2}(2\pi\sigma^2)^{-1/4} e^{-x^2/4\sigma^2} 2\sigma^2 \left(\frac{x}{\sigma\sqrt{2}}\right)^2 H_n\left(\frac{x}{\sigma\sqrt{2}}\right) =$$

$$= (2^n n!)^{-1/2}(2\pi\sigma^2)^{-1/4} e^{-x^2/4\sigma^2} 2\sigma^2 \times$$

$$\times \left[n(n-1)H_{n-2}\left(\frac{x}{\sigma\sqrt{2}}\right) + \left(n + \frac{1}{2}\right)H_n\left(\frac{x}{\sigma\sqrt{2}}\right) + \frac{1}{4}H_{n+2}\left(\frac{x}{\sigma\sqrt{2}}\right)\right] =$$

$$= (2^n n!)^{-1/2}(2^{n-2}(n-2)!)^{1/2} 2\sigma^2 n(n-1)\chi_{n-2}(x)+$$

48

$$+2\sigma^2 \left(n + \frac{1}{2}\right) \chi_n(x) + (2^n n!)^{-1/2} (2^{n+2}(n+2)!)^{1/2} 2\sigma^2 \frac{1}{4} \chi_{n+2}(x) =$$

$$= \sigma^2 \sqrt{n(n-1)} \chi_{n-2}(x) + 2\sigma^2 \left(n + \frac{1}{2}\right) \chi_n(x) +$$

$$+ \sigma^2 \sqrt{(n+1)(n+2)} \chi_{n+2}(x).$$

which implies

$$< \chi_\ell x^2 \chi_n > = \begin{cases} \sigma^2 \sqrt{n(n-1)}, & \text{if } \ell = n - 2; \\[2mm] 2\sigma^2 \left(n + \frac{1}{2}\right), & \text{if } \ell = n; \\[2mm] \sigma^2 \sqrt{(n+1)(n+2)}, & \text{if } \ell = n + 2; \\[2mm] 0, & \text{elsewhere.} \end{cases}$$

Using this equality and (40), we obtain

$$< \chi \hat{H} \chi > = \tag{41}$$

$$= \left(\frac{\hbar^2}{4\sigma^2 m} + k\sigma^2\right) \sum_{n=1}^{N} c_n^2 \left(n + \frac{1}{2}\right) - \frac{1}{2} \left(\frac{\hbar^2}{4\sigma^2 m} - k\sigma^2\right) \times$$

$$\times \left(\sum_{n=3}^{N} c_{n-2} c_n \sqrt{n(n-1)} + \sum_{n=1}^{N-2} c_n c_{n+2} \sqrt{(n+1)(n+2)}\right).$$

Introducing (41) and (39) into

$$< \hat{E} > = \frac{< \chi \hat{H} \chi >}{< \chi^2 >} \tag{42}$$

we get

$$\sum_{n=1}^{N} \left[\left(\frac{\hbar^2}{4\sigma^2 m} + k\sigma^2\right) \left(n + \frac{1}{2}\right) - < \hat{E} >\right] c_n^2 - \frac{1}{2} \left(\frac{\hbar^2}{4\sigma^2 m} - k\sigma^2\right) \times$$

$$\times \left(\sum_{n=3}^{N} c_{n-2} c_n \sqrt{n(n-1)} + \sum_{n=1}^{N-2} c_n c_{n+2} \sqrt{(n+1)(n+2)}\right) = 0. \tag{43}$$

The expected value $< \hat{E} >$ of energy depends on the probability wave function χ, i.e., it depends both on the generalized coordinates $\{H_n; n = 0, 1, \ldots\}$ and on the coefficients $\{c_n; n = 1, 2, \ldots\}$ associated to χ. The stationary values of the mean energy are obtained from the system of equations

$$\frac{\partial < \hat{E} >}{\partial c_n} = 0, \quad (n = 1, 2, \ldots, N).$$

Using these equations and taking in (43) partial derivatives with respect to each c_n, we get

$$2A_n c_n - \frac{1}{2}Bc_{n+2}\sqrt{(n+1)(n+2)} = 0, \quad (n = 1, 2);$$

$$2A_n c_n - \frac{1}{2}B(c_{n-2}\sqrt{n(n-1)} + c_{n+2}\sqrt{(n+1)(n+2)}) = 0,$$

$$(n = 3, 4, \ldots, N - 2);$$

$$2A_n c_n - \frac{1}{2}Bc_{n-2}\sqrt{n(n-1)} = 0, \quad (n = N - 1, N).$$

where

$$A_n = \left(\frac{\hbar^2}{4\sigma^2 m} + k\sigma^2\right)\left(n + \frac{1}{2}\right) - < \hat{E} >,$$

$$B = \frac{\hbar^2}{4\sigma^2 m} - k\sigma^2.$$

This homogeneous system consists of N linear equations with N unknows $c_1, \ldots, c_N$. Either $c_1 = \ldots = c_N = 0$, which corresponds to the statistical equilibrium described by the probability distribution $N(0, \sigma^2)$, or $B = 0$ and $A_n = 0$ for the values of n for which $c_n \neq 0$. But $B = 0$ implies

$$\sigma^2 = \frac{\hbar}{2\sqrt{mk}}, \tag{44}$$

which is a mathematical expression of the correspondence principle: If we neglect $\hbar$ (i.e., $\hbar \to 0$), or if the relative mass m is large (i.e., $m \to +\infty$), or if the force k is strong (i.e., $k \to +\infty$), then the variance σ^2 becomes negligible (i.e., $\sigma^2 \to 0$) and we obtain the classical harmonic oscillator whose behavior has nothing random in it and for which equilibrium means having the deviation from origin equal to zero. As for the equalities $A_n = 0$ if $c_n \neq 0$,

50

they show the possible values of the energy. Thus, if $\chi = \chi_n$, then $A_n = 0$ becomes

$$< \hat{E} >= E_n = \left(\frac{\hbar^2}{4\sigma^2 m} + k\sigma^2 \right) \left(n + \frac{1}{2} \right),$$

which, taking into account (44), becomes

$$E_n = \hbar \sqrt{\frac{k}{m}} \left(n + \frac{1}{2} \right). \tag{45}$$

Introducing (39), (41) and (44) into (42) and taking (45) into account, we get

$$< \hat{E} >= \sum_{n=1}^{N} E_n \frac{c_n^2}{\sum_{\ell=1}^{N} c_\ell^2},$$

which shows that

$$\frac{c_n^2}{\sum_{\ell=1}^{N} c_\ell^2}$$

may be interpreted as being the probability that the value of energy is E_n.

7. APPLICATION 5: THE HYDROGEN ATOM

The hydrogen atom consists of a proton fixed at the origin and an electron of reduced mass m interacting with the proton through a Coulomb potential:

$$U(r) = -\frac{e^2}{4\pi\varepsilon_0 r},$$

where e is the charge on the proton, ε_0 is the permitivity of free space, and r is the distance between the electron and the proton, namely, $r = \sqrt{x^2 + y^2 + z^2}$.

The Hamiltonian is

$$\hat{H} = -\frac{\hbar^2}{2m} \nabla^2 + U(r).$$

where ∇^2 is the Laplacian.

1. *The Ground State.* As the Coulomb potential depends only on

$$r(x, y, z) = \sqrt{x^2 + y^2 + z^2},$$

the statistical equilibrium should also depend only on $r(x, y, z)$. As the range of $r(x, y, z)$ is $[0, +\infty)$, according to Proposition 2, the most unbiased probability distribution on $[0, +\infty)$ subject to the mean value μ is the exponential distribution $\mathbf{E}(\mu)$, with the density

$$g(x, y, z) = M \frac{1}{\mu} e^{-r(x,y,z)/\mu},$$

where M is a positive constant. The model itself suggests that we use a spherical coordinate system with the proton at the origin, namely,

$$x = r \sin\theta \cos\omega, \quad y = r \sin\theta \sin\omega, \quad z = r \cos\theta.$$

Taking $s = \cos\theta$, the element of volume in the three-dimensional Euclidean space becomes

$$dx\, dy\, dz = r^2 \sin\theta\, dr\, d\theta\, d\omega = r^2 dr\, ds\, d\omega.$$

The three variables r, s, ω are independent and their ranges are

$$0 \le r < +\infty, \quad -1 \le s \le 1, \quad 0 \le \omega \le 2\pi.$$

Therefore, the statistical equilibrium of the hydrogen atom is described by the probability density function

$$g(x, y, z)\, dx\, dy\, dz = M \frac{1}{\mu} e^{-\sqrt{x^2+y^2+z^2}/\mu}\, dx\, dy\, dz =$$

$$= M \frac{1}{\mu} e^{-r/\mu} r^2\, dr\, ds\, d\omega,$$

where the positive constant M is determined from

$$1 = \int_{\mathbf{R}^3} g(x, y, z)\, dx\, dy\, dz =$$

$$= M \int_0^{+\infty} \frac{1}{\mu} r^2 e^{-r/\mu}\, dr \int_{-1}^1 ds \int_0^{2\pi} d\omega =$$

$$= M(2\mu^2)(2)(2\pi) = 8\pi\mu^2 M,$$

which implies $M = 1/(8\pi\mu^2)$, and

$$g(x, y, z) = \frac{1}{8\pi\mu^3} e^{-\sqrt{x^2+y^2+z^2}/\mu} = \frac{1}{8\pi\mu^3} e^{-r(x,y,z)/\mu}.$$

Denoting by

$$u(r) = \frac{1}{2\mu^3} r^2 e^{-r/\mu}, \qquad (0 \le r < +\infty);$$

$$v(s) = \frac{1}{2}, \quad (-1 \le s \le 1); \quad w(\omega) = \frac{1}{2\pi}, \quad (0 \le \omega \le 2\pi),$$

we can write

$$g(x, y, z)\, dx\, dy\, dz = \frac{1}{8\pi\mu^3} e^{-\sqrt{x^2+y^2+z^2}/\mu}\, dx\, dy\, dz = \tag{46}$$

$$= \frac{1}{8\pi\mu^3} e^{-r/\mu} r^2\, dr\, ds\, d\omega = u(r)v(s)w(\omega)\, dr\, ds\, d\omega.$$

Statistical equilibrium is therefore described in the space

$$[0, +\infty) \times [-1, +1] \times [0, 2\pi]$$

by the uniform probability distribution $\mathbf{U}(0, 2\pi)$ on $[0, 2\pi]$ for ω, the uniform probability distribution $\mathbf{U}(-1, +1)$ on $[-1, 1]$, for s, and the Gamma probability distribution $\mathbf{G}(\mu, 3)$ with parameters $1/\mu$ and 3, on $[0, +\infty)$, for the radial variable r. The ground probability wave function induced by the probability density function $g(x, y, z)$ is

$$\psi_0(x, y, z) = \sqrt{g(x, y, z)} = \frac{1}{\sqrt{\pi}} \left(\frac{1}{2\mu}\right)^{3/2} e^{-r(x,y,z)/(2\mu)}. \tag{47}$$

As ψ_0 depends only on the radial distance $r(x, y, z)$,

$$\nabla^2 \psi_0 = \frac{d^2\psi_0}{dr^2} + \frac{2}{r} \frac{d\psi_0}{dr} = \left(\frac{1}{4\mu^2} - \frac{1}{\mu r}\right) \psi_0. \tag{48}$$

In general, if the function g depends only on

$$r = \sqrt{\sum_{i=1}^{n} x_i^2}$$

53

in $\mathbf{R}^n$, then g is integrable in $\mathbf{R}^n$ if and only if $g(r)r^{n-1}$ is integrable on $[0, +\infty)$, and we have

$$\int_{\mathbf{R}^n} g(x_1, \ldots, x_n)dx_1 \ldots dx_n = S_n \int_0^{+\infty} g(r)r^{n-1}dr,$$

where S_n is the area of the sphere in $\mathbf{R}^n$ with radius 1, namely, $S_1 = 2, S_2 = 2\pi, S_3 = 4\pi$, etc. Consequently, using (47) and (48), we get

$$\int_{\mathbf{R}^3} \psi_0 \nabla^2 \psi_0 dx dy dz = -\frac{1}{4\mu^2},$$

and, similarly,

$$< \hat{E} >= \int_{\mathbf{R}^3} \psi_0 \hat{H} \psi_0 dx dy dz = \frac{\hbar^2}{8m\mu^2} - \frac{e^2}{8\pi\varepsilon_0\mu}. \qquad (49)$$

Using $\alpha = 1/\mu$, the mean energy becomes:

$$< \hat{E} >= \frac{\hbar^2}{8m}\alpha^2 - \frac{e^2}{8\pi\varepsilon_0}\alpha.$$

Looking for a value of α for which the mean energy $< \hat{E} >$ is stationary, we have

$$\frac{\partial < \hat{E} >}{\partial \alpha} = 0$$

if and only if

$$\alpha = \frac{e^2 m}{2\hbar^2 \pi\varepsilon_0},$$

or, equivalently, if and only if

$$\mu = \frac{2\hbar^2 \pi\varepsilon_0}{me^2} = \frac{a_0}{2},$$

where

$$a_0 = 4\pi\varepsilon_0\hbar^2/(me^2)$$

is the Bohr radius. Introducing this value of μ into (49), we get

$$< \hat{E} >= -\frac{1}{2}\left(\frac{e^2}{4\pi\varepsilon_0}\right)^2 \frac{m}{\hbar^2}.$$

The mean location of the electron, i.e., the mean distance from nucleus, is

$$\int_{\mathbf{R}^3} r(x,y,z)\, g(x,y,z)\, dx\, dy\, dz =$$

$$= \int_{[0,+\infty)\times[-1,1]\times[0,2\pi]} r \frac{1}{8\pi\mu^3} e^{-r/\mu} r^2 dr\, ds\, d\omega =$$

$$= \int_0^{+\infty} \frac{1}{8\pi\mu^3} r^3 e^{-r/\mu} dr \int_{-1}^{1} ds \int_0^{2\pi} d\omega = 3\mu = \frac{3}{2}a_0.$$

Denoting by $u(r)$ the radial probability density, we have

$$u(r) = \int_{[-1,1]\times[0,2\pi]} \frac{1}{8\pi\mu^3} e^{-r/\mu} r^2 ds d\omega =$$

$$= \frac{1}{8\pi\mu^3} r^2 e^{-r/\mu} \int_{-1}^{1} ds \int_0^{2\pi} d\omega = \frac{1}{2\mu^3} r^2 e^{-r/\mu}$$

which is the Gamma probability distribution with parameters $1/\mu$ and 3. Its mean value and variance are 3μ and $3\mu^2 = \frac{3}{4}a_0^2$, respectively. The most probable location of the electron, i.e., the most probable distance from the nucleus, is given by the root of the equation:

$$u'(r) = \frac{1}{2\mu^3} r e^{-r/\mu} \left(2 - \frac{r}{\mu}\right) = 0,$$

which gives $r = 2\mu = a_0$, i.e., the Bohr radius.

2. *Jumps from the Ground State.* As we have just seen, the probability density corresponding to statistical equilibrium is

$$g(x,y,z)\, dx\, dy\, dz = \frac{1}{8\pi\mu^3} e^{-r(x,y,z)/\mu}\, dx\, dy\, dz =$$

$$= \frac{1}{8\pi\mu^3} e^{-\sqrt{x^2+y^2+z^2}/\mu}\, dx\, dy\, dz =$$

$$= g(x(r,\theta,\omega), y(r,\theta,\omega), z(r,\theta,\omega)) r^2 \sin\theta dr d\theta d\omega =$$

$$= g(r\sin\theta\cos\omega, r\sin\theta\sin\omega, r\cos\theta) r^2 \sin\theta dr d\theta d\omega =$$

$$= \frac{1}{8\pi\mu^3} r^2 e^{-r/\mu} \sin\theta dr d\theta d\omega =$$

$$= \frac{1}{2\mu^3}r^2 e^{-r/\mu}\frac{1}{2}\frac{1}{2\pi}dr\,ds\,d\omega = u(r)v(s)w(\omega)\,dr\,ds\,d\omega.$$

According to (5),(6),(9), and (17), any probability density function q in the space

$$[0, +\infty) \times [-1, 1] \times [0, 2\pi]$$

may be written as

$$q(r, s, \omega)\,dr\,ds\,d\omega = u(r)v(s)w(\omega) \times \qquad (50)$$

$$\times \left[1 + \sum_{\substack{n=0 \\ (n,\ell,k)\neq(0,0,0)}}^{+\infty}\sum_{\ell=0}^{+\infty}\sum_{k=0}^{+\infty} c_{n\ell k}\, U_n(r)V_\ell(s)W_k(\omega)\right]dr\,ds\,d\omega,$$

where: $U_n(r)$ is the generalized Laguerre polynomial $L_n^{(2)}(r/\mu)$, $V_\ell(s)$ is the Legendre polynomial $P_\ell(s)$, and $W_k(\omega)$ is the trigonometric function $\sqrt{2}\sin\frac{k}{2}\omega$, with $L_0^{(2)} \equiv 1, P_0 \equiv 1$, and $W_0 \equiv 1$. The marginal probability density corresponding to the radial variable r is

$$\int_{-1}^{1}\int_{0}^{2\pi} q(r, s, \omega)\,ds\,d\omega =$$

$$= \frac{1}{2\mu^3}r^2 e^{-r/\mu}\left[1 + \sum_{n=1}^{+\infty}c_{n00}L_n^{(2)}\left(\frac{r}{\mu}\right)\right].$$

Let us denote by

$$f(x, y, z) = \frac{1}{8\pi\mu^3}e^{-r(x,y,z)/\mu}\left[1 + \sum_{n=1}^{+\infty}c_{n00}L_n^{(2)}\left(\frac{r(x,y,z)}{\mu}\right)\right]. \qquad (51)$$

Focusing on the radial variable r, the probability wave function induced by the deviation of the probability density $f(x, y, z)$ from the ground probability density $g(x, y, z)$ which describes the statistical equilibrium is

$$\chi(r) = \chi(r(x, y, z)) = \frac{f(x, y, z) - g(x, y, z)}{\sqrt{g(x, y, z)}} =$$

$$= \frac{1}{\sqrt{8\mu^3\pi}}e^{-r/(2\mu)}\sum_{n=1}^{+\infty}c_{n00}L_n^{(2)}(r/\mu) = \sum_{n=1}^{+\infty}c_{n00}\chi_n.$$

56

The ground probability wave function, depending on $r = r(x, y, z)$, is

$$\psi_0(r) = \frac{1}{\sqrt{8\pi\mu^3}} e^{-r/(2\mu)}$$

For $\beta = 2$ equality (10) becomes:

$$L_n^{(2)}(x) - L_{n-1}^{(2)}(x) = L_n^{(1)}(x). \tag{52}$$

The simple jump from the ground state, as a function of $r = r(x, y, z)$, is

$$\psi_1(r) = \chi_1 - \chi_0 = \frac{1}{\sqrt{8\pi\mu^3}} e^{-r/(2\mu)} [L_1^{(2)}(r/\mu) - L_0^{(2)}(r/\mu)],$$

which, taking into account (52), gives

$$\psi_1(r) = \frac{1}{\sqrt{8\pi\mu^3}} e^{-r/(2\mu)} L_1^{(1)}(r/\mu) = \psi_0(r) L_1^{(1)}(r/\mu).$$

Similarly, the simple jump from the elementary probability wave function χ_{n-1} to the next elementary probability wave function χ_n is

$$\psi_n(r) = \chi_n - \chi_{n-1} = \frac{1}{\sqrt{8\pi\mu^3}} e^{-r/(2\mu)} [L_n^{(2)}(r/\mu) - L_{n-1}^{(2)}(r/\mu)],$$

which, based on (52), as a function of $r = r(x, y, z)$, becomes

$$\psi_n(r) = \frac{1}{\sqrt{8\pi\mu^3}} e^{-r/(2\mu)} L_n^{(1)}(r/\mu) = \psi_0(r) L_n^{(1)}(r/\mu). \tag{53}$$

Computer tip: Here is a MATHEMATICA session for getting the plot of the radial wave function $\psi_3(r)$ (for $\mu = 1$), given in Figure 12, and the plot of the corresponding probability density function $r^2\psi_3^2(r)$, shown in Figure 13:

```
math
f[r_]:=((8*Pi)^(−1/2))*
Exp[−r/2]*LaguerreL[3,1,r]
c=Plot[f[r],{r,0,25}];
Display["terry1",c]
ff[r_]:=(r^2)*(f[r]^2)
d=Plot[ff[r],{r,0,25}];
```

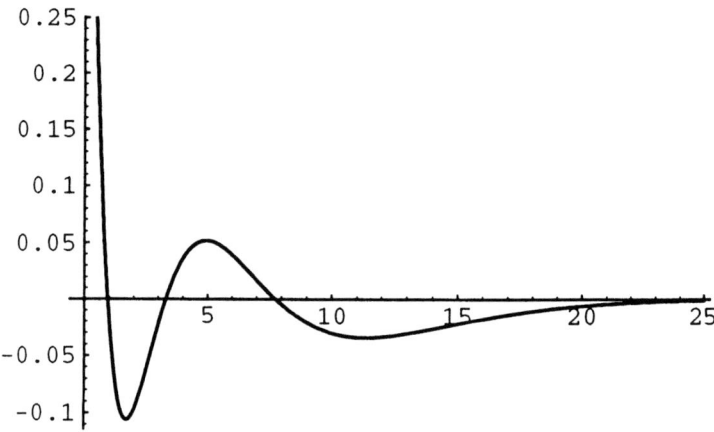

Figure 12: The plot of the radial wave function $\psi_3(r)$ for $\mu = 1$.

Display["terry2",d]
Quit
hardcopy terry1
hardcopy terry2

For $\beta = 1$ equality (11) becomes:

$$x\frac{d^2}{dx^2}L_n^{(1)}(x) + (2 - x)\frac{d}{dx}L_n^{(1)}(x) + nL_n^{(1)}(x) = 0. \tag{54}$$

From (53), we get

$$\frac{d}{dr}\psi_n(r) = \psi_0(r)\left[-\frac{1}{2\mu}L_n^{(1)}(r/\mu) + \frac{d}{dr}L_n^{(1)}(r/\mu)\right], \tag{55}$$

and, also,

$$\frac{d^2}{dr^2}\psi_n(r) =$$

$$= \psi_0(r)\left[\frac{1}{4\mu^2}L_n^{(1)}(r/\mu) - \frac{1}{\mu}\frac{d}{dr}L_n^{(1)}(r/\mu) + \frac{d^2}{dr^2}L_n^{(1)}(r/\mu)\right]. \tag{56}$$

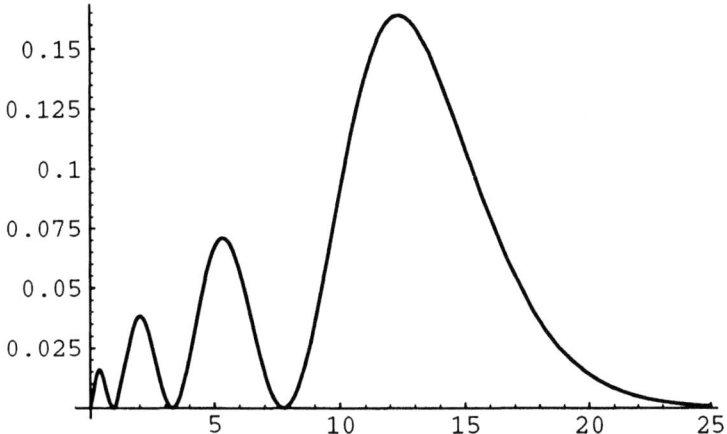

Figure 13: The plot of the radial probability density function $r^2\psi_3^2(r)$ for $\mu = 1$.

Replacing x by r/μ in (54), we get

$$\frac{r}{\mu}\frac{d^2}{d(r/\mu)^2}L_n^{(1)}(r/\mu) + \left(2 - \frac{r}{\mu}\right)\frac{d}{d(r/\mu)}L_n^{(1)}(r/\mu) + n\,L_n^{(1)}(r/\mu) = 0,$$

which, dividing by $r/\mu > 0$ and taking into account that

$$d(r/\mu) = \frac{1}{\mu}dr, \quad \text{and} \quad d(r/\mu)^2 = \frac{1}{\mu^2}dr^2,$$

becomes

$$\frac{d^2}{dr^2}L_n^{(1)}(r/\mu) + \left(\frac{2}{r} - \frac{1}{\mu}\right)\frac{d}{dr}L_n^{(1)}(r/\mu) = -\frac{n}{r\mu}L_n^{(1)}(r/\mu) \qquad (57)$$

Introducing (56) and (55) into

$$\nabla^2\psi_n = \frac{d^2\psi_n}{dr^2} + \frac{2}{r}\frac{d\psi_n}{dr},$$

59

using (57), and taking (53) into account, we get

$$\nabla^2 \psi_n = \psi_0(r) \left(\frac{1}{4\mu^2} - \frac{1}{\mu r} - \frac{n}{\mu r} \right) L_n^{(1)}(r/\mu) =$$

$$= \left(\frac{1}{4\mu^2} - \frac{n+1}{\mu r} \right) \psi_n(r).$$

Taking this result into account, the Hamiltonian

$$\hat{H} = -\frac{\hbar^2}{2m} \nabla^2 - \frac{e^2}{4\pi\varepsilon_0 r}$$

applied to the probability wave function ψ_n gives

$$\hat{H}\psi_n(r) = \left[-\frac{\hbar^2}{8m\mu^2} + \left(\frac{\hbar^2(n+1)}{2m\mu} - \frac{e^2}{4\pi\varepsilon_0} \right) \frac{1}{r} \right] \psi_n(r). \qquad (58)$$

The probability wave function ψ_n is an eigenfunction of the Hamiltonian $\hat{H}$ if the coefficient of ψ_n in (58) does not depend on $1/r$, which happens if

$$\mu = \frac{4\pi\varepsilon_0 \hbar^2 (n+1)}{2me^2}. \qquad (59)$$

Introducing (59) into (58), we get

$$\hat{H}\psi_n(r) = -\frac{me^4}{32\pi^2\varepsilon_0^2\hbar^2(n+1)^2} \psi_n(r). \qquad (60)$$

Therefore, the value of energy corresponding to ψ_n is

$$E_n = -\frac{1}{2} \left(\frac{e^2}{4\pi\varepsilon_0} \right)^2 \frac{m}{\hbar^2(n+1)^2}, \qquad (n = 0, 1, \ldots). \qquad (61)$$

Remark 1: In this section we have used the spherical coordinates (r, θ, ω) in the space

$$[0, +\infty) \times [0, \pi] \times [0, 2\pi],$$

or, equivalently, the coordinates (r, s, ω) in the space

$$[0, +\infty) \times [-1, 1] \times [0, 2\pi],$$

where $s = \cos\theta$. As shown in (46), in obtaining a probabilistic model for the behavior of the hydrogen atom we have started from a statistical equilibrium described by the Gamma distribution $\mathbf{G}(\mu, 3)$ for the radial variable r on $[0, +\infty)$, the uniform distribution for s on $[-1, 1]$, and the uniform distribution for ω on $[0, 2\pi]$. In its general form, the probability wave function has the form

$$\chi(r, s, \omega) = \tag{62}$$

$$= \psi_0(r) \ \sum_{n=0}^{+\infty} \sum_{\ell=0}^{+\infty} \sum_{k=0}^{+\infty} \ c_{n\ell k} \, U_n(r) V_\ell(s) W_k(\omega),$$
$$\scriptstyle (n,\ell,k) \neq (0,0,0)$$

where: $\psi_0(r)$ is the probability wave function (47) of the ground state, $U_n(r)$ is the generalized Laguerre polynomial $L_n^{(2)}(r/\mu)$, $V_\ell(s)$ is the Legendre polynomial $P_\ell(s)$, and $W_k(\omega)$ is the trigonometric function $\sqrt{2} \sin \frac{k}{2}\omega$, as discussed in Section 2 of this part. In all the considerations made in this section we focused on the radial variable r and, instead of (62), we dealt with the simpler probability wave function

$$\chi(r) = \psi_0(r) \sum_{n=1}^{+\infty} c_{n00} \, U_n(r) V_0(s) W_0(\omega),$$

where $V_0 \equiv 1, W_0 \equiv 1$. In the general context, the jumps (53) should be replaced by

$$\psi_{mjq,n\ell k}(r, s, \omega) =$$
$$= \psi_0(r)[U_n(r)V_\ell(s)W_k(\omega) - U_m(r)V_j(s)W_q(\omega)].$$

Remark 2: The probabilistic model for the hydrogen atom, just discussed, started from the statistical equilibrium described by the exponential distribution

$$g(x, y, z) = M \frac{1}{\mu} e^{-r(x,y,z)/\mu},$$

which is the solution of the maximum entropy principle

$$\max_g \mathbf{H}(g),$$

subject to the mean value μ or, equivalently, the solution of the principle of minimum relative entropy

$$\min_g \mathbf{H}(g \mid 1)$$

on $[0, +\infty)$, where $1 = r^0(x, y, z)$, in which case the radial probability density on $0 \le r < +\infty$ is

$$u(r) = \frac{1}{2\mu^3} r^2 e^{-r/\mu},$$

i.e., the Gamma probability distribution $\mathbf{G}(\mu, 3)$, and the system of orthogonal polynomials with the weight u is the sequence of generalized Laguerre polynomials

$$\{L_n^{(2)}(r/\mu); \; n = 0, 1, \ldots\}.$$

This proved to be enough for getting the entire energy spectrum (61) of the hydrogen atom. The approach can be generalized, if we start from the statistical equilibrium described by the solution of the principle of minimum relative entropy

$$\min_g \mathbf{H}(g \mid r^{2\ell}(x, y, z)),$$

which is the Gamma probability distribution $\mathbf{G}(\mu, 2\ell + 1)$, whose density is

$$g(x, y, z) = \frac{1}{\Gamma(2\ell + 1)} \left(\frac{1}{\mu}\right)^{2\ell+1} r^{2\ell}(x, y, z) e^{-r(x,y,z)/\mu},$$

in which case the radial probability density on $0 \le r < +\infty$ is

$$u(r) = M \left(\frac{1}{\mu}\right)^{2\ell+3} r^{2\ell+2} e^{-r/\mu}, \qquad (M \text{ constant}),$$

i.e., the Gamma distribution $\mathbf{G}(\mu, 2\ell + 3)$, and the system of orthogonal polynomials with the weight u is the sequence of generalized Laguerre polynomials:

$$\{L_n^{(2\ell+2)}(r/\mu); \; n = 0, 1, \ldots\}.$$

The simple jumps are in this case described by the probability wave functions

$$\psi_n(r) = \sqrt{u(r)}[L_n^{(2\ell+2)}(r/\mu) - L_{n-1}^{(2\ell+2)}(r/\mu)] =$$

$$= \sqrt{u(r)} L_n^{(2\ell+1)}(r/\mu).$$

The hydrogen atom is the only atom for which the corresponding Schrödinger equation may be solved exactly. For more complex atoms approximative methods have been used. There is no indication that the shell structure of the atoms consisting of more than two electrons could be obtained from the corresponding Schrödinger equations.

62

8. APPLICATION 6: THE GROUND STATE OF THE HELIUM ATOM

As it is not possible to solve the Schrödinger equation exactly for any atom or molecule more complicated than the hydrogen atom, approximate methods, as perturbation theory and the variational method, are used instead. For helium atom, for instance, the corresponding Schrödinger equation cannot be solved exactly.

The helium atom has a nucleus and two electrons. Fixing the nucleus at the origin of a three-dimensional Euclidean coordinate system (x, y, z), the Hamiltonian operator has the form

$$\hat{H} = -\frac{\hbar^2}{2m}\left(\nabla_1^2 + \nabla_2^2\right) - \frac{2e^2}{4\pi\varepsilon_0}\left(\frac{1}{r_1} + \frac{1}{r_2}\right) + \frac{e^2}{4\pi\varepsilon_0}\frac{1}{r_{12}},$$

where

$$\nabla_i^2 = \frac{\partial^2}{\partial x_i^2} + \frac{\partial^2}{\partial y_i^2} + \frac{\partial^2}{\partial z_i^2}$$

is the Laplacian operator for electron $i, (i = 1, 2)$, r_i is the radial distance from the nucleus (the origin) to electron i, r_{12} is the distance between the two electrons, m is the electronic mass, and $\hbar, e, \varepsilon_0$ are constants. To be more specific,

$$m = 9.1091 \times 10^{-31} \ kg. \ [\text{electron (rest) mass}];$$

$$\hbar = 1.05459 \times 10^{-34} \ J.s. \ [\text{Planck's constant}];$$

$$e = 1.60219 \times 10^{-19} \ C \ [\text{proton charge}];$$

$$\varepsilon_0 = 8.854188 \times 10^{-12} \ C^2.s^2.kg^{-1}.m^{-3} \ [\text{permittivity of a vacuum}].$$

Using atomic units, i.e., taking

$$\hbar = 1, \quad m = 1, \quad e = 1, \quad 4\pi\varepsilon_0 = 1,$$

the Hamiltonian becomes

$$\hat{H} = -\frac{1}{2}\left(\nabla_1^2 + \nabla_2^2\right) - \left(\frac{2}{r_1} + \frac{2}{r_2}\right) + \frac{1}{r_{12}}, \tag{63}$$

For such a quantum system the variational theorem (see Appendix 4) provides an upper bound to its ground state energy. The ground state wave function ψ_0 and energy E_0 satisfy

$$\hat{H}\psi_0 = E_0\psi_0,$$

where $\hat{H}$ is the Hamiltonian operator of the system. According to the variational theorem (see Appendix 4), for any other function ψ we have $E_\psi \geq E_0$, with

$$E_\psi = \frac{<\psi^*\hat{H}\psi>}{<\psi^*\psi>},$$

where ψ^* is the complex conjugate of ψ. The trial function ψ is chosen such that it depends on some arbitrary parameters $\alpha, \beta, \gamma, \ldots$ in which case E_ψ also will depend on these variational parameters,

$$E_\psi(\alpha, \beta, \gamma, \ldots) \geq E_0.$$

The variational parameters are determined in order to minimize

$$E_\psi(\alpha, \beta, \gamma, \ldots).$$

There is no unique or optimal way of selecting the trial function ψ in the present literature. The first trial function of interest was introduced long ago by E.A. Hylleraas (1929), namely,

$$\psi(r_1, r_2, r_{12}) = e^{-\alpha r_1} e^{-\alpha r_2}[1 + P(r_1, r_2, r_{12})], \tag{64}$$

where P was a polynomial in r_1, r_2, r_{12} whose coefficients, together with α, were taken as variational parameters. No theoretical justification of this choice has been given but it proved to give a value of the ground energy of the helium atom in a very accurate agreement with the experimental value. Reminiscences from early quantum mechanics of two-electron atoms may be found in the very well-written paper Hylleraas (1963). Since 1929, many generalizations of Hylleraas' approach have been proposed but again, without a sound theoretical justification. The trial functions of the form (64) proposed in the literature contained more and more variational parameters in the expression of the polynomial $P(r_1, r_2, r_{12})$ namely, 39 in Kinoshita (1957) and even 1078 in Pekeris (1959). Details may be found in McQuarrie (1983, p.291).

In his paper containing reminiscences from early quantum mechanics of two-electron atoms, Hylleraas (1963, p.425) gives the following motivation of his taking a trial function of the form (64). "The Bohr theory, applying ... a model of atom with its two electrons in strictly opposite positions with respect to the nucleus, led to a numerical value of about 28 eV for the ionization energy of the first electron. On the other hand, a simple perturbation treatment of the Schrödinger equation, as given by Unsöld, led to a much lower value of about 20.3 eV. The true value of 24.46 eV, as known from spectroscopic measurement, was about in the middle in between. Hence, there was a broad gap of about 4 eV to be filled up. The reason for the bad results is easily seen. In the Bohr picture, with the electrons held strictly at the largest possible mutual distance, the interelectronic energy came out too low. In the wave mechanics picture with independent spherical electronic charge distributions, the interelectronic energy would be unreasonably high. About half of the gap might have been filled at once, by letting the two electronic distributions expand to a reasonable degree such that interelectronic repulsion and nuclear attraction would be more balanced. This means a minimization of the total energy ... Finally, if we would think of displacing the electronic charge distributions relative to each other, a bit to either side of the nucleus, remembering the Bohr picture, we would have what is called a polarization effect or correlation energy." What clearly transpires from the above text is the need for a probabilistic model of correlated electrons replacing both the strictly deterministic model of rigidly connected electrons and the too liberal model of randomly but independent electrons. But back in 1929, when Hylleraas introduced his trial function (64), no probabilistic justification was mentioned.

The objective of this section is to show that the general formalism discussed in this paper may be applied for approximating the ground state energy of the helium atom. The case of one electron was already discussed in Section 7. For the helium atom, we deal with a system of two interdependent electrons with random behavior. Switching from the Cartesian coordinates $(x_1, y_1, z_1, x_2, y_2, z_2)$ to the spherical coordinate system

$$x_i = r_i \sin \theta_i \cos \omega_i,$$

$$y_i = r_i \sin \theta_i \sin \omega_i,$$

$$z_i = r_i \cos \theta_i,$$

$$0 \leq r_i < +\infty, \quad 0 \leq \theta_i \leq \pi,$$

$$0 \leq \omega_i \leq 2\pi, \quad (i = 1, 2),$$

and eventually changing to variables

$$s_i = \cos\theta_i, \quad -1 \leq s_i \leq 1, \quad (i = 1, 2),$$

the joint probability density of the system formed by the two electrons of the helium atom is

$$g(x_1, y_1, z_1; x_2, y_2, z_2)\, dx_1\, dy_1\, dz_1\, dx_2\, dy_2\, dz_2 =$$

$$= f(r_1, s_1, \omega_1; r_2, s_2, \omega_2) r_1^2\, r_2^2\, dr_1\, ds_1\, d\omega_1\, dr_2\, ds_2\, d\omega_2 =$$

$$= f_1(r_1, s_1, \omega_1)\, f_2(r_2, s_2, \omega_2) \times \tag{65}$$

$$\times [1 + \overset{*}{\sum} c_{n\ell k, mjq} U_n(r_1) V_\ell(s_1) W_k(\omega_1) U_m(r_2) V_j(s_2) W_q(\omega_2)] \times$$

$$\times r_1^2 r_2^2 dr_1 ds_1 d\omega_1 dr_2 ds_2 d\omega_2,$$

where

$$\overset{*}{\sum} = \sum_{(n,\ell,k,m,j,q)\neq(0,0,0,0,0,0)}$$

and

$$f_i(r_i, s_i, \omega_i) = M h(r_i) v(s_i) w(\omega_i), \quad (i = 1, 2),$$

where:

$$h(r) = \frac{1}{\mu} e^{-r/\mu}$$

is the radial exponential probability density with parameter μ on $[0, +\infty)$;

$$v(s) = \frac{1}{2}$$

is the uniform probability density on $[-1, 1]$;

$$w(\omega) = \frac{1}{2\pi}$$

is the uniform probability density on $[0, 2\pi]$. Therefore,

$$f_i(r_i, s_i, \omega_i) = f_i(r_i) = M \frac{1}{4\pi\mu} e^{-r_i/\mu},$$

where M is a constant which may be found using the useful formula

$$\int_0^{+\infty} r^n e^{-r/\mu} dr = n!\, \mu^{n+1}, \qquad (66)$$

from the condition

$$1 = < f_i > = \int_0^{+\infty} \int_{-1}^{+1} \int_0^{2\pi} f_i(r_i, s_i, \omega_i) r_i^2 dr_i ds_i d\omega_i = 2M\mu^2,$$

implying $M = 1/(2\mu^2)$, which gives

$$f_i(r_i, s_i, \omega_i) = f_i(r_i) = \frac{1}{8\pi\mu^3} e^{-r_i/\mu}, \qquad (i = 1, 2). \qquad (67)$$

Thus,

$$u(r) = Mr^2 h(r) = \frac{1}{2\mu^2} r^2 h(r) = \frac{1}{2\mu^3} r^2 e^{-r/\mu}$$

is the radial Gamma probability distribution with parameters μ and 3 on $[0, +\infty)$. Also,

$$U_n(r) = L_n^{(2)}(r/\mu)$$

is the second order Laguerre polynomial of degree n, with $U_0 \equiv 1$;

$$V_\ell(s) = (2\ell + 1)^{1/2} P_\ell(s)$$

is the Legendre polynomial of degree ℓ, with $V_0 \equiv 1$;

$$W_k(\omega) = \sqrt{2} \sin \frac{k}{2}\omega,$$

is the trigonometric function, with $W_0 \equiv 1$.

The corresponding joint and marginal probability wave functions are

$$\psi = \sqrt{f}, \quad \psi_i = \sqrt{f_i}, \quad (i = 1, 2),$$

respectively. Thus,

$$\psi_i(r_i, s_i, \omega_i) = \psi_i(r_i) = \frac{1}{\sqrt{8\pi\mu^3}} e^{-r_i/(2\mu)},$$

depending only on r_i. Using the approximation

$$\sqrt{1+t} \approx 1 + \frac{1}{2}t,$$

which gives good results for small values of t, an approximation of the joint probability wave function of the system of two electrons of the helium atom is

$$\psi(r_1, s_1, \omega_1; r_2, s_2, \omega_2) = \psi_1(r_1, s_1, \omega_1)\, \psi_2(r_2, s_2, \omega_2) \times \qquad (68)$$

$$\times [1 + \sum^{*} \frac{1}{2} c_{n\ell k, mjq} U_n(r_1) V_\ell(s_1) W_k(\omega_1) U_m(r_2) V_j(s_2) W_q(\omega_2)],$$

where

$$\sum^{*} = \sum_{(n,\ell,k,m,j,q) \neq (0,0,0,0,0,0)},$$

which gives good results when the generalized correlation coefficients $c_{n\ell k, mjq}$ have relatively small values. Let us take now some simple joint probability wave functions of type (68) into account and see how accurately they can describe the ground state of the helium atom.

1) *The ground energy using the first-order radial generalized correlation.* We assume in this subsection that the two electrons are linearly dependent in terms of the radial variable r and denote by $c = \frac{1}{2}c_{100,100}$ assuming that all the other generalized correlation coefficients $c_{n\ell k, mjq}$ are equal to zero. The joint probability wave function (68) becomes

$$\psi(r_1, r_2) = \psi_1(r_1)\psi_2(r_2)[1 + cL_1^{(2)}(r_1/\mu)L_1^{(2)}(r_2/\mu)] = \qquad (69)$$

$$= \psi_1(r_1)\psi_2(r_2)\left[1 + c\left(3 - \frac{r_1}{\mu}\right)\left(3 - \frac{r_2}{\mu}\right)\right] =$$

$$= \frac{1}{8\pi\mu^3}e^{-(r_1+r_2)/(2\mu)}\left[1 + c\left(3 - \frac{r_1}{\mu}\right)\left(3 - \frac{r_2}{\mu}\right)\right].$$

As ψ_1 depends only on the radial variable r_1, the corresponding Laplacian has the form

$$\nabla_1^2\psi_1 = \frac{d^2\psi_1}{dr_1^2} + \frac{2}{r_1}\frac{d\psi_1}{dr_1} = \left(\frac{1}{4\mu^2} - \frac{1}{\mu r_1}\right)\psi_1,$$

and from (69) we get

$$\nabla_1^2 \psi = \frac{d^2\psi}{dr_1^2} + \frac{2}{r_1}\frac{d\psi}{dr_1} =$$

$$= \left[\left(\frac{1}{4\mu^2} - \frac{1}{\mu r_1} \right) + c \left(\frac{33}{4\mu^2} - \frac{15}{\mu r_1} - \frac{3r_1 + 11r_2}{4\mu^3} + \frac{5r_2}{\mu^2 r_1} + \frac{r_1 r_2}{4\mu^4} \right) \right] \times$$

$$\times \psi_1(r_1)\psi_2(r_2).$$

Using (66) several times, in particular the integrals

$$\int_0^{+\infty} \frac{1}{r_i}\psi_i^2(r_i)r_i^2 dr_i = \frac{1}{8\pi\mu}, \quad (i = 1,2),$$

$$\int_0^{+\infty} r_i\psi_i^2(r_i)r_i^2 dr_i = \frac{3\mu}{4\pi}, \quad (i = 1,2),$$

$$\int_0^{+\infty} r_i^2\psi_i^2(r_i)r_i^2 dr_i = \frac{3\mu^2}{\pi}, \quad (i = 1,2),$$

and taking into account the expression (69), we obtain

$$< \psi \nabla_1^2 \psi >=$$

$$= \int_0^{+\infty} \int_{-1}^{+1} \int_0^{2\pi} \int_0^{+\infty} \int_{-1}^{+1} \int_0^{2\pi} \psi \nabla_1^2 \psi \, r_1^2 \, r_2^2 \, dr_1 ds_1 d\omega_1 \, dr_2 ds_2 d\omega_2 =$$

$$= -\frac{1}{4\mu^2}(1 + 21\,c^2). \tag{70}$$

Similarly, as the wave function ψ is symmetric in r_1 and r_2, we get

$$< \psi \nabla_2^2 \psi >= -\frac{1}{4\mu^2}(1 + 21\,c^2). \tag{71}$$

From (69), using again (66) several times, we obtain

$$< \psi \frac{1}{r_i} \psi >=$$

$$= \int_0^{+\infty} \int_{-1}^{+1} \int_0^{2\pi} \int_0^{+\infty} \int_{-1}^{+1} \int_0^{2\pi} \frac{1}{r_i}\psi^2 \, r_1^2 \, r_2^2 \, dr_1 ds_1 d\omega_1 \, dr_2 ds_2 d\omega_2 =$$

$$= \frac{1}{2\mu}(1 + 9\,c^2), \quad (i = 1,2). \tag{72}$$

69

The following series representation of $1/r_{12}$ in terms of Legendre polynomials is given without proof in McQuarrie (1983, p.340):

$$\frac{1}{r_{12}} = \sum_{\ell=0}^{+\infty} \frac{r_<^\ell}{r_>^{\ell+1}} P_\ell(s_1) \, P_\ell(s_2), \tag{73}$$

where

$$r_< = \min\{r_1, r_2\}, \qquad r_> = \max\{r_1, r_2\}.$$

Let θ be the angle between r_1 and r_2. If we choose one of the radius vectors, say r_1, to be the z-axis, then $\theta_1 = 0$ and $\theta_2 = \theta$. Using the law of cosines,

$$r_{12} = (r_1^2 + r_2^2 - 2r_1 r_2 \cos\theta)^{1/2},$$

and we have

$$\int_0^\pi \frac{1}{r_{12}} \sin\theta d\theta = \int_0^\pi \frac{\sin\theta d\theta}{(r_1^2 + r_2^2 - 2r_1 r_2 \cos\theta)^{1/2}} =$$

$$= \int_{-1}^1 \frac{dx}{(r_1^2 + r_2^2 - 2r_1 r_2 x)^{1/2}} = \begin{cases} 2/r_1, & \text{if } r_1 > r_2; \\ 2/r_2, & \text{if } r_1 < r_2, \end{cases}$$

As $P_\ell(\cos 0) = P_\ell(1) = 1$, taking $\theta_1 = 0$ and $\theta_2 = \theta$, the equality (73) becomes

$$\frac{1}{r_{12}} = \sum_{\ell=0}^{+\infty} \frac{r_<^\ell}{r_>^{\ell+1}} P_\ell(\cos\theta),$$

which means

$$\frac{1}{r_{12}} = \frac{1}{r_2} \sum_{\ell=0}^{+\infty} \left(\frac{r_1}{r_2}\right)^\ell P_\ell(\cos\theta), \qquad \text{if } r_1 < r_2;$$

$$\frac{1}{r_{12}} = \frac{1}{r_1} \sum_{\ell=0}^{+\infty} \left(\frac{r_2}{r_1}\right)^\ell P_\ell(\cos\theta), \qquad \text{if } r_1 > r_2.$$

The first terms of this series are:

$$\frac{1}{r_{12}} = \frac{1}{r_>} + \frac{r_<}{r_>^2} \cos\theta + \ldots$$

70

The above expansion of $1/r_{12}$ into spherical harmonics of the angle θ between the position vectors r_1 and r_2 of the electrons is mentioned without any proof in Bethe and Salpeter (1957, p.221, and p.224), Hylleraas (1963, p.427), Anderson (1971, p.381), and Flügge, (1974, vol.2, p.60). As this formula and (73) play an important role in what follows, a proof of them is given in Appendix 5.

Using (69),(73), the orthogonality of Legendre polynomials, and the useful formulas

$$\int_a^b x e^{\alpha x} dx = \left(\frac{1}{\alpha^2} + \frac{a}{\alpha} \right) e^{-\alpha a} - \left(\frac{1}{\alpha^2} + \frac{b}{\alpha} \right) e^{-\alpha b};$$

$$\int_a^{+\infty} x e^{-\alpha x} dx = \left(\frac{1}{\alpha^2} + \frac{a}{\alpha} \right) e^{-\alpha a};$$

$$\int_a^b x^2 e^{-\alpha x} dx = \left(\frac{2}{\alpha^3} + \frac{2a}{\alpha^2} + \frac{a^2}{\alpha} \right) e^{-\alpha a} - \left(\frac{2}{\alpha^3} + \frac{2b}{\alpha^2} + \frac{b^2}{\alpha} \right) e^{-\alpha b};$$

$$\int_0^b x^2 e^{-\alpha x} dx = \frac{2}{\alpha^3} - \left(\frac{2}{\alpha^3} + \frac{2b}{\alpha^2} + \frac{b^2}{\alpha} \right) e^{-\alpha b};$$

$$\int_a^{+\infty} x^2 e^{-\alpha x} dx = \left(\frac{2}{\alpha^3} + \frac{2a}{\alpha^2} + \frac{a^2}{\alpha} \right) e^{-\alpha a};$$

$$\int_0^b x^3 e^{-\alpha x} dx = \frac{6}{\alpha^4} - \left(\frac{6}{\alpha^4} + \frac{6b}{\alpha^3} + \frac{3b^2}{\alpha^2} + \frac{b^3}{\alpha} \right) e^{-\alpha b};$$

$$\int_a^{+\infty} x^3 e^{-\alpha x} dx = \left(\frac{6}{\alpha^4} + \frac{6a}{\alpha^3} + \frac{3a^2}{\alpha^2} + \frac{a^3}{\alpha} \right) e^{-\alpha a};$$

$$\int_0^b x^4 e^{-\alpha x} dx = \frac{24}{\alpha^5} - \left(\frac{24}{\alpha^5} + \frac{24b}{\alpha^4} + \frac{12b^2}{\alpha^3} + \frac{4b^3}{\alpha^2} + \frac{b^4}{\alpha} \right) e^{-\alpha b},$$

we get

$$< \psi \frac{1}{r_{12}} \psi > = \tag{74}$$

$$= \int_0^{+\infty} \int_{-1}^{+1} \int_0^{2\pi} \int_0^{+\infty} \int_{-1}^{+1} \int_0^{2\pi} \psi^2(r_1, r_2) \times$$

$$\times \sum_{\ell=0}^{+\infty} \frac{r_<^\ell}{r_>^{\ell+1}} P_\ell(s_1) P_\ell(s_2) r_1^2 r_2^2 dr_1 ds_1 d\omega_1 dr_2 ds_2 d\omega_2 =$$

$$= 4\pi^2 \int_0^{+\infty} \int_0^{+\infty} \psi^2(r_1, r_2) \frac{2}{r_>} r_1^2 r_2^2 dr_1 dr_2 =$$

$$= 8\pi^2 \int_0^{+\infty} \left[\frac{1}{r_1} \int_0^{r_1} \psi^2(r_1, r_1) r_2^2 dr_2 + \int_{r_1}^{+\infty} \psi^2(r_1, r_2) r_2 dr_2 \right] r_1^2 dr_1 =$$

$$= \frac{1}{32\mu}(10 + 12c + 63c^2).$$

From (69), applying again (66), we get

$$< \psi\, \psi > = \tag{75}$$

$$= \int_0^{+\infty} \int_{-1}^{+1} \int_0^{2\pi} \int_0^{+\infty} \int_{-1}^{+1} \int_0^{2\pi} \psi^2(r_1, r_2) r_1^2 r_2^2 dr_1 ds_1 d\omega_1 dr_2 ds_2 d\omega_2 =$$

$$= 1 + 9c^2.$$

Taking (63) into account, the mean energy is

$$< \hat{E} > = \frac{< \psi\, \hat{H}\psi >}{< \psi\, \psi >} = \tag{76}$$

$$= \left[-\frac{1}{2}(< \psi \nabla_1^2 \psi > + < \psi \nabla_1^2 \psi >) - \right.$$

$$\left. -2\,(< \psi \frac{1}{r_1}\psi > + < \psi \frac{1}{r_2}\psi >) + < \psi \frac{1}{r_{12}}\psi > \right] / < \psi\, \psi >,$$

and from (70)-(72), (74), and (75), we obtain

$$< \hat{E} > = \frac{1 + 21c^2}{4\mu^2(1 + 9c^2)} - \frac{54 - 12c + 513c^2}{32\mu(1 + 9c^2)}. \tag{77}$$

Introducing $\alpha = 1/\mu$, the mean energy $< \hat{E} >$ becomes a rational function depending on two variational parameters α and c. Looking for stationary points of $< \hat{E} >$, the system of equations

$$\frac{\partial < \hat{E} >}{\partial \alpha} = 0, \qquad \frac{\partial < \hat{E} >}{\partial c} = 0,$$

gives

$$\alpha = \frac{54 - 12c + 513c^2}{16(1 + 21c^2)} = \frac{-6 + 27c + 54c^2}{96c}. \tag{78}$$

The second equality implies

$$378c^4 - 837c^3 - 99c - 2 = 0.$$

It is easy to approximate the real roots of this polynomial in c. For instance, using the package MATHEMATICA, Version 2.1 (Wolfram (1991)), the necessary commands are:

```
math
f[c_]:=378*c^4-837*c^3-99*c-2
NSolve[f[c]==0,c]
Quit
```

The output gives the two real roots

$$c = -0.0201324043753751, \qquad c = 2.26575777385865,$$

the other two roots being complex numbers. The first value,

$$c = -0.020132404,$$

gives the minimum value of $< \hat{E} >$. From (78) we obtain $\alpha = 3.37437$, and therefore

$$\mu = 1/\alpha = 0.296351 \text{ bohr},$$

and the minimum value of the mean energy (77) is

$$\min < \hat{E} >= -2.860394 \text{ a.u.} = -77.8359 \text{ eV}.$$

Computer tip: Here is a MATHEMATICA session for getting:
(a) the plot (Figure 14), the contour plot (Figure 15), and the density plot (Figure 16) of the radial wave function $\psi(r_1, r_2)$ given by (69) with

$$\mu = 0.296351, \quad c = -0.0201324043753751;$$

(b) the plot (Figure 17), the contour plot (Figure 18), and the density plot (Figure 19) of the radial joint probability density function

$$r_1^2 r_2^2 \psi^2(r_1, r_2).$$

```
math
g[r1_,r2_]:=(1/(8*Pi*0.296351^3))*
Exp[-(r1+r2)/(2*0.296351)]*
(1-0.0201324043753751*(3-(r1/0.296351))
*(3-(r2/0.296351)))
a1=Plot3D[g[r1,r2],{r1,0,1},{r2,0,1},PlotPoints->30];
Display["john1",a1]
a2=ContourPlot[g[r1,r2],{r1,0,1},{r2,0,1}];
Display["john2",a2]
a3=DensityPlot[g[r1,r2],{r1,0,1},{r2,0,1},Mesh->False];
Display["john3",a3]
gg[r1_,r2_]:=(r1^2)*
(r2^2)*(g[r1,r2]^2)
b1=Plot3D[gg[r1,r2],{r1,0,2},{r2,0,2},PlotPoints->30];
Display["mary1",b1]
b2=ContourPlot[gg[r1,r2],{r1,0,2},{r2,0,2}];
Display["mary2",b2]
b3=DensityPlot[gg[r1,r2],{r1,0,2},{r2,0,2},Mesh->False];
Display["mary3",b3]
Quit
hardcopy john1
hardcopy john2
hardcopy john3
hardcopy mary1
hardcopy mary2
hardcopy mary3
```

Remark 1: From experimental point of view, there is a variety of results mentioned for the true value of the double ionization potential of the helium atom, i.e., the energy value needed to dissociate both electrons from the helium nucleus. Thus, the following experimental values were mentioned: -78.6 eV (Messiah (1964)), -78.62 eV (Anderson (1971)), -78.8882 eV and -78.8932 eV (Harnwell and Livingood (1961)), -78.9832 eV (Born (1969)), -78.99673525 eV (Dean (1992)), -79.0 eV (Striganov and Sventitskii (1968), Fano (1969), Böhm (1979)), -79.0052 eV (Lide (2000)). Some experimental physicists (Fano (1969), Böhm (1979)) consider the value -79.0 eV as being

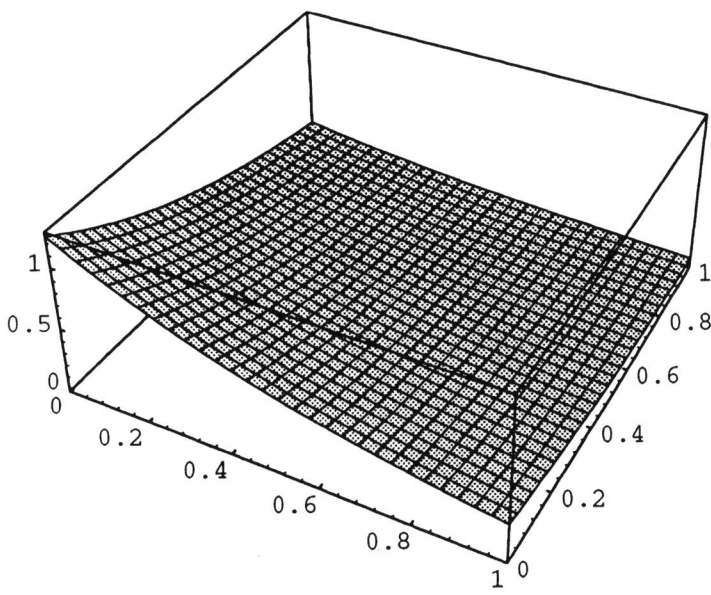

Figure 14: The plot of the radial wave function $\psi(r_1, r_2)$.

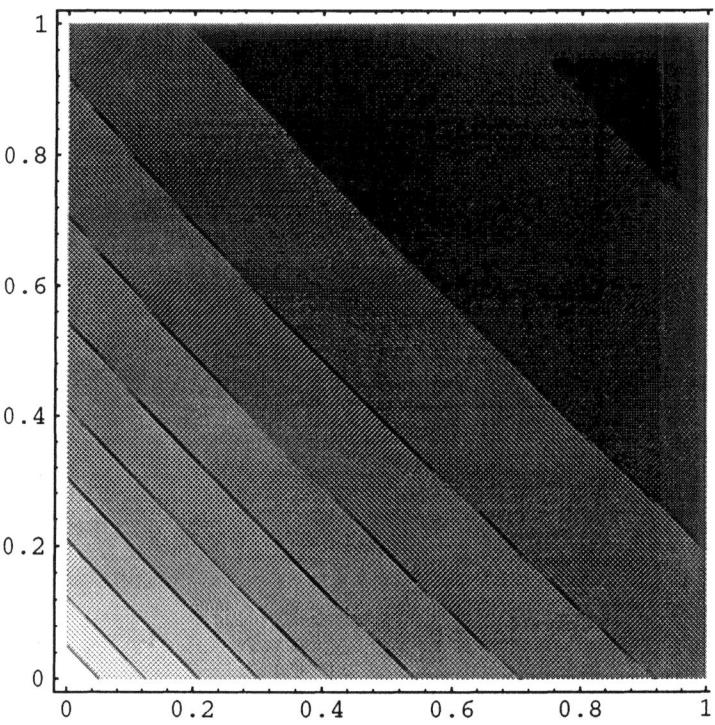

Figure 15: The contour plot of the radial wave function $\psi(r_1, r_2)$.

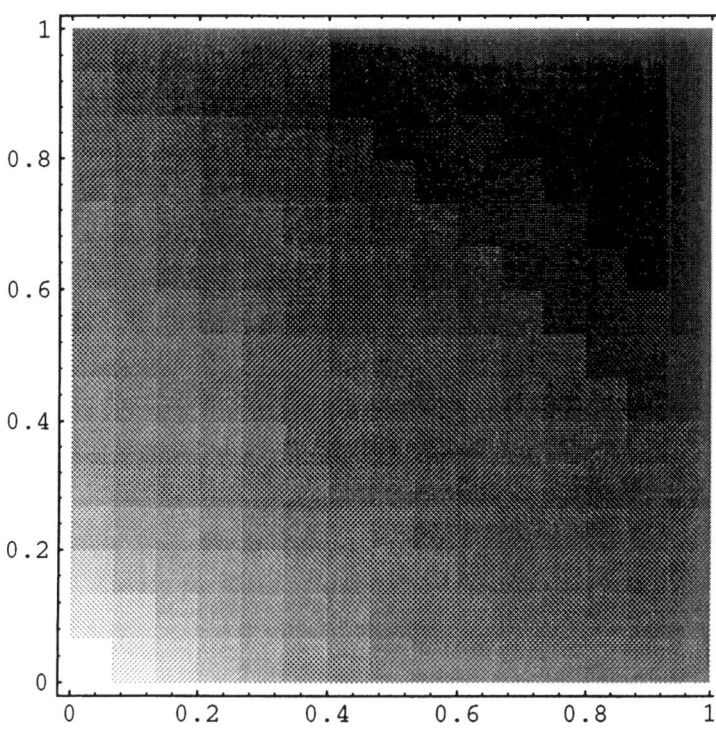

Figure 16: The density plot of the radial wave function $\psi(r_1, r_2)$.

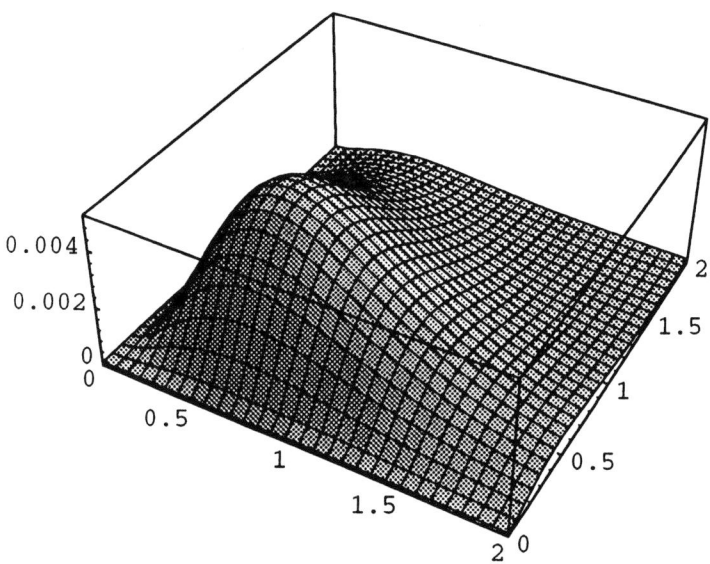

Figure 17: The plot of the radial joint probability density function $r_1^2 r_2^2 \psi^2(r_1, r_2)$.

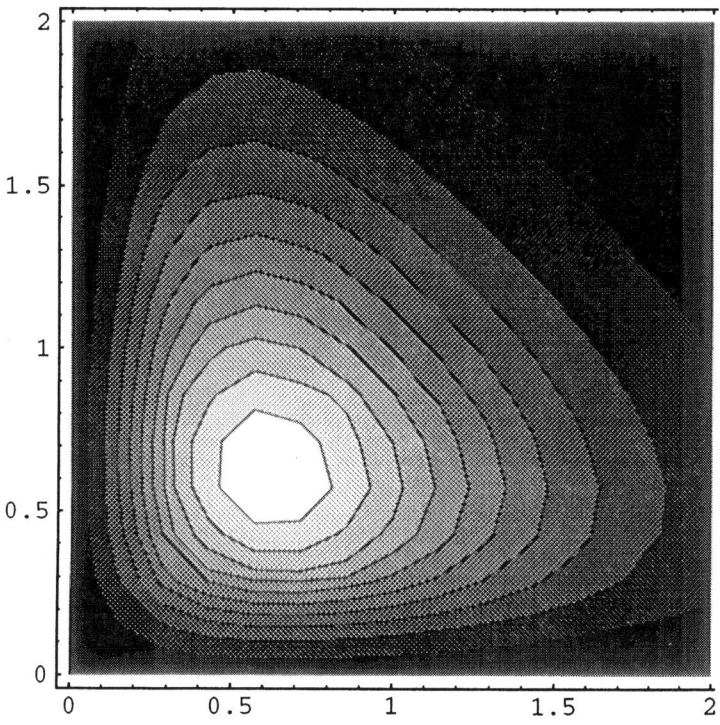

Figure 18: The contour plot of the radial joint probability density function $r_1^2 r_2^2 \psi^2(r_1, r_2)$.

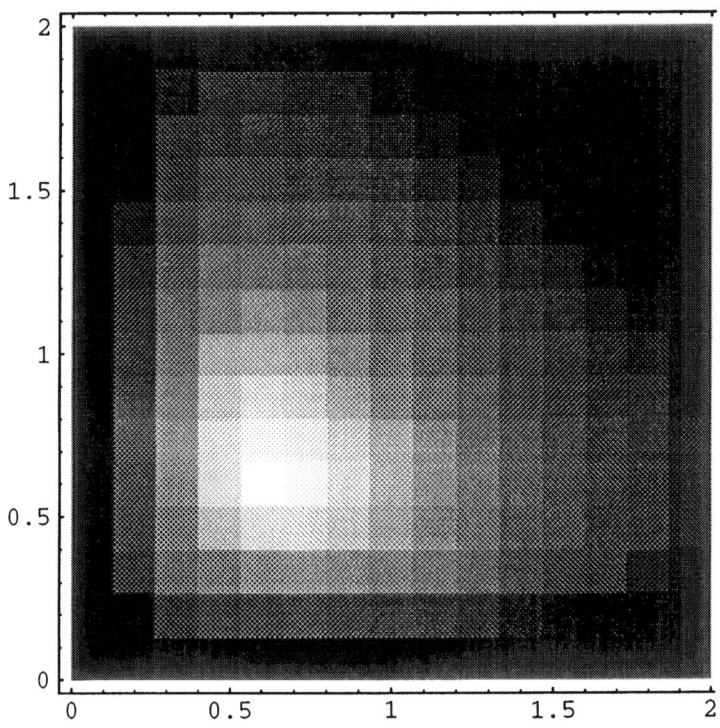

Figure 19: The density plot of the radial joint probability density function $r_1^2 r_2^2 \psi^2(r_1, r_2)$.

a lower bound, or a *threshold* for the double ionization potential of the helium atom.

Remark 2: If the two electrons of the helium atom are assumed to be statistically independent, which is equivalent to taking $c = 0$, the probability wave function (69) becomes

$$\psi(r_1, r_2) = \psi_1(r_1)\,\psi(r_2) = \frac{1}{\sqrt{8\pi\mu^3}}e^{-r_1/(2\mu)}\frac{1}{\sqrt{8\pi\mu^3}}e^{-r_2/(2\mu)}.$$

Going through the same steps as above, we obtain

$$< \psi\,\nabla_i^2\,\psi >= -\frac{1}{4\mu^2}, \quad < \psi\frac{1}{r_i}\psi >= \frac{1}{2\mu}, \quad (i = 1, 2),$$

$$< \psi\frac{1}{r_{12}}\psi >= \frac{5}{16\mu},$$

and, introducing these expressions into (76), we get

$$< \hat{E} >= \frac{1}{4\mu^2} - \frac{27}{16\mu}.$$

Using again $\alpha = 1/\mu$ and taking

$$\frac{\partial < \hat{E} >}{\partial\alpha} = 0,$$

we get $\alpha = 27/8 = 3.375$, which gives $\mu = 1/\alpha = 0.2962962$ bohr and

$$\min < \hat{E} >= -2.84765625 \text{ a.u.} = -77.48928281 \text{ eV}.$$

This last case, dealing with statistically independent electrons, is generally treated in textbooks. It gives the approximation -77.4893 eV for the ground energy of the helium atom. As seen before, the more general assumption that the electrons are statistically linearly dependent gave a better approximation, namely, -77.8359 eV. Moreover, the two variational parameters from the expression (69) of the probability wave function ψ may be easily interpreted. Thus, $\mu = 0.296351$ bohr gives the mean radial distance of the electrons from the nucleus whereas the correlation coefficient $c = -0.020132404$ shows the intensity of the statistical linear dependence between the two electrons. As the two electrons have both a negative electric charge, they repel each other and consequently it is normal to have a negative linear correlation coefficient that reflects such a behavior. Due to the approximation

$$\sqrt{1+t} \approx 1 + \frac{t}{2}$$

used in getting the probability wave function ψ from the joint probability density of the two electrons, a more accurate evaluation of the linear correlation coefficient between the two electrons of the helium atom would be

$$\rho = \frac{2c}{1+9c^2} = -0.040118463,$$

where the denominator is the norm of the probability wave function

$$< \psi^2 >= 1 + 9c^2.$$

The difference of the two approximations obtained for the ground energy, namely,

$$(-77.4893 \text{ eV}) - (-77.8359 \text{ eV}) = 0.3466 \text{ eV},$$

could be taken as an energy measure of the linear radial dependence between the two electrons of the helium atom, whereas the difference between -77.8359 eV and the experimental value -79.0 eV , namely 1.1641 eV, is an energy measure of the nonlinear dependence between the two electrons.

Remark 3: Referring to (69), let us notice that the probability wave function should by antisymmetric, as required by Pauli's exclusion principle. If σ_i is the spin variable of electron $i, (i = 1, 2)$, let α and β be the spin eigenfunctions. The orthonormality conditions imply

$$\int \alpha^*(\sigma_i)\alpha(\sigma_i) \, d\sigma_i = \int \beta^*(\sigma_i)\beta(\sigma_i) \, d\sigma_i = 1, \quad (i = 1, 2),$$

$$\int \alpha^*(\sigma_i)\beta(\sigma_i) \, d\sigma_i = \int \alpha(\sigma_i)\beta^*(\sigma_i) \, d\sigma_i = 0, \quad (i = 1, 2),$$

because two electrons with the same spatial quantum numbers must have opposite spin. The antisymmetric probability wave function is

$$\Psi = \psi(r_1, r_2)[\alpha(\sigma_1)\beta(\sigma_2) - \alpha(\sigma_2)\beta(\sigma_1)].$$

Taking the orthonormality conditions into account and the fact that

$$\Psi^* = \psi(r_1, r_2)[\alpha^*(\sigma_1)\beta^*(\sigma_2) - \alpha^*(\sigma_2)\beta^*(\sigma_1)],$$

we obtain

$$< \Psi^* \hat{H} \Psi > = \int \ldots \int \Psi^* \hat{H} \Psi \, dr_1 dr_2 d\sigma_1 d\sigma_2 =$$

$$= \int \int \psi(r_1, r_2) \, \hat{H}\psi(r_1, r_2) \, dr_1 dr_2 = < \psi \, \hat{H} \psi >,$$

because the Hamiltonian $\hat{H}$ does not depend on spin. Similarly,

$$< \Psi^* \Psi > = \int \ldots \int \Psi^* \Psi \, dr_1 dr_2 d\sigma_1 d\sigma_2 = \int \int \psi^2(r_1, r_2) \, dr_1 dr_2 = < \psi^2 > .$$

Thus, we obtain

$$< \hat{E} > = \frac{< \Psi^* \hat{H} \Psi >}{< \Psi^* \Psi >} = \frac{< \psi \, \hat{H} \psi >}{< \psi^2 >}.$$

The last ratio has been used to calculate min $< \hat{E} >$.

2) *The ground energy using higher-order radial generalized correlations.*
Let us take a probability wave function (68) that contains five generalized
radial correlations, replacing (69), which contains only one linear radial correlation, by

$$\psi(r_1, r_2) = \frac{1}{8\pi\mu^3} e^{-(r_1+r_2)/(2\mu)} \left\{ 1 + a \left[L_1^{(2)} \left(\frac{r_1}{\mu} \right) + L_1^{(2)} \left(\frac{r_2}{\mu} \right) \right] + \right.$$

$$+ b L_1^{(2)} \left(\frac{r_1}{\mu} \right) L_1^{(2)} \left(\frac{r_2}{\mu} \right) + c \left[L_2^{(2)} \left(\frac{r_1}{\mu} \right) + L_2^{(2)} \left(\frac{r_2}{\mu} \right) \right] +$$

$$+ q \left[L_1^{(2)} \left(\frac{r_1}{\mu} \right) L_2^{(2)} \left(\frac{r_2}{\mu} \right) + L_2^{(2)} \left(\frac{r_1}{\mu} \right) L_1^{(2)} \left(\frac{r_2}{\mu} \right) \right] +$$

$$\left. + g L_2^{(2)} \left(\frac{r_1}{\mu} \right) L_2^{(2)} \left(\frac{r_2}{\mu} \right) \right\}.$$

Appendix 6 contains a MATHEMATICA program that does in this more involved case what we did analytically in the previous simpler case that dealt with the probability wave function (69). We obtain in this case:

$$\mu = 0.260945353 \text{ bohr}, \quad a = -0.0857747, \quad b = -0.0149716,$$

$$c = 0.0255457, \quad q = -0.0000732054, \quad g = -0.00114415,$$

$$\min < \hat{E} >= -2.8783 \text{ a.u.} = -78.32315 \text{ eV}.$$

It may be seen that the value obtained for the ground energy is closer to the experimental threshold -79.0 eV.

3) *The ground energy using radial and angular generalized correlations.* The dependence between the two electrons of the helium atom may be expressed not only in terms of radial correlations but of angular correlations as well. Thus, let us take the probability wave function (68) of the form:

$$\psi(r_1, r_2, s_1, s_2) = \tag{79}$$

$$= \frac{1}{8\pi\mu^3} e^{-(r_1+r_2)/(2\mu)} \left\{ 1 + a \left[L_1^{(2)}\left(\frac{r_1}{\mu}\right) + L_1^{(2)}\left(\frac{r_2}{\mu}\right) \right] + \right.$$

$$+ b L_1^{(2)}\left(\frac{r_1}{\mu}\right) L_1^{(2)}\left(\frac{r_2}{\mu}\right) + c \left[L_2^{(2)}\left(\frac{r_1}{\mu}\right) + L_2^{(2)}\left(\frac{r_2}{\mu}\right) \right] +$$

$$\left. + g L_2^{(2)}\left(\frac{r_1}{\mu}\right) L_2^{(2)}\left(\frac{r_2}{\mu}\right) + f P_1(s_1) P_1(s_2) \right\},$$

where $L_n^{(2)}$ is the second order Laguerre polynomial of degree n, and P_ℓ is the Legendre polynomial of degree ℓ.

We use here the Laplacian

$$\nabla_i^2 \psi = \frac{\partial^2 \psi}{\partial r_i^2} + \frac{2}{r_i}\frac{\partial \psi}{\partial r_i} - \frac{2s_i}{r_i^2}\frac{\partial \psi}{\partial s_i} + \frac{1-s_i^2}{r_i^2}\frac{\partial^2 \psi}{\partial s_i^2}, \quad (i=1,2), \tag{80}$$

and the series expansion

$$\frac{1}{r_{12}} = \sum_{\ell=0}^{+\infty} \frac{r_<^\ell}{r_>^{\ell+1}} P_\ell(\cos\theta_1) P_\ell(\cos\theta_2) = \sum_{\ell=0}^{+\infty} \frac{r_<^\ell}{r_>^{\ell+1}} P_\ell(s_1) P_\ell(s_2),$$

where

$$r_< = \min\{r_1, r_2\}, \quad r_> = \max\{r_1, r_2\},$$
$$s_i = \cos\theta_i, \quad (i = 1, 2).$$

Appendix 7 contains a MATHEMATICA program that finds the minimum mean energy $\min < \hat{E} >$, the mean distance μ of the electrons from the nucleus, and the values of the generalized radial (a, b, c, g) and angular (f) correlations. Running the program interactively, we obtain

$$\mu = 0.269261 \text{ bohr}, \quad a = -0.0703336, \quad b = -0.0174889,$$

$$c = 0.0241668, \quad g = -0.00116699, \quad f = -0.0455481,$$

$$\min < \hat{E} > = -2.88134 \text{ a.u.} = -78.40587154 \text{ eV}.$$

The value of f shows that the linear angular correlation between θ_1 and θ_2 is not negligible and its inclusion into the probability wave function (79) has improved the corresponding approximation of the ground energy.

4) *A generalization of Hylleraas' trial function.* Long ago, Hylleraas (1929) had the idea of using r_1, r_2, and r_{12} as basic variables in dealing with the two electrons of the helium atom instead of the Cartesian or spherical coordinates. As mentioned in Hylleraas (1963, p.425), "A systematic attack on the ground state problem of the helium atom had been planned by Max Born in cooperation with a pupil, Dr. Biemüller, since Born himself had no preference for numerical work. However, the enterprise came to a stop by the failing health of Dr. Biemüller before his work became particularly useful ... Professor Born first suggested to me that – as he said – I was the right one to go on with the helium problem ... One thing which I noticed fairly soon was that solutions must exist which depend only on three coordinates, instead of the full number of six, and these were the coordinates r_1, r_2, φ, defining the shape of the electron-nucleus triangle, leaving the orientation in space out of interest." But the angle φ between r_1 and r_2 may be obviously connected to the interdistance r_{12} between the two electrons, because

$$2r_1 r_2 \cos\varphi = r_1^2 + r_2^2 - r_{12}^2.$$

According to Hylleraas (1963, p.427), "What I really invented was rather $u = r_{12}$, together with the $s = r_1 + r_2$ and $t = -r_1 + r_2$, forming the triple s,t,u of which I am really proud ... The triple is forever reserved for atomic research ... This change of coordinates had, to my astonishment and to my great satisfaction as well, almost the effect of a miracle."

Let r_1, θ_1, ω_1 be the spherical coordinates of the first electron with respect to the Cartesian coordonate system x_1, y_1, z_1, i.e.,

$$x_1 = r_1 \sin \theta_1 \cos \omega_1, \quad y_1 = r_1 \sin \theta_1 \sin \omega_1, \quad z_1 = r_1 \cos \theta_1,$$

$$0 \le r_1 < +\infty, \quad 0 \le \theta_1 \le \pi, \quad 0 \le \omega_1 \le 2\pi,$$

and let r_2, θ_2, ω_2 be the spherical coordinates of the second electron with respect to the Cartesian coordonate system x_2, y_2, z_2, i.e.,

$$x_2 = r_2 \sin \theta_2 \cos \omega_2, \quad y_2 = r_2 \sin \theta_2 \sin \omega_2, \quad z_2 = r_2 \cos \theta_2,$$

$$0 \le r_2 < +\infty, \quad 0 \le \theta_2 \le \pi, \quad 0 \le \omega_2 \le 2\pi,$$

where the z-axis is in the direction of r_1, in which case θ_2 is just the angle between r_1 and r_2, and we have

$$r_{12}^2 = r_1^2 + r_2^2 - 2r_1 r_2 \cos \theta_2.$$

Differentiating the last equality with respect to θ_2, we get

$$r_{12} dr_{12} = r_1 r_2 \sin \theta_2 d\theta_2.$$

Taking this into account, the volume element in the six dimensional space of the Cartesian coordinates $x_1, y_1, z_1, x_2, y_2, z_2$, becomes

$$d\tau = dx_1 dy_1 dz_1 dx_2 dy_2 dz_2 =$$

$$= r_1^2 \sin \theta_1 dr_1 d\theta_1 d\omega_1 r_2^2 \sin \theta_2 dr_2 d\theta_2 d\omega_2 =$$

$$= r_1 r_2 r_{12} dr_1 dr_2 dr_{12} \sin \theta_1 d\theta_1 d\omega_1 d\omega_2.$$

Integrating $d\tau$ with respect to θ_1, ω_1, ω_2, and taking into account that

$$\int_0^{2\pi} d\omega_1 \int_0^{2\pi} d\omega_2 \int_0^{\pi} \sin \theta_1 d\theta_1 = 4\pi^2 \int_{-1}^{+1} ds = 8\pi^2,$$

we obtain $8\pi^2\,d\hat{\tau}$, where the element of volume with respect to the remaining coordinates r_1, r_2, r_{12} is

$$d\hat{\tau} = r_1 r_2 r_{12} dr_1 dr_2 dr_{12}.$$

Taking a trial function of the form (64) with the polynomial $P(r_1, r_2, r_{12}) = c\, r_{12}$, namely,

$$\psi(r_1, r_2, r_{12}) = e^{-\alpha r_1} e^{-\alpha r_2} (1 + c\, r_{12}), \qquad (81)$$

with the volume element $r_1 r_2 r_{12} dr_1 dr_2 dr_{12}$, and using only two variational parameters α and c, he obtained

$$\min < \hat{E} > = -2.89112 \text{ a.u.} = -78.672 \text{ eV},$$

a result which, as he mentioned (Hylleraas (1963, p.427)), "was greatly admired and thought of as almost a proof of the validity of wave mechanics, also, in the strict numerical sense." Hylleraas' approach, based on the variation-perturbation method, was refined in numerous subsequent papers, taking more and more coefficients of the polynomial $P(r_1, r_2, r_{12})$ in (64) as variational parameters, culminating with the work done by Pekeris (1959), who obtained the value

$$\min < \hat{E} > = -2.903724375 \text{ a.u.} = -79.0149862 \text{ eV},$$

by solving a determinant of order 1078. In fact this theoretical approximation is smaller than the smallest experimental value (-79.00519 eV) mentioned in the literature (Lide (2000, p.10-175)), slightly contradicting the variational theorem. Details on the computations made by Hylleraas and his followers may be found in Hylleraas (1929, 1963), Bethe and Salpeter (1957), and Kinoshita (1957). Summarizing their approach, McQuarrie (1983, p.294) writes: "Although these calculations do show that one can obtain essentially exact energies by using the variational method with r_{12} in the trial function explicitely, these calculations are quite difficult computationally and do not readily lend themselves to large atoms and molecules. Furthermore, [in following this line of thought] we have abandoned the orbital concept altogether. The orbital concept has been of great use to chemists and so the trend nowadays is to find the Hartree-Fock orbitals ... and to correct these by perturbation theory." Let us notice that the starting point of the Hartree-Fock procedure for helium atom consists in writing the two-electron wave function as a product of orbitals, i.e.

$$\psi(r_1, r_2) = \psi_1(r_1)\,\psi_2(r_2) \tag{82}$$

which, according to our probabilistic model, is equivalent to taking a probability wave function (68) with all generalized correlation coefficients equal to zero, which means that the electrons are supposed to be statistically independent. Let us mention that, back in 1929, Hylleraas did not give any probabilistic justification for his trial function (81). Remarkably enough, he chose an exponential distribution for the marginals of the joint probability density, perhaps without being aware of the statistical significance of his choice, and introduced the interdistance r_{12} between electrons as a measure of their correlation. Using a wave function of the form (81), or (64), or more general (68), does not mean to abandon orbitals. It simply means to take *statistically dependent* orbitals into account instead of *statistically independent* ones as expressed by (82). It is said (McQuarrie, 1983, p.293) that starting from (82) and applying perturbation theory "it turns out that one reaches a limit [i.e., min $< \hat{E} >= -2.8617$ a.u. $= -77.87143572$ eV] which is the best value of the energy that can be obtained using a trial function of the form of a product of one-electron wave functions (82)." Or, we have seen that using orbitals with radial linear dependence (69), the simplest case of statistical dependence between electrons, we have directly obtained a comparable value, -77.8359 eV, for the ground energy, without applying perturbation theory, whereas taking higher-order correlations between electrons into account, the corresponding theoretical values of the ground energy approach the experimental threshold -79.0 eV.

Let us start from Hylleraas' trial function (81), which takes the interelectron distance r_{12} as an indicator of the interdependence between the two electrons, and include also the linear radial correlation between electrons, as in (69), namely,

$$\psi(r_1, r_2, r_{12}) = \tag{83}$$

$$= e^{-r_1/(2\mu)} e^{-r_2/(2\mu)} \left[1 + c\, r_{12} + b\, L_1^{(1)}\left(\frac{r_1}{\mu}\right) L_1^{(1)}\left(\frac{r_2}{\mu}\right) \right],$$

with the volume element

$$r_1 r_2 r_{12} dr_1 dr_2 dr_{12},$$

where

$$L_1^{(1)}(x) = 2 - x.$$

Let us notice that the choice of the Laguerre polynomial $L_1^{(1)}$ is justified by the fact that the radial probability density is $r_i e^{-r_i/\mu}$, i.e., the density of the Gamma distribution $\mathbf{G}(\mu, 2)$. We take into account that, with respect to the variables r_1, r_2, and r_{12}, the Hamiltonian (63) has the form

$$\hat{H} = -\frac{1}{2}\frac{\partial^2 \psi}{\partial r_1^2} - \frac{1}{r_1}\frac{\partial \psi}{\partial r_1} - \frac{1}{2}\frac{\partial^2 \psi}{\partial r_2^2} - \tag{84}$$

$$-\frac{1}{r_2}\frac{\partial \psi}{\partial r_2} - \frac{\partial^2 \psi}{\partial r_{12}^2} - \frac{2}{r_{12}}\frac{\partial \psi}{\partial r_{12}} - \frac{r_1^2 - r_2^2 + r_{12}^2}{2r_1 r_{12}}\frac{\partial^2 \psi}{\partial r_1 \partial r_{12}} -$$

$$-\frac{r_2^2 - r_1^2 + r_{12}^2}{2r_2 r_{12}}\frac{\partial^2 \psi}{\partial r_2 \partial r_{12}} - \left(\frac{2}{r_1} + \frac{2}{r_2} - \frac{1}{r_{12}}\right)\psi.$$

We obtain in this case

$$\mu = 0.279794 \text{ bohr}, \quad c = 0.310042, \quad b = -0.016188,$$

$$\min < \hat{E} >= -2.89511 \text{ a.u.} = -78.7806 \text{ eV},$$

a very good result, obtained by using only three variational parameters.

The most surprising fact comes up when an impressive improvement is obtained by taking into account a second-order radial correlation as well. Thus, starting from the probability wave function

$$\psi(r_1, r_2, r_{12}) = \tag{85}$$

$$= e^{-r_1/(2\mu)}e^{-r_2/(2\mu)}\left\{1 + c\,r_{12} + b\,L_1^{(1)}\left(\frac{r_1}{\mu}\right)L_1^{(1)}\left(\frac{r_2}{\mu}\right) + \right.$$

$$\left. +a\left[L_2^{(1)}\left(\frac{r_1}{\mu}\right) + L_2^{(1)}\left(\frac{r_2}{\mu}\right)\right]\right\},$$

and using again the Hamiltonian (84), we obtain

$$\mu = 0.281424 \text{ bohr}, \quad c = 0.281515, \quad , b = -0.0171302,$$

$$a = 0.0204191, \quad \min < \hat{E} > = -2.90252 \text{ a.u.} = -78.9822 \text{ eV},$$

using only four variational parameters. At the same time, the probability wave function (85) and its coefficients have a clear interpretation: the distance of each electron from the nucleus is exponentially distributed with the mean μ which, taking the volume element into account, gives a Gamma radial probability distribution with parameters μ and 2, and the coefficients $2c, 2b, 2a$ are the interdistance and first- and second-order generalized correlation coefficients, respectively.

Appendix 8 gives the MATHEMATICA program which implements the computation of the ground energy for (85), program which runs interactively in approximately one hour on a very common personal computer IBM PS/2, Model 55 SX, with only 8MB RAM and 60 MB hard disk. Let us notice that after calculating

$$\min < \hat{E} > = \frac{< \psi \hat{H} \psi >}{< \psi^2 >}$$

as a function of the variational parameters μ, c, b, and a, the last command in the program asks for the minimum value of $< \hat{E} >$. Such a command needs arbitrary starting values of the variables μ, c, b, a. It is reasonable to choose $\mu = 1$ bohr and $c = b = a = 0$, which means that we start from the first average orbit and initially assume that the two electrons are statistically independent. Minimizing the mean energy of the system, the two electrons prove to interact statistically, moving on average closer to the nucleus, as an interelectron repelling effect, with a negative first-order (linear) correlation and positive interdistance and radial second-order (quadratic) generalized correlations.

9. APPLICATION 7: THE GROUND STATE OF THE LITHIUM ATOM.

Lithium atom represents an interesting system, where the three electrons randomly moving around the nucleus do not belong to the same energy shell. A shell picture of the atom has been in force since the time of the old quantum theory. In fact Pauli's exclusion principle ("Two electrons cannot be in exactly the same state") emerged along with spin ("If the spatial quantum numbers are the same then the two electrons must have opposite spin") to explain the picture before coming of quantum mechanics. The energy shells were originally defined to be the quantized electron orbits of Bohr's model.

If we (wrongly!) allow all three electrons of the lithium atom to interact freely inside the same energy shell, let us take the probability wave function

$$\psi(r_1, r_2, r_3, s_1, s_2, s_3) = \tag{86}$$

$$= \frac{1}{\sqrt{8\pi\mu^3}} e^{-r_1/(2\mu)} \frac{1}{\sqrt{8\pi\mu^3}} e^{-r_2/(2\mu)} \frac{1}{\sqrt{8\pi\mu^3}} e^{-r_3/(2\mu)} [1 +$$

$$+ b_3 L_1^{(2)}\left(\frac{r_1}{\mu}\right) L_1^{(2)}\left(\frac{r_2}{\mu}\right) + b_2 L_1^{(2)}\left(\frac{r_1}{\mu}\right) L_1^{(2)}\left(\frac{r_3}{\mu}\right) +$$

$$+ b_1 L_1^{(2)}\left(\frac{r_2}{\mu}\right) L_1^{(2)}\left(\frac{r_3}{\mu}\right) +$$

$$+ f_3 P_1(s_1) P_1(s_2) + f_2 P_1(s_1) P_1(s_3) + f_1 P_1(s_2) P_1(s_3)],$$

which is a direct generalization to three electrons of the probability wave function (79), with $a = c = g = 0$, where

$$s_i = \cos\theta_i, \quad (i = 1, 2, 3).$$

Using the Hamiltonian

$$\hat{H} = -\frac{1}{2}(\nabla_1^2 + \nabla_2^2 + \nabla_3^2) - \tag{87}$$

$$- \left(\frac{3}{r_1} + \frac{3}{r_2} + \frac{3}{r_3}\right) + \left(\frac{1}{r_{12}} + \frac{1}{r_{13}} + \frac{1}{r_{23}}\right),$$

with the Laplacian ∇_i^2 given by (80), for $i = 1, 2, 3$, and

$$\frac{1}{r_{ij}} = \sum_{\ell=0}^{+\infty} \frac{(\min\{r_i, r_j\})^\ell}{(\max\{r_i, r_j\})^{\ell+1}} P_\ell(s_i) P_\ell(s_j),$$

91

we obtain the values

$$\mu = 0.210549 \text{ bohr}, \quad b_1 = b_2 = b_3 = -0.0138212,$$

$$f_1 = f_2 = f_3 = -0.0380186,$$

$$\min < \hat{E} >= -8.50913 \text{ a.u.} = -231.547 \text{ eV},$$

a much smaller value than the experimental value -203.48619 eV (Lide (2000, p.**10**-175)), which contradicts the variational theorem. This result shows that indeed, the three electrons of the lithium atom are not interacting freely inside the same energy shell. Also, if the electrons are supposed to be statistically independent inside the same energy shell, supposition which is equivalent to taking

$$b_i = f_i = 0, \quad (i = 1, 2, 3)$$

in (86), then we get

$$\min < \hat{E} >= -230.236 \text{ eV},$$

a value which is also much below the experimental value -203.48619 eV.

Therefore, let us take the structure of the lithium atom as consisting of two interdependent electrons in the first closed energy shell, at a mean radial distance μ from the nucleus, and an independent electron in the second, open energy shell, at a mean radial distance ν from the nucleus. Let us emphasize that the two shells are not rigidly separated; the old orbits from Bohr's planetary model of the atom correspond to the mean radial distances μ and ν mentioned above. Because there is interdependence between the two electrons of the first shell, the mean radial distance μ is kept undetermined, as a variational parameter, starting from the initial value $\mu = 1$ bohr (as we did for the helium atom in the previous section), whereas the valence electron is the only one in the second energy shell whose mean radial distance is taken to be $\nu = 2$ bohr. The probability wave function which assumes a linear radial dependence between the two electrons of the closed energy shell, with the linear correlation coefficient $2c$, an undetermined mean radial distance μ of the closed shell from the nucleus (starting with the initial value $\mu = 1$ bohr), and an independent valence electron in the open shell at a mean radial distance $\nu = 2$ bohr from the nucleus, is

$$\psi(r_1, r_2, r_3) = \qquad\qquad\qquad\qquad (88)$$

$$= \left\{ \frac{1}{\sqrt{8\pi\mu^3}} e^{-r_1/(2\mu)} \frac{1}{\sqrt{8\pi\mu^3}} e^{-r_2/(2\mu)} \left[1 + c\, L_1^{(2)}\left(\frac{r_1}{\mu}\right) L_1^{(2)}\left(\frac{r_2}{\mu}\right) \right] \right\} \times$$

$$\times \frac{1}{\sqrt{8\pi\nu^3}} e^{-r_3/(2\nu)},$$

with $\nu = 2$ bohr, and $\mu < \nu$, where

$$L_1^{(2)}(x) = 3 - x, \quad (i = 1, 2),$$

and the volume element is

$$r_1^2 r_2^2 r_3^2 dr_1 dr_2 dr_3 ds_1 ds_2 ds_3 d\omega_1 d\omega_2 d\omega_3,$$

with

$$s_i = \cos\theta_i, \quad (i = 1, 2, 3).$$

Computer tip: Here is a MATHEMATICA session for getting:

(a) the plot (Figure 20), the contour plot (Figure 21), and the density plot (Figure 22) of the radial wave function $\psi(r_1, r_2, r_3)$ for

$$r_2 = 0.5, \quad \mu = 0.18622, \quad \nu = 2, \quad c = -0.0122754;$$

(b) the plot (Figure 23), the contour plot (Figure 24), and the density plot (Figure 25) of the radial joint probability density function

$$r_1^2 r_2^2 r_3^2 \psi^2(r_1, r_2, r_3)$$

for

$$r_2 = 0.5, \quad \mu = 0.18622, \quad \nu = 2, \quad c = -0.0122754.$$

```
math
h[r1_,r2_,r3_,mu_,nu_,c_]:=(1/(8*Pi*
mu^3))*Exp[−(r1+r2)/(2*mu)]*
(1−c*LaguerreL[1,2,r1/mu]*
LaguerreL[1,2,r2/mu])*((8*Pi*nu^3)
^(−1/2))*Exp[−r3/(2*nu)]
a1=Plot3D[h[r1,0.5,r3,0.18622,2,−0.0122754],
{r1,0,2},{r3,0,9},PlotPoints− >30];
```

93

```
Display["john1",a1]
a2=ContourPlot[h[r1,0.5,r3,0.18622,2,−0.0122754],
{r1,0,2},{r3,0,9}];
Display["john2",a2]
a3=DensityPlot[h[r1,0.5,r3,0.18622,2,−0.0122754],
{r1,0,2},{r3,0,9}, Mesh− >False];
Display["john3",a3]
hh[r1_,r2_,r3_,mu_,nu_,c_]:=(r1^2)⋆
(r2^2)⋆(r3^2)⋆(h[r1,r2,r3,
mu,nu,c])^2
b1=Plot3D[hh[r1,0.5,r3,0.18622,2,−0.0122754],
{r1,0,2},{r3,0,9},PlotPoints− >30];
Display["mary1",b1]
b2=ContourPlot[hh[r1,0.5,r3,0.18622,2,−0.0122754],
{r1,0,2},{r3,0,9}];
Display["mary2",b2]
b3=DensityPlot[hh[r1,0.5,r3,0.18622,2,−0.0122754],
{r1,0,2},{r3,0,9},Mesh− >False];
Display["mary3",b3]
Quit
hardcopy john1
hardcopy john2
hardcopy john3
hardcopy mary1
hardcopy mary2
hardcopy mary3
```

Using the Hamiltonian (87) with the Laplacian

$$\nabla_i^2 \, \psi = \frac{\partial^2 \psi}{\partial r_i^2} + \frac{2}{r_i}\frac{\partial \psi}{\partial r_i}, \qquad (i = 1, 2, 3), \qquad (89)$$

we obtain the values

$$\mu = 0.18622 \text{ bohr}, \qquad c = -0.0122754,$$

$$\min < \hat{E} >= -7.46071 \text{ a.u.} = -203.018 \text{ eV},$$

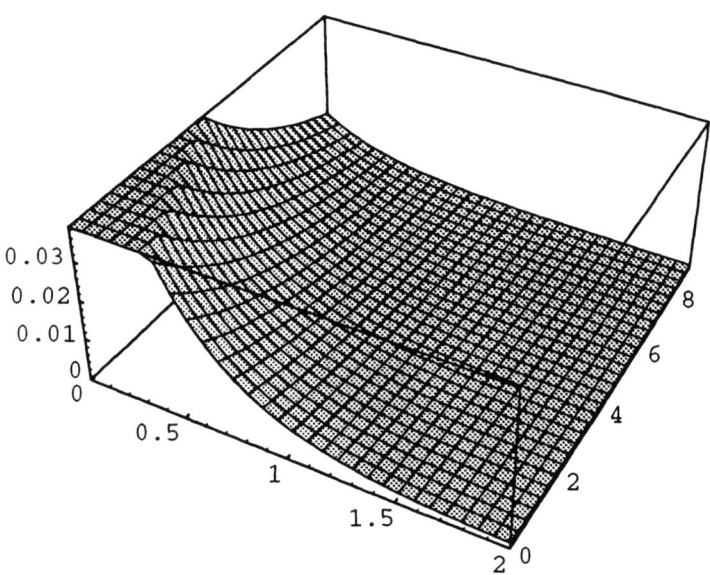

Figure 20: The plot of the radial wave function $\psi(r_1, r_2, r_3)$ for $r_2 = 0.5, \mu = 0.18622, \nu = 2, c = -0.0122754$.

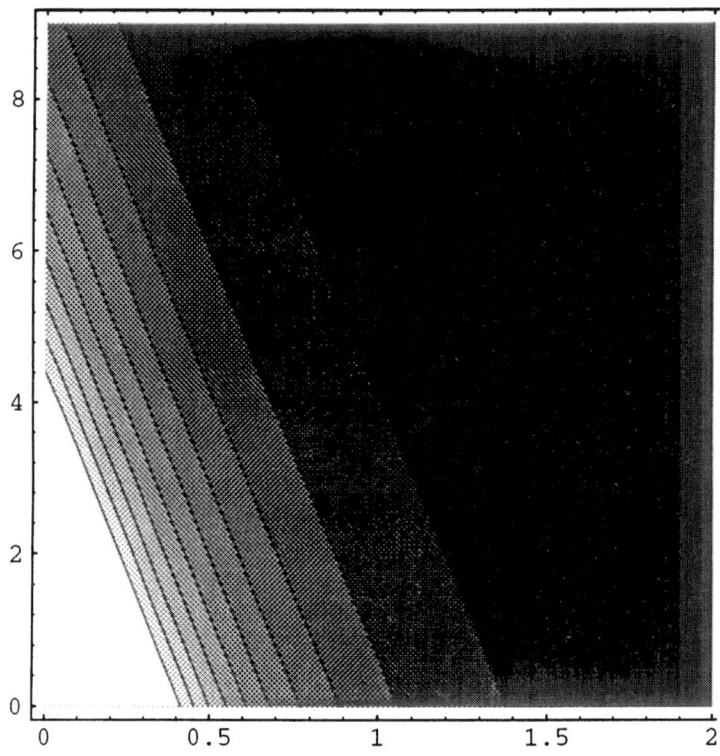

Figure 21: The contour plot of the radial wave function $\psi(r_1, r_2, r_3)$ for $r_2 = 0.5, \mu = 0.18622, \nu = 2, c = -0.0122754$.

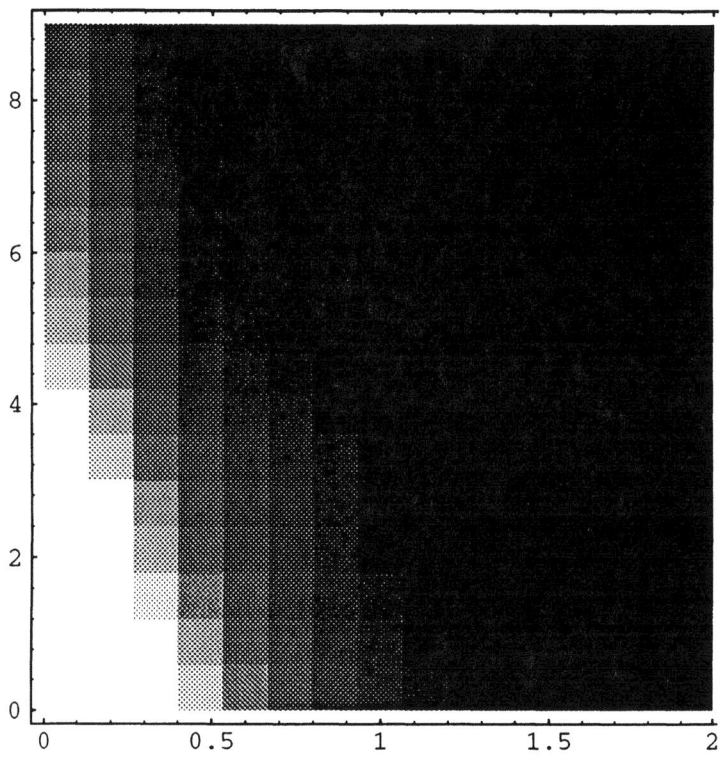

Figure 22: The density plot of the radial wave function $\psi(r_1, r_2, r_3)$ for $r_2 = 0.5, \mu = 0.18622, \nu = 2, c = -0.0122754$.

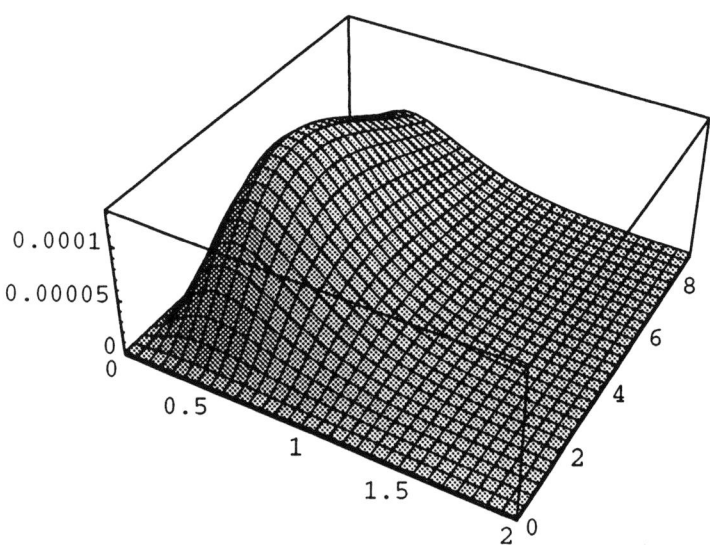

Figure 23: The plot of the radial joint probability density function $r_1^2 r_2^2 r_3^2 \psi^2(r_1, r_2, r_3)$ for $r_2 = 0.5, \mu = 0.18622, \nu = 2, c = -0.0122754$.

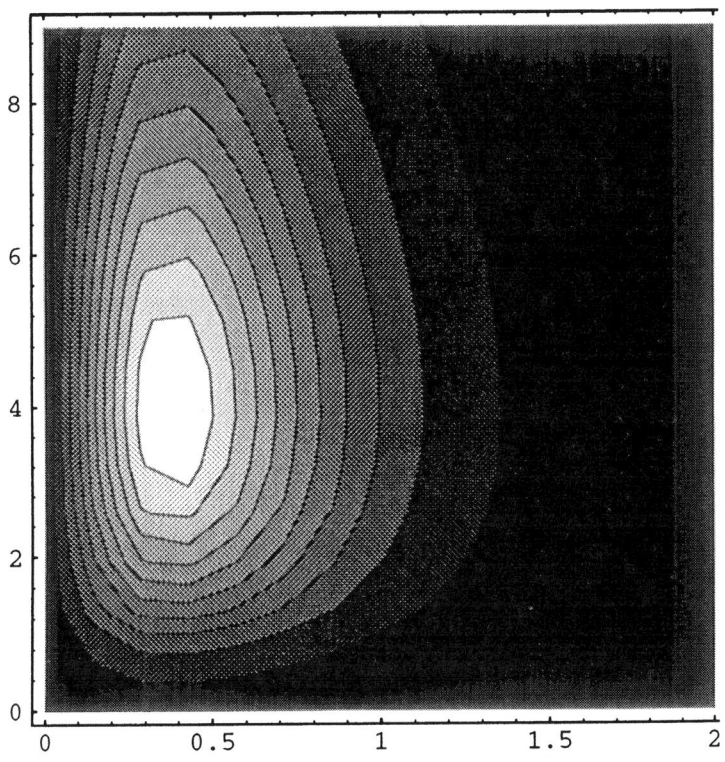

Figure 24: The contour plot of the radial joint probability density function $r_1^2 r_2^2 r_3^2 \psi^2(r_1, r_2, r_3)$ for $r_2 = 0.5, \mu = 0.18622, \nu = 2, c = -0.0122754$.

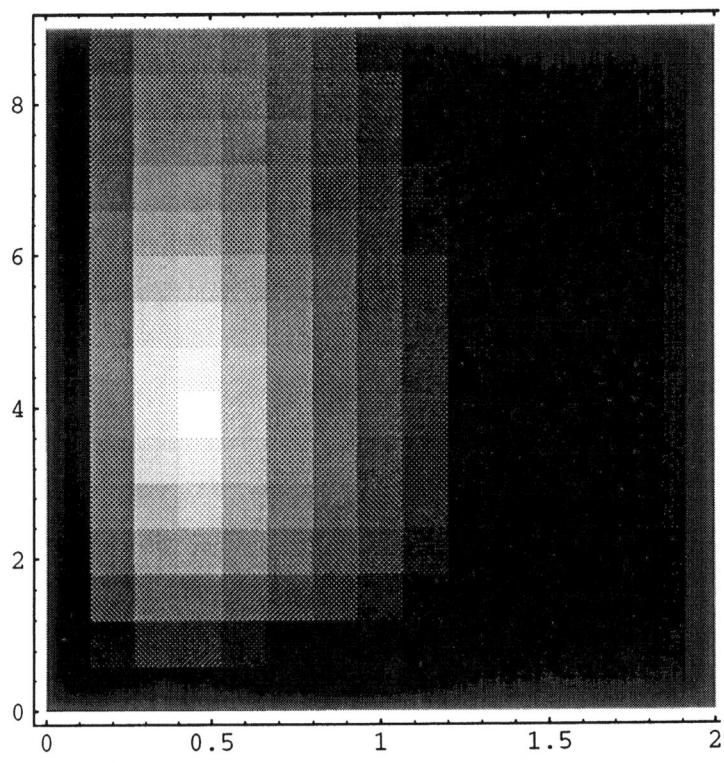

Figure 25: The density plot of the radial joint probability density function $r_1^2 r_2^2 r_3^2 \psi^2(r_1, r_2, r_3)$ for $r_2 = 0.5, \mu = 0.18622, \nu = 2, c = -0.0122754$.

a value of the ground energy amazingly close to the experimental value -203.486 eV.

Appendix 9 contains a MATHEMATICA program for minimizing the mean energy min $< \hat{E} >$, using the probability wave function (88).

Remark 1: Compared with the helium atom, the two electrons making up the closed energy shell of the lithium atom are influenced by the existence of the open shell in the sense that on average they move closer to the nucleus and their random behavior is more regular, being adequately described by a linear radial statistical interdependence.

Remark 2: The value of min $< \hat{E} >$ has been calculated by introducing ψ, given by (88), into

$$< \hat{E} >= \frac{< \psi \hat{H} \psi >}{< \psi^2 >},$$

and minimizing this expression with respect to the variational parameters μ and c. The probability wave function (88) is not antisymmetric, as required by the Pauli exclusion principle. Let us denote

$$u(i,j) = \frac{1}{\sqrt{8\pi\mu^3}} e^{-r_i/(2\mu)} \frac{1}{\sqrt{8\pi\mu^3}} e^{-r_j/(2\mu)} \left[1 + c\, L_1^{(2)} \left(\frac{r_i}{\mu} \right) L_1^{(2)} \left(\frac{r_j}{\mu} \right) \right],$$

$$v(k) = \frac{1}{\sqrt{8\pi 2^3}} e^{-r_k/4},$$

for $i, j, k = 1, 2, 3$. If σ_i is the spin variable of electron $i, (i = 1, 2, 3)$, let α and β be the spin eigenfunctions of the electrons belonging to the closed energy shell and γ be the spin eigenfunction of the valence electron. We have the normality conditions

$$\int \alpha^*(\sigma_i)\alpha(\sigma_i)\, d\sigma_i = \tag{90}$$

$$= \int \beta^*(\sigma_i)\beta(\sigma_i)\, d\sigma_i = \int \gamma^*(\sigma_i)\gamma(\sigma_i)\, d\sigma_i = 1, \quad (i = 1, 2, 3),$$

and the orthogonality conditions

$$\int \alpha^*(\sigma_i)\beta(\sigma_i)\, d\sigma_i = \int \alpha(\sigma_i)\beta^*(\sigma_i)\, d\sigma_i = \tag{91}$$

$$= \int \alpha^*(\sigma_i)\gamma(\sigma_i)\, d\sigma_i = \int \alpha(\sigma_i)\gamma^*(\sigma_i)\, d\sigma_i =$$

$$= \int \beta^*(\sigma_i)\gamma(\sigma_i)\, d\sigma_i = \int \beta(\sigma_i)\gamma^*(\sigma_i)\, d\sigma_i = 0,$$

for $i = 1, 2, 3$, because two electrons belonging to the closed energy shell cannot have the same spin and an electron cannot belong simultaneously both to the open energy shell and to the closed energy shell. The antisymmetric probability wave function corresponding to (88) is

$$\Psi = u(1,2)v(3)[\alpha(\sigma_1)\beta(\sigma_2) - \beta(\sigma_1)\alpha(\sigma_2)]\gamma(\sigma_3) - \tag{92}$$

$$-u(1,3)v(2)[\alpha(\sigma_1)\beta(\sigma_3) - \beta(\sigma_1)\alpha(\sigma_3)]\gamma(\sigma_2) +$$

$$+u(2,3)v(1)[\alpha(\sigma_2)\beta(\sigma_3) - \beta(\sigma_2)\alpha(\sigma_3)]\gamma(\sigma_1) =$$

$$= \begin{vmatrix} \alpha(\sigma_1) & \beta(\sigma_1) & u(2,3)v(1)\gamma(\sigma_1) \\ \alpha(\sigma_2) & \beta(\sigma_2) & u(1,3)v(2)\gamma(\sigma_2) \\ \alpha(\sigma_3) & \beta(\sigma_3) & u(1,2)v(3)\gamma(\sigma_3) \end{vmatrix}$$

Taking into account the orthonormality conditions (90), (91), and the fact that

$$\Psi^* = u(1,2)v(3)[\alpha^*(\sigma_1)\beta^*(\sigma_2) - \beta^*(\sigma_1)\alpha^*(\sigma_2)]\gamma^*(\sigma_3) -$$

$$-u(1,3)v(2)[\alpha^*(\sigma_1)\beta^*(\sigma_3) - \beta^*(\sigma_1)\alpha^*(\sigma_3)]\gamma^*(\sigma_2) +$$

$$+u(2,3)v(1)[\alpha^*(\sigma_2)\beta^*(\sigma_3) - \beta^*(\sigma_2)\alpha^*(\sigma_3)]\gamma^*(\sigma_1),$$

we obtain

$$< \Psi^*\hat{H}\Psi > = \int \ldots \int \Psi^*\hat{H}\Psi \, dr_1 dr_2 dr_3 d\sigma_1 d\sigma_2 d\sigma_3 =$$

$$= 2\int \ldots \int u(1,2)v(3)\hat{H}u(1,2)v(3)\, dr_1 dr_2 dr_3 +$$

$$+2\int \ldots \int u(2,3)v(1)\hat{H}u(2,3)v(1)\, dr_1 dr_2 dr_3 +$$

$$+2\int \ldots \int u(1,3)v(2)\hat{H}u(1,3)v(2)\, dr_1 dr_2 dr_3,$$

because the Hamiltonian $\hat{H}$ does not depend on spin. As $\hat{H}$ is symmetric,

$$< \Psi^*\hat{H}\Psi > = 6\int \ldots \int u(1,2)v(3)\hat{H}u(1,2)v(3)\, dr_1 dr_2 dr_3 =$$

$$= 6 < \psi\hat{H}\psi > .$$

Similarly, we get

$$< \Psi^* \Psi > = \int \ldots \int \Psi^* \Psi \, dr_1 dr_2 dr_3 d\sigma_1 d\sigma_2 d\sigma_3 =$$

$$= 6 \int \ldots \int [u(1,2)v(3)]^2 \, dr_1 dr_2 dr_3 = 6 < \psi^2 > .$$

Thus, we finally get

$$< \hat{E} > = \frac{< \Psi^* \hat{H} \Psi >}{< \Psi^* \Psi >} = \frac{< \psi \hat{H} \psi >}{< \psi^2 >},$$

as used by us to calculate min $< \hat{E} >$.

The antisymmetrization of the probability wave function of the lithium atom will be discussed in detail in Part II of the monograph.

10. CONCLUSION

In the standard nonrelativistic quantum mechanics, the Schrödinger equation is taken as a postulate and the squared absolute value of its solution is interpreted as being a probability density function used for making predictions about the behavior of quantum systems. This paper deals with a non-standard approach. Given a quantum system, we determine the probability wave function whose corresponding probability distribution (i.e., the square of its absolute value) is the closest one to statistical equilibrium subject to generalized correlation coefficients whose values are obtained by looking for the stationary points of the mean energy of the system. Statistical equilibrium, determined by using the principle of maximum entropy, gives the most unbiased probability distribution on the possible states of the system subject to given mean values. The closest probability distribution to statistical equilibrium is obtained by minimizing Pearson's mean deviation subject to given generalized correlation coefficients whose values are obtained, as said before, by looking for stationary points of the mean energy of the system.

Whether Nature is acting according to the principle of minimum mean deviation from statistical equilibrium and independence is a matter of philosophy. This monograph uses this principle only as a tool for constructing a mathematical model for the behavior of quantum systems. Surprisingly enough, this variational method recovers, in a unitary way, the exact solutions for the harmonic oscillator, the free particle in a box, two independent particles in a box, the hydrogen atom, and gives excellent simple approximations for the ground state of the helium and lithium atoms. Such a variational approach avoids the use of the Schrödinger equation and is in agreement with William of Ockham's 'razor' (it is futile to employ many principles when it is possible to employ fewer) and Ernst Mach's economy of thought principle. The method used in this monograph may be briefly called the *miniminimax* model, which is an abbreviation for 'minimizing the mean (ground or transition) energy corresponding to the probability wave function obtained by minimizing the mean deviation from maximum entropy condition subject to generalized correlations.' According to McQuarrie (1983, p.297), "the inclusion of electron correlations in atomic and molecular wave functions is a problem of current and active interest."

The main conclusion of this monograph is that whereas the maximum entropy principle is a widely accepted way for describing the state of statistical equilibrium, the minimization of the mean deviation from statistical equilibrium and independence proves to be a suitable mathematical tool for describing the natural evolution of a system. It also shows that the old 'planetary' model of the atom reflects only what happens with the mean values of the random variables that describe the behavior of the electrons in a probabilistic model of the atom. The rigid orbits and shelves from the old theory are replaced by different levels of probabilistic global interdependence between electrons.

In dealing with various systems, what is generally noticeable is not the equilibrium state but deviations from it due to fluctuations, not the independence but changes of it due to interactions. In order to reach some kind of stability and continue to exist and function in a distinct and somehow predictable way, the systems have to remain close to equilibrium and regain as much independence as possible. But there is no room for rigidity here. The fluctuations are random and both equilibrium, independence, and interdependence are essentially statistical. The study of some quantum systems shows that minimizing the mean deviations from statistical equilibrium and independence due to random fluctuations induced by internal and external connections seems to be a sound methodology in building models of reality and maybe even a way in which Nature is acting in order to maintain its statistical stability.

APPENDIX 1. Entropy as a Measure of Uncertainty.

Let

$$A = \begin{pmatrix} a_1 & \cdots & a_n \\ p_1 & \cdots & p_n \end{pmatrix},$$

be a probabilistic experiment, whose possible outcomes are labeled by

$$a_1, \ldots, a_n,$$

and have the corresponding probabilities satisfying

$$p_k \geq 0, \qquad \sum_{k=1}^{n} p_k = 1.$$

Every time the experiment A is performed, only one of the possible outcomes does occur. Before performing the experiment A, there is an uncertainty on what particular outcome will occur. The amount of uncertainty contained by the probabilistic experiment A is measured by Shannon entropy

$$\mathbf{H}_n(A) = \mathbf{H}_n(p_1, \ldots, p_n) = -\sum_{k=1}^{n} p_k \log p_k, \qquad (93)$$

introduced (Shannon, 1948) by analogy (the discrete case) with Boltzmann's function $\mathbf{E}$ discussed at the beginning of Section 2. The following propositions show that the function (93) is justified to be used as a global measure of the amount of uncertainty contained by a probabilistic experiment.

Proposition A1: $\mathbf{H}_n(p_1, \ldots, p_n) \geq 0$.

Proposition A2: $\mathbf{H}_n(p_1, \ldots, p_n)$ *is continuous and symmetric in all variables.*

Proposition A3: $\mathbf{H}_{n+1}(p_1, \ldots, p_n, 0) = \mathbf{H}_n(p_1, \ldots, p_n)$.

These propositions are obvious if we take into account that the concave function $-x \log x$, shown in Figure 26, is extended by continuity at the origin $x = 0$ to be equal to 0, i.e. $-0 \log 0 = 0$. The first two propositions tell us that the entropy is nonnegative, as we expect from any measure, and it

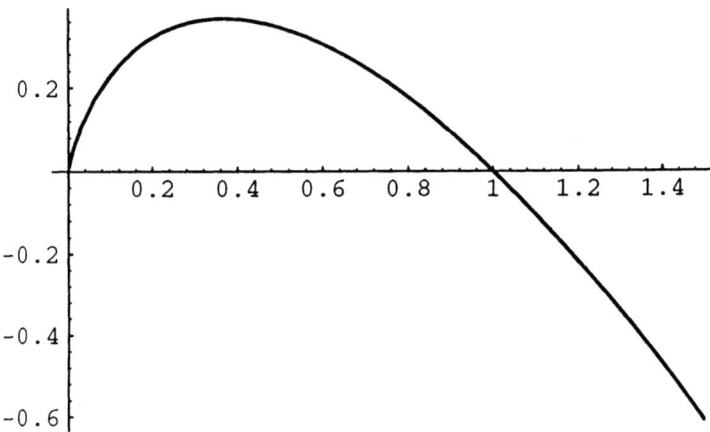

Figure 26: The plot of the function $-x \log x$.

does not depend on how the possible outcomes are labeled. Proposition A3 illustrates the obvious fact that if an outcome having the probability equal to zero is included into the set of outcomes of a probabilistic experiment, then the amount of global uncertainty remains the same.

Proposition A4: $\mathbf{H}_n(p_1, \ldots, p_n) \le \mathbf{H}_n(1/n, \ldots, 1/n) = \log n.$

Proof: If $f : [a, b] \longleftarrow \mathbf{R}^1$ is a concave function, $x_1, \ldots, x_n$ are arbitrary numbers from $[a, b]$, and $\lambda_1, \ldots, \lambda_n$ are nonnegative numbers of sum 1, then

$$\sum_{k=1}^{n} \lambda_k f(x_k) \le f \left(\sum_{k=1}^{n} \lambda_k x_k \right). \tag{94}$$

From (94), taking

$$a = 0, \quad b = 1, \quad x_k = p_k,$$

$$\lambda_k = \frac{1}{n}, \quad f(x) = -x \log x,$$

we obtain

$$-\sum_{k=1}^{n} \frac{1}{n} p_k \log p_k \le - \left(\sum_{k=1}^{n} \frac{1}{n} p_k \right) \log \left(\sum_{k=1}^{n} \frac{1}{n} p_k \right),$$

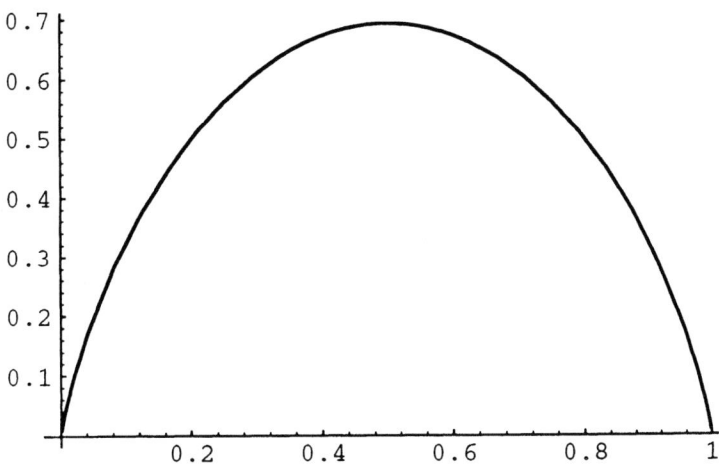

Figure 27: The plot of the entropy $\mathbf{H}_2(p, 1-p)$.

which means

$$\frac{1}{n}\mathbf{H}_n(p_1, \ldots, p_n) \leq -\frac{1}{n}\log\frac{1}{n},$$

or

$$\mathbf{H}_n(p_1, \ldots, p_n) \leq \log n = \mathbf{H}_n\left(\frac{1}{n}, \ldots, \frac{1}{n}\right).$$

Proposition A4 tells us that among all possible probabilistic experiments with n possible outcomes, the maximum uncertainty occurs when these outcomes are equally likely. The plot of the entropy

$$\mathbf{H}_2(p, 1-p) = -p\log p - (1-p)\log(1-p),$$

as a function of $p \in [0, 1]$, is given in Figure 27. Let us consider a *joint* or compound probabilistic experiment

$$(A, B) = \left(\begin{array}{ccc} (a_1, b_1) & \cdots & (a_n, b_m) \\ \pi_{11} & \cdots & \pi_{nm} \end{array}\right),$$

where $\pi_{k\ell}$ is the joint probability that the outcome a_k occurs in the experiment A and the outcome b_ℓ occurs in the experiment B, i.e.

$$\pi_{k\ell} \geq 0, \quad \sum_{k=1}^{n}\sum_{\ell=1}^{m}\pi_{k\ell} = 1.$$

Let us take the marginal probability distributions:

$$p_k = \sum_{\ell=1}^{m} \pi_{k\ell}, \quad (k = 1, \ldots, n),$$

$$q_\ell = \sum_{k=1}^{n} \pi_{k\ell}, \quad (\ell = 1, \ldots, m),$$

and the conditional probability distributions:

$$p_{k|\ell} = \frac{\pi_{k\ell}}{q_\ell}, \quad q_{\ell|k} = \frac{\pi_{k\ell}}{p_k}.$$

Here p_k is the probability of the outcome a_k in the experiment A, regardless of what happens in the experiment B, q_ℓ is the probability of the outcome b_ℓ in the experiment B, regardless of what happens in the experiment A, $p_{k|\ell}$ is the probability of the outcome a_k in the experiment A *if* the outcome b_ℓ occurs in the experiment B, and $q_{\ell|k}$ is the probability of the outcome b_ℓ in the experiment B *if* the outcome a_k occurs in the experiment A. We have

$$p_k \geq 0, \quad q_\ell \geq 0, \quad \sum_{k=1}^{n} p_k = \sum_{\ell=1}^{m} q_\ell = 1,$$

$$p_{k|\ell} \geq 0, \quad \sum_{k=1}^{n} p_{k|\ell} = 1, \quad (\ell = 1, \ldots, m),$$

$$q_{\ell|k} \geq 0, \quad \sum_{\ell=1}^{m} q_{\ell|k} = 1, \quad (k = 1, \ldots, n).$$

Several entropies may be introduced, corresponding to the above probability distributions. Thus,

$$\mathbf{H}_{nm}(A, B) = -\sum_{k=1}^{n} \sum_{\ell=1}^{m} \pi_{k\ell} \log \pi_{k\ell}, \tag{95}$$

is the entropy of the joint probabilistic experiment (A, B), measuring the amount of uncertainty on the occurrence of the outcomes (a_k, b_ℓ),

$$\mathbf{H}_n(A) = -\sum_{k=1}^{n} p_k \log p_k, \tag{96}$$

is the entropy of the experiment A, measuring the amount of uncertainty on the occurrence of the outcomes a_k, regardless of what happens in the experiment B,

$$\mathbf{H}_m(B) = -\sum_{\ell=1}^{m} q_\ell \log q_\ell, \tag{97}$$

is the entropy of the experiment B, measuring the amount of uncertainty on the occurrence of the outcomes b_ℓ, regardless of what happens in experiment A,

$$\mathbf{H}_n(A \mid b_\ell) = -\sum_{k=1}^{n} p_{k|\ell} \log p_{k|\ell},$$

is the entropy of the experiment A *if* the outcome b_ℓ occurs in the experiment B, measuring the amount of uncertainty on the occurrence of the outcomes $a_1, \ldots, a_n$ in A given that the outcome b_ℓ occurs in the experiment B,

$$\mathbf{H}_m(B \mid a_k) = -\sum_{\ell=1}^{m} q_{\ell|k} \log q_{\ell|k},$$

is the entropy of the experiment B *if* the outcome a_k occurs in the experiment A, measuring the amount of uncertainty on the occurrence of the outcomes $b_1, \ldots, b_m$ in B given that the outcome a_k occurs in the experiment A,

$$\mathbf{H}_n(A \mid B) = \sum_{\ell=1}^{m} \mathbf{H}_n(A \mid b_\ell) q_\ell = -\sum_{k=1}^{n}\sum_{\ell=1}^{m} \pi_{k\ell} \log p_{k|\ell}, \tag{98}$$

is the entropy of the experiment A given the experiment B, measuring the amount of uncertainty on the outcome of the experiment A conditioned by all possible outcomes of the experiment B,

$$\mathbf{H}_m(B \mid A) = \sum_{k=1}^{n} \mathbf{H}_m(B \mid a_k) p_k = -\sum_{k=1}^{n}\sum_{\ell=1}^{m} \pi_{k\ell} \log q_{\ell|k}, \tag{99}$$

is the entropy of the experiment B given the experiment A, measuring the amount of uncertainty on the outcome of the experiment B conditioned by all possible outcomes of the experiment A.

Proposition A5: *We have*

$$\mathbf{H}_{nm}(A, B) = \mathbf{H}_n(A) + \mathbf{H}_m(B \mid A) = \tag{100}$$

$$= \mathbf{H}_n(A \mid B) + \mathbf{H}_m(B).$$

Proof: The two equalities are obtained replacing $\pi_{k\ell}$ in (95) by $p_k q_{\ell|k}$ and $p_{k|\ell} q_\ell$, respectively, and taking into account the expressions (96)-(99).

Proposition A6: We have the inequalities

$$\mathbf{H}_m(B \mid A) \le \mathbf{H}_m(B), \quad \mathbf{H}_n(A \mid B) \le \mathbf{H}_n(A). \tag{101}$$

Proof: We apply inequality (94) for

$$a = 0, \quad b = 1, \quad f(x) = -x \log x,$$

$$\lambda_k = p_k, \quad , x_k = q_{\ell|k}, \quad (k = 1, \dots, n).$$

where ℓ is an arbitrary integer from $\{1, \dots, m\}$. We get

$$-\sum_{k=1}^n p_k q_{\ell|k} \log q_{\ell|k} \le -\left(\sum_{k=1}^n p_k q_{\ell|k}\right) \log \left(\sum_{k=1}^n p_k q_{\ell|k}\right),$$

or, equivalently,

$$-\sum_{k=1}^n \pi_{k\ell} \log q_{\ell|k} \le -q_\ell \log q_\ell.$$

Applying this inequality to all the values $\ell = 1, \dots, m$, and summing up, we get

$$-\sum_{k=1}^n \sum_{\ell=1}^m \pi_{k\ell} \log q_{\ell|k} \le -\sum_{\ell=1}^m q_\ell \log q_\ell,$$

which is just the first inequality in (101). The second inequality in (101) may be obtained similarly, by applying inequality (94), with k and n replaced by ℓ and m, respectively, taking

$$a = 0, \quad b = 1, \quad f(x) = -x \log x,$$

$$\lambda_\ell = q_\ell, \quad , x_\ell = p_{k|\ell}, \quad (\ell = 1, \dots, m).$$

Proposition A6 is in agreement with our intuition that the amount of uncertainty on the possible outcomes of a probabilistic experiment can only decrease if we know what happens in another probabilistic experiment. From Propositions A5 and A6 we get:

Proposition A7: For any two probabilistic experiments A and B with n and m possible outcomes, respectively, the following inequality holds

$$\mathbf{H}_{nm}(A, B) \leq \mathbf{H}_n(A) + \mathbf{H}_m(B). \tag{102}$$

Based on inequality (102), we can introduce

$$\mathbf{W}(A, B) = \mathbf{H}_n(A) + \mathbf{H}_m(B) - \mathbf{H}_{nm}(A, B), \tag{103}$$

called the entropic measure of interdependence between probabilistic experiments A and B. This name is justified by the following proposition:

Proposition A8: We have $\mathbf{W}(A, B) \geq 0$, with equality if and only if the probabilistic experiments A and B are independent.

Proof: Taylor's formula applied to the twice differentiable function φ, having continuous second derivative, gives

$$\varphi(t) = \varphi(1) + \varphi'(1)(t - 1) + \frac{1}{2!}\varphi''(\tau)(t - 1)^2, \tag{104}$$

where τ is a value between t and 1. Taking $\varphi(t) = t \log t$, we have

$$\varphi(1) = 0, \quad \varphi'(t) = \log t + 1,$$

$$\varphi'(1) = 1, \quad \varphi''(\tau) = \frac{1}{\tau},$$

and (104) becomes

$$t \log t = (t - 1) + \frac{1}{2\tau}(t - 1)^2 \geq t - 1, \tag{105}$$

with equality if and only if $t = 1$. Using (105) and taking into account (95)-(97), from (103) we get

$$\mathbf{W}(A, B) = \sum_{k=1}^{n} \sum_{\ell=1}^{m} \pi_{k\ell} \log \frac{\pi_{k\ell}}{p_k q_\ell} =$$

$$= \sum_{k=1}^{n} \sum_{\ell=1}^{m} p_k q_\ell \frac{\pi_{k\ell}}{p_k q_\ell} \log \frac{\pi_{k\ell}}{p_k q_\ell} \geq$$

$$\geq \sum_{k=1}^{n} \sum_{\ell=1}^{m} p_k q_\ell \left(\frac{\pi_{k\ell}}{p_k q_\ell} - 1 \right) = \sum_{k=1}^{n} \sum_{\ell=1}^{m} \pi_{k\ell} - \sum_{k=1}^{n} p_k \sum_{\ell=1}^{m} q_\ell = 1 - 1 = 0,$$

with equality if and only if

$$\pi_{k\ell} = p_k q_\ell, \quad (k = 1, \dots, n; \ell = 1, \dots, m), \tag{106}$$

which means that A and B are independent probabilistic experiments.

Propositions A8 and A5 imply that the inequalities from Propositions A6 and A7 become equalities if and only if the equalities (106) hold, i.e., if and only if A and B are independent probabilistic experiments.

Remark 1: The measure of interdependence $\mathbf{W}(A, B)$ between two probabilistic experiments may be generalized to an arbitrary finite number of probabilistic experiments $A_1, \dots, A_s$ by

$$\mathbf{W}(A_1, \dots, A_s) = \sum_{r=1}^{s} \mathbf{H}(A_r) - \mathbf{H}(A_1, \dots, A_s).$$

Remark 2: Is entropy $\mathbf{H}$ the only function that satisfies the above mentioned Properties A1-A7? The answer is yes. In fact, the uniqueness theorem says that if only Propositions A1-A5 are taken as starting properties of an arbitrary measure of uncertainty $\mathbf{H}_n(p_1, \dots, p_n)$, then it must have the expression

$$\mathbf{H}_n(p_1, \dots, p_n) = -\lambda \sum_{k=1}^{n} p_k \log p_k,$$

where λ is an arbitrary positive constant. Therefore, the entropy (93), as a measure of uncertainty is unique up to an arbitrary multiplicative constant, which allows us to choose arbitrarily the base of the logarithm in its expression without altering its properties. The proof of the uniqueness of entropy may be found in Appendix 3.

Let $p = \{p_1, \ldots, p_n\}$ and $\tilde{p} = \{\tilde{p}_1, \ldots, \tilde{p}_n\}$ be two probability distributions such that p is absolutely continuous with respect to $\tilde{p}$, which means that $\tilde{p}_k = 0$ implies $p_k = 0$. The *divergence of p from $\tilde{p}$*, or Kullback-Leibler (1951) *indicator*, is

$$\mathbf{I}(p : \tilde{p}) = \sum_{k=1}^{n} p_k \log \frac{p_k}{\tilde{p}_k}. \tag{107}$$

If p and $\tilde{p}$ are equivalent probability distributions, which means that $p_k = 0$ if and only if $\tilde{p}_k = 0$, then the *divergence* between p and $\tilde{p}$ is defined by

$$\mathbf{J}(p, \tilde{p}) = \mathbf{I}(p : \tilde{p}) + \mathbf{I}(\tilde{p} : p).$$

In the particular case when $\tilde{p}$ is the uniform probability distribution, i.e.,

$$\tilde{p}_k = \frac{1}{n}, \quad (k = 1, \ldots, n),$$

we have

$$\mathbf{I}(p : \tilde{p}) = \log n - \mathbf{H}_n(p_1, \ldots, p_n), \tag{108}$$

which shows that the entropy of any probability distribution p may be viewed as being a divergence from the uniform probability distribution.

In the continuous case, if f and g are two probability densities on the Lebesgue measurable set $D \subseteq \mathbf{R}^s$, such that $g(x) = 0$ on a subset of D of positive Lebesgue measure implies $f(x) = 0$, then the divergence of f from g is defined by

$$\mathbf{I}(f : g) = \int_D f(x) \log \frac{f(x)}{g(x)} \, dx,$$

and the entropy of f is

$$\mathbf{H}(f) = -\int_D f(x) \log f(x) \, dx,$$

provided that the two integrals exist. If $D = [a, b] \subset \mathbf{R}$, and

$$g(x) = \frac{1}{b - a},$$

then

$$\mathbf{I}(f : g) = \log(b - a) - \mathbf{H}(f),$$

113

which is the continuous analogue of equality (108), showing that in the continuous case too, the entropy of any probability density function f on an arbitrary interval $[a, b]$ of the real line, i.e.,

$$\mathbf{H}(f) = -\int_a^b f(x) \log f(x)\, dx,$$

may be viewed as being the divergence of f from the uniform probability distribution on $[a, b]$.

*Application: The **H**-theorem.* The famous **H**-theorem states the increase of entropy as a measure of uncertainty. It plays an important role in statistical mechanics. As an application of the properties of entropy proved in this appendix, we can easily show that for a class of Markov type stochastic chains the **H**-theorem holds.

Lemma 1: If

$$\pi_{ij} \geq 0, \quad p_i \geq 0, \quad (i, j = 1, \ldots, m),$$

$$\sum_{i=1}^m \pi_{ij} = 1, \quad (j = 1, \ldots, m), \qquad \sum_{i=1}^m p_i = 1,$$

we have

$$\prod_{i=1}^m (p_i)^{\pi_{ij}} \leq \sum_{i=1}^m \pi_{ij} p_i, \quad (j = 1, \ldots, m).$$

Proof: Taking

$$n = m, \quad \lambda_k = q_k, \quad q_k \geq 0, \quad \sum_{k=1}^m q_k = 1,$$

$$f(x) = \log x, \quad a = 0, \quad b = 1,$$

in inequality (94), we obtain

$$\sum_{i=1}^m q_i \log x_i \leq \log \left(\sum_{i=1}^m q_i x_i \right),$$

or

$$\prod_{i=1}^m (x_i)^{q_i} \leq \sum_{i=1}^m q_i x_i,$$

114

from which the inequality from Lemma 1 is obtained by simply taking

$$x_i = p_i, \quad q_i = \pi_{ij}.$$

Lemma 2: If $p_i \geq 0, q_i > 0, (i = 1, \ldots, m)$, and

$$\sum_{i=1}^{m} p_i = \sum_{i=1}^{m} q_i = 1,$$

then

$$-\sum_{i=1}^{m} q_i \log q_i \leq -\sum_{i=1}^{m} q_i \log p_i.$$

Proof: It is enough to take

$$x_1 = \frac{p_1}{q_1}, \ldots, x_m = \frac{p_m}{q_m}$$

in the inequality proved in Lemma 1.

Now, let Ω be a finite set. An element $\omega \in \Omega$ will be interpreted as the macroscopic state of an arbitrary system. Let also

$$t_0 < t_1 < t_2 < \ldots < t_n < t_{n+1} < \ldots$$

be an increasing sequence of positive real numbers. Here t_n will denote an arbitrary instant of time. In order to simplify the writing, we suppose that $t_n = n$. Let us give an initial probability distribution on the set Ω, i.e.,

$$p_0(\omega) \geq 0, \quad \sum_{\omega \in \Omega} p_0(\omega) = 1,$$

and a transition stochastic matrix family

$$p_{i,i+1}(\omega' \mid \omega) > 0, \quad \sum_{\omega' \in \Omega} p_{i,i+1}(\omega' \mid \omega) = 1,$$

for $(\omega' \in \Omega, \omega \in \Omega; i = 0, 1, \ldots)$. Here $p_0(\omega)$ represents the probability that the system is in the macroscopic state ω at instant 0, whereas $p_{i,i+1}(\omega' \mid \omega)$ represents the transition probability from the macroscopic state ω at instant i to the macroscopic state ω' at instant $i+1$. Let us suppose that the successive probabilities of the different macroscopic states at different instants of time are connected according to the Markov type evolution

$$p_{n+1}(\omega') = \sum_{\omega \in \Omega} p_n(\omega)\, p_{n,n+1}(\omega' \mid \omega), \quad (\omega' \in \Omega).$$

At every instant n, the average amount of uncertainty on the set of macroscopic states Ω is given by the entropy

$$\mathbf{H}_n = -\sum_{\omega \in \Omega} p_n(\omega) \log p_n(\omega).$$

*The **H**-Theorem: If the transition stochastic matrix*

$$[p_{n,n+1}(\omega' \mid \omega)]$$

is doubly stochastic, i.e.,

$$\sum_{\omega \in \Omega} p_{n,n+1}(\omega' \mid \omega) = 1, \quad \text{for every } \omega' \in \Omega,$$

then

$$\mathbf{H}_n \leq \mathbf{H}_{n+1}.$$

Proof: Applying Lemma 1 and Lemma 2, and taking into account that

$$\sum_{\omega \in \Omega} \left(\sum_{\omega' \in \Omega} p_{n+1}(\omega') p_{n,n+1}(\omega' \mid \omega) \right) =$$

$$= \sum_{\omega' \in \Omega} p_{n+1}(\omega') \left(\sum_{\omega \in \Omega} p_{n,n+1}(\omega' \mid \omega) \right) = 1,$$

we have

$$\mathbf{H}_{n+1} = -\sum_{\omega' \in \Omega} p_{n+1}(\omega') \log p_{n+1}(\omega') =$$

$$= -\sum_{\omega' \in \Omega} \sum_{\omega \in \Omega} p_n(\omega)\, p_{n,n+1}(\omega' \mid \omega) \log p_{n+1}(\omega') =$$

$$= -\sum_{\omega \in \Omega} p_n(\omega) \log \prod_{\omega' \in \Omega} (p_{n+1}(\omega'))^{p_{n,n+1}(\omega' \mid \omega)} \geq$$

$$\geq -\sum_{\omega \in \Omega} p_n(\omega) \log \left(\sum_{\omega' \in \Omega} p_{n+1}(\omega')\, p_{n,n+1}(\omega' \mid \omega) \right) \geq$$

$$\geq -\sum_{\omega \in \Omega} p_n(\omega) \log p_n(\omega) = \mathbf{H}_n.$$

As a consequence of the **H**-theorem, we can write

$$\mathbf{H}_0 \leq \mathbf{H}_1 \leq \ldots \mathbf{H}_n \leq \mathbf{H}_{n+1} \leq \ldots \leq \log \gamma,$$

where γ denotes the number of elements of the set Ω. Therefore, a stabilization of entropy occurs, i.e., there exists the limit

$$\lim_{n \to +\infty} \mathbf{H}_n = \mathbf{H}_\infty \leq \log \gamma.$$

Let us notice that a strictly deterministic flow (defined by a system of differential equations, for instance)

$$T_{n,n+1}(\theta' \mid \theta) : \Theta \to \Theta, \quad (n = 0, 1, \ldots),$$

in the infinite space Θ of possible microscopic states of a physical system induces a stochastic flow

$$p_{n,n+1}(\omega' \mid \omega) : \Omega \to \Omega,$$

in the finite space of macroscopic states Ω, where Ω is the quotient space $\Omega = \Theta/R$, with respect to the equivalence relation R defined by: 'Two microscopic states θ and θ' are R-equivalent, i.e., $\theta R \theta'$, if they belong to the same macroscopic state ω defined by R. For instance, the macroscopic energy could define such an equivalence relation, in which case two microscopic states were equivalent if they would correspond to the same macroscopic value of energy, assuming a finite spectrum for the possible values of macroscopic energy. Mathematically, the macroscopic space Ω is a partition (or dissection) of the microscopic space Θ in a finite number of disjoint subsets representing the equivalence classes with respect to the equivalence relation R. If the strictly deterministic flow

$$\{T_{n,n+1}(\theta' \mid \theta); \ n = 0, 1, \ldots\}$$

in the microscopic space Θ is time reversible, i.e., if there are the inverse transformations

$$T_{n,n+1}^{-1}, \quad (n = 0, 1, \ldots),$$

117

defining the inverse flow

$$T_{n+1,n} = T_{n,n+1}^{-1},$$

then the stochastic flow

$$\{p_{n,n+1}(\omega' \mid \omega);\ n = 0, 1, \ldots\}$$

in the macroscopic space Ω is doubly stochastic and the **H**-theorem holds. This shows that microscopic reversibility could imply macroscopic irreversibility. This variant of the **H**-theorem was proved in Guiasu (1965) and Watanabe (1969). Let us notice that the first example of Markov chain was given in statistical mechanics by Paul and Tatiana Ehrenfest (1907).

At the end of this appendix, it is useful to mention that, generally, the measures used in applied mathematics referred to entities directly accessible to our senses (like length, area, volume, etc.) or to our instruments (like speed, temperature, pressure, altitude, etc.). Entropy is a very special new type of measure, assigning a number to the amount of uncertainty contained by a probability distribution. Any probabilistic experiment contains a prior uncertainty about its outcome and Shannon's entropy, inspired by the old Boltzmann's **H** function, allows us to measure how large or small this amount of uncertainty is. Appendix 3 will show that, under very reasonable assumptions, in agreement with our intuition, this probabilistic measure of uncertainty is unique. The probabilistic entropy may be used, as shown convincingly by Watanabe (1969), to obtain a more sophisticated measure of the global interdependence among the subsystems of a system, as mentioned in Remark 1. If a system is the union of a finite number of disjoint subsystems, then the amount of internal interdpendence is equal to the sum of the entropies of the respective subsystems minus the entropy of the entire system. This global measure of interdependence among the disjoint subsystems making up a system is a super-additive set function, unless the sub-additive set measures that are studied in the standard measure theory books, reflecting the fact that a real system, of any kind, is something more complex than the simple juxtaposition of its parts and this "something more" is just the interdependence among its components. The new entropic measures of uncertainty and global interdependence have been recently used, with surprising results, in pattern recognition and classification. Here, they are used for getting the most unbiased probability distributions subject to constraints given by generalized mean values, correlations, and mixed moments.

APPENDIX 2. Maximum Entropy Principle.

As seen in Appendix 1, uncertainty is measured in terms of the probability of events. To determine the value of a particular probability, we may use the classical approach, the relative frequency method, and the subjective probability approach. The classical definition for the probability of event E occurring assumes that the experiment has n equally likely outcomes and event E occurs in m of the n outcomes with a resulting occurrence probability equal to m/n. When using the relative frequency method, the experiment is observed n times. Letting m represent the number of times event E occurs out of n times, then the resulting probability of event E occurring in the future is m/n. A subjective approach is a measure of our belief that a particular event E will occur, and like all probabilities, ranges from zero to one, inclusive.

Quite often in mathematical modelling, the corresponding probability distribution is not known and there is no way of repeating a probabilistic experiment under absolutely the same conditions which would allow us to evaluate relative frequencies of possible outcomes as a substitute for their true but unknown occurrence probabilities. This is the case in statistical mechanics, where the only information available to us is provided by some mean values of some random variables associated with the respective probabilistic experiment. Unfortunately, there are infinitely many probability distributions compatible with such given mean values. How to select one of them and which one would be the most reliable? Entropy, as a global measure of the amount of uncertainty contained by a probabilistic experiment offers a solution to the problem.

The uncertainty is maximum when the outcomes are equally likely. As shown by Proposition A4, the uniform probability distribution maximizes the entropy; the uniform probability distribution contains the largest amount of uncertainty. But this is just Laplace's principle of insufficient reason, (Laplace (1812, 1814)), according to which, if there is no reason to discriminate between two or several events, the best strategy is to consider them as being equally likely. Of course, for Laplace this was a subjective point of view, based on prudence and common sense. Indeed, without knowing anything about entropy, we apply Laplace's principle of insufficient reason in everyday life, even in analyzing the simplest probabilistic experiments. Indeed, in tossing a coin we usually attach equal probabilities to the two possible outcomes (head or tail) not after a long series of repetitions of this simple experiment followed by a careful analysis of the stability of the relative frequencies of the possible outcomes, as required by the relative frequency method, but simply because we apply, without knowing, just Laplace's principle and realize that we have no good reasons for discriminating between the two outcomes. But, as we have already seen, if we accept the Shannon entropy as a measure of uncertainty, then Proposition A4 from Appendix 1 is just the mathematical justification of the maximum entropy principle, which generalizes Laplace's principle of insufficient reason and asserts that entropy is maximized by the uniform distribution when no constraint is imposed on the probability distribution. In such a case, our intuition, based on our past experience, gives the right solution. But what happens when there are some constraints imposed on the probability distribution?

Before answering this question let us see again what kinds of constraints may be imposed. Quite often in applications, we have at our disposal one or several mean values of one or several random variables. Thus, in statistical mechanics the state functions are random variables because the state space is a probability space and we can measure only some mean values of such state functions. For instance, to each microscopic state there corresponds a well-defined value of the energy of the system. But we cannot determine with certainty the real, unique, microscopic state of the system at some instant t, and so we construct instead a probability distribution on the possible states of the system. Then the energy becomes a random variable and what we can really measure, at the macroscopic level, is the mean value of this random variable, i.e., the macroscopic energy. The macroscopic level is the level of mean values and some of these mean values can be measured. But we need a probabilistic model at the microscopic level, i.e., a probability distribution on the possible microscopic states of the system. In general, there are many probability distributions (even an infinity!) compatible with the known mean values. Hence the question: What probability distribution is "the best" and with respect to what criterion?

E.T. Jaynes (1957) gave a very natural criterion of choice by introducing explicitly the maximum entropy principle, which could be found in an implicit form in the last part of von Neumann's (1932) book of quantum mechanics and later in Kullback's (1959) book on applications of information theory in statistical inference : From the set of all probability distributions compatible with one or several mean values of one or several random variables, choose the one that maximizes the corrseponding entropy. Such a probability distribution is the most unbiased one; it will ignore no possibility, being the most random one, subject to the given constraints. Introduced for solving a problem in statistical mechanics, the maximum entropy principle has become a widely applied tool for constructing the most random probability distribution subject to given mean values of random variables. This variational principle may be viewed as being the most objective subjective criterion for constructing a probability distribution describing the statistical equilibrium corresponding to given mean values. In what follows, we will analyze the solution of the maximum entropy principle describing statistical equilibrium for some important cases in applications.

1. Discrete finite case. To see how the maximum entropy principle works, let us take the simplest possible case, i.e., the case in which we know the mean value $\mathbf{E}(X)$ of a random variable X whose possible values are $x_1, \ldots, x_n$. We need a probability distribution $p = \{p_1, \ldots, p_n\}$,

$$p_k > 0, \quad (k = 1, \ldots, n); \qquad \sum_{k=1}^{n} p_k = 1, \qquad (109)$$

satisfying the constraint

$$\mathbf{E}(X) = \sum_{k=1}^{n} x_k p_k. \qquad (110)$$

In the trivial case $n = 2$, the mean value $\mathbf{E}(X)$ uniquely defines the corresponding probability distribution from the linear equation

$$x_1 p_1 + x_2 (1 - p_1) = \mathbf{E}(X).$$

But for any $n \geq 3$ there are infinitely many probability distributions (109) satisfying (110). Applying the maximum entropy principle, we choose the most uncertain (i.e., the most random, the most unbiased) probability distribution, i.e., the probability distribution that maximizes the entropy subject to the constraints (109) and (110).

Proposition A9: The probability distribution that maximizes

$$\mathbf{H}_n(p_1, \ldots, p_n) = - \sum_{k=1}^{n} p_k \log p_k,$$

subject to the constraints (109) and (110) is

$$p_k = \frac{1}{\Phi(\beta)} e^{-\beta x_k}, \quad (k = 1, \ldots, n), \tag{111}$$

where

$$\Phi(\beta) = \sum_{k=1}^{n} e^{-\beta x_k}, \tag{112}$$

and β is the unique solution of the equation

$$\frac{d \log \Phi(\beta)}{d\beta} = -\mathbf{E}(X). \tag{113}$$

Proof: Let us notice that H_n is a concave and continuous function defined in the convex domain characterized by (109) and (110). There is only one global maximum point belonging to the set

$$\{(p_1, \ldots, p_n); \ p_k > 0, (k = 1, \ldots, n)\},$$

satisfying the constraints

$$\sum_{k=1}^{n} p_k - 1 = 0,$$

$$\sum_{k=1}^{n} x_k p_k - \mathbf{E}(X) = 0.$$

Taking the Lagrange function

$$L = \mathbf{H}_n(p_1, \ldots, p_n) - \alpha \left(\sum_{k=1}^{n} p_k - 1 \right) - \beta \left(\sum_{k=1}^{n} x_k p_k - \mathbf{E}(X) \right),$$

where α and β are the Lagrange multipliers corresponding to the two constraints, and putting the first order partial derivatives equal to zero, we get

$$\frac{\partial L}{\partial p_k} = -\log p_k - 1 - \alpha - \beta f_k = 0, \quad (k = 1, \ldots, n),$$

$$\frac{\partial L}{\partial \alpha} = 1 - \sum_{k=1}^{n} p_k = 0,$$

$$\frac{\partial L}{\partial \beta} = \mathbf{E}(X) - \sum_{k=1}^{n} x_k p_k = 0.$$

The solution of this system is

$$p_k = \frac{e^{-\beta x_k}}{\sum_{r=1}^{n} e^{-\beta x_r}}, \quad (k = 1, \ldots, n), \tag{114}$$

where β is the solution of the exponential equation

$$\sum_{k=1}^{n} [x_k - \mathbf{E}(X)] e^{-\beta [x_k - \mathbf{E}(X)]} = 0. \tag{115}$$

If the random variable X is nondegenerate (i.e., if X takes on at least two different values with positive probabilities), such a solution exists and is unique because the function

$$G(\beta) = \sum_{k=1}^{n} [x_k - \mathbf{E}(X)] e^{-\beta [x_k - \mathbf{E}(X)]} \tag{116}$$

is strictly decreasing, and

$$\lim_{\beta \to -\infty} G(\beta) = +\infty, \quad \lim_{\beta \to +\infty} G(\beta) = -\infty.$$

But (114) is just (111) and, using the notation (112), the equation (115) is just (113).

Remark 1: Let us notice that (111) is just Gibbs', or canonical, distribution encountered in almost all books on statistical mechanics and, more recently, in some books on decision theory and simulation too. Now we can see why the canonical distribution is so useful in many applications; it is the most uncertain (i.e., the most random, or the most unbiased) probability distribution subject to the only constraint given by the mean value $\mathbf{E}(X)$.

Remark 2: Since (115) is an exponential equation, its solution β may be a transcendental number. The fact, however, that the function G given by (116) is strictly decreasing allows us to approximate the solution β with great accuracy.

2. Discrete countable case. Let us take now a random variable X whose range is countable, namely,

$$\{ku;\; u > 0, k = 0, 1, 2, \ldots\}.$$

This is the case of energy in quantum mechanics, for instance, in which case u is the quantum of energy – or of many discrete functions in operations research – in which case u is the unit. Let us mention an unusual property of entropy that allows us to maximize it even when the unknown is an entire sequence satisfying

$$p_k > 0, \quad \sum_{k=0}^{+\infty} p_k = 1.$$

In such a case, instead of computing the partial derivatives with respect to a countable set of variables ($p_0, p_1, \ldots$, and the Lagrange multipliers corresponding to the constraints), it is enough to take into account again the simple equality (obtained from the Taylor expansion of $t \log t$ about 1)

$$t \log t = (t - 1) + \frac{1}{2\tau}(t - 1)^2, \tag{117}$$

true for any $t > 0$, where τ, depending on t, is a positive number located somewhere between 1 and t. We want to maximize the countable entropy

$$\mathbf{H} = -\sum_{k=0}^{+\infty} p_k \log p_k,$$

subject to the constraints

$$\sum_{k=0}^{+\infty} p_k = 1,$$

$$\mathbf{E}(X) = \sum_{k=0}^{+\infty} ku\, p_k < +\infty.$$

Applying the equality (117), we have for $\alpha > 0$, $\beta > 0$,

$$\mathbf{H} - \alpha.1 - \beta \mathbf{E}(X) = -\sum_{k=0}^{+\infty} p_k \log \left(p_k e^{\alpha + \beta ku} \right) =$$

$$= -\sum_{k=0}^{+\infty} e^{-\alpha - \beta ku} \left(p_k e^{\alpha + \beta ku} \right) \log \left(p_k e^{\alpha + \beta ku} \right) \leq$$

124

$$\leq -\sum_{k=0}^{+\infty} e^{-\alpha-\beta ku}\left(p_k e^{\alpha+\beta ku} - 1\right) = -1 + \sum_{k=0}^{+\infty} e^{-\alpha-\beta ku};$$

here the upper bound is independent of the probability distribution

$$\{p_k;\ k = 0, 1, \dots\},$$

and we have equality if and only if

$$p_k = e^{-\alpha-\beta ku}, \quad (k = 0, 1, \dots).$$

From the first constraint

$$\sum_{k=0}^{+\infty} e^{-\alpha-\beta ku} = 1,$$

we obtain

$$e^{-\alpha} = 1 - e^{-\beta u},$$

and from the second constraint

$$\mathbf{E}(X) = \sum_{k=0}^{+\infty} ku(1 - e^{-\beta u})e^{-\beta ku},$$

we obtain the solution

$$p_k = \frac{u[\mathbf{E}(X)]^k}{[u + \mathbf{E}(X)]^{k+1}}, \quad (k = 0, 1, \dots).$$

We see that the unit u and the mean value $\mathbf{E}(X)$ completely determine the solution of the entropy maximization in this case. The importance of this probability distribution in atomic physics was stressed by Born (1969).

3. Continuous case. Four continuous probability distributions are very important. Let $D = [a, b]$ be an interval on the real line and u a probability density on D, i.e.,

$$u(x) \geq 0, \quad \int_a^b u(x)\, dx = 1. \tag{118}$$

125

Proposition A10: The probability density function u on [a, b] that maximizes the entropy

$$\mathbf{H}(u) = -\int_a^b u(x) \log u(x)\, dx$$

subject to the only constraint (118) is the uniform probability density function

$$u(x) = \frac{1}{b-a}, \quad x \in [a, b]. \tag{119}$$

Proof: For $\alpha > 0$, we have

$$-\mathbf{H}(u) + \alpha.1 = \int_a^b u(x) \log u(x)\, dx + \alpha \int_a^b u(x)\, dx =$$

$$= \int_a^b e^{-\alpha} [u(x)e^{\alpha}] \log [u(x)e^{\alpha}]\, dx \geq$$

$$\geq \int_a^b e^{-\alpha} [u(x)e^{\alpha} - 1]\, dx = 1 - e^{-\alpha}(b-a),$$

with equality if and only if

$$u(x) = e^{-\alpha}, \quad (a \leq x \leq b), \tag{120}$$

in which case $\mathbf{H}(u) = \alpha$. Introducing (120) into the constraint (118) we get

$$1 = \int_a^b u(x)\, dx = e^{-\alpha} \int_a^b dx = e^{-\alpha}(b-a),$$

which gives

$$e^{-\alpha} = \frac{1}{b-a},$$

and (120) becomes (119), in which case $\mathbf{H}(u) = \log(b-a)$.

Proposition A11: If $D = [0, +\infty)$, then the solution of the nonlinear program

$$\max_u \mathbf{H}(u)$$

subject to the constraints

$$< u >= 1, \tag{121}$$

$$< x u(x) >= \mu, \tag{122}$$

is the exponential distribution $\mathbf{E}(\mu)$ *whose density is*

$$u(x) = \frac{1}{\mu} e^{-x/\mu}, \qquad x \in [0, +\infty). \tag{123}$$

Proof: For $\alpha > 0$ and $\beta > 0$, we have

$$-\mathbf{H}(u) + \alpha.1 + \beta\mu = \int_0^{+\infty} u(x) \log u(x) \, dx +$$

$$+\alpha \int_0^{+\infty} u(x) \, dx + \beta \int_0^{+\infty} x \, u(x) \, dx =$$

$$= \int_0^{+\infty} u(x) \log\left[u(x)e^{\alpha+\beta x}\right] \, dx =$$

$$= \int_0^{+\infty} e^{-\alpha-\beta x} \left[u(x)e^{\alpha+\beta x}\right] \log\left[u(x)e^{\alpha+\beta x}\right] \, dx \geq$$

$$\geq \int_0^{+\infty} e^{-\alpha-\beta x} \left[u(x)e^{\alpha+\beta x} - 1\right] =$$

$$= \int_0^{+\infty} u(x) \, dx - \int_0^{+\infty} e^{-\alpha-\beta x} \, dx = 1 - \int_0^{+\infty} e^{-\alpha-\beta x} \, dx,$$

with equality if and only if

$$u(x) = e^{-\alpha-\beta x}, \tag{124}$$

in which case

$$\mathbf{H}(u) = \alpha + \beta \, \mu. \tag{125}$$

127

Introducing (124) into constraint (121), we get

$$1 = \int_0^{+\infty} u(x)\,dx = e^{-\alpha} \int_0^{+\infty} e^{-\beta x} dx = e^{-\alpha} \frac{1}{\beta},$$

which implies

$$e^{-\alpha} = \beta,$$

giving

$$u(x) = \beta e^{-\beta x}, \quad (x > 0). \tag{126}$$

Introducing this function into constraint (122), we get

$$\mu = \int_0^{+\infty} x\, u(x)\,dx = \int_0^{+\infty} \beta x e^{-\beta x} dx =$$

$$= -[x e^{-\beta x}]_0^{+\infty} + \int_0^{+\infty} e^{-\beta x} dx = \left[-\frac{1}{\beta} e^{-\beta x} \right]_0^{+\infty} = \frac{1}{\beta},$$

which implies $\beta = 1/\mu$. Introducing this value of β into (126), we get (123). Also, using

$$\beta = \frac{1}{\mu} \text{ and } \alpha = -\log \beta = \log \mu,$$

we get from (125)

$$\mathbf{H}(u) = 1 + \log \mu.$$

Proposition A12: If $D = (-\infty, +\infty)$, then the solution of the nonlinear program

$$\max_u \mathbf{H}(u)$$

subject to the constraints

$$< u > = \int_{-\infty}^{+\infty} u(x)\,dx = 1, \tag{127}$$

$$< x\, u(x) > = \int_{-\infty}^{+\infty} x\, u(x)\,dx = \mu, \tag{128}$$

$$< (x - \mu)^2\, u(x) > = \int_{-\infty}^{+\infty} (x - \mu)^2 u(x)\,dx = \sigma^2, \tag{129}$$

is the normal distribution $\mathbf{N}(\mu, \sigma^2)$ with parameters μ and σ^2, whose density is

$$u(x) = \frac{1}{\sigma\sqrt{2\pi}}e^{-(x-\mu)^2/(2\sigma^2)}, \qquad x \in (-\infty, +\infty). \tag{130}$$

Proof: Let α, β, and γ be the Lagrange multipliers corresponding to the three constraints (127)-(129), respectively. Applying again the inequality (105), we get

$$-\mathbf{H}(u) + \alpha.1 + \beta\mu + \gamma\sigma^2 =$$

$$= \int_{-\infty}^{+\infty} u(x)[\log u(x) + \alpha + \beta x + \gamma(x - \mu)^2]\, dx =$$

$$= \int_{-\infty}^{+\infty} u(x) \log\left[u(x)e^{\alpha+\beta x+\gamma(x-\mu)^2}\right]\, dx =$$

$$= \int_{-\infty}^{+\infty} e^{-\alpha-\beta x-\gamma(x-\mu)^2} \left[u(x)e^{\alpha+\beta x+\gamma(x-\mu)^2}\right] \times$$

$$\times \log\left[u(x)e^{\alpha+\beta x+\gamma(x-\mu)^2}\right]\, dx \geq$$

$$\geq \int_{-\infty}^{+\infty} e^{-\alpha-\beta x-\gamma(x-\mu)^2} \left[u(x)e^{\alpha+\beta x+\gamma(x-\mu)^2} - 1\right]\, dx =$$

$$= 1 - \int_{-\infty}^{+\infty} e^{-\alpha-\beta x-\gamma(x-\mu)^2}\, dx,$$

the lower bound being independent of $u(x)$, with equality if and only if

$$u(x) = e^{-\alpha-\beta x-\gamma(x-\mu)^2}, \qquad (-\infty < x < +\infty). \tag{131}$$

In the subsequent computations we will frequently use the following elementary integrals:

$$\frac{1}{\sqrt{2\pi}} \int_{-\infty}^{+\infty} e^{-y^2/2}\, dy = 1, \qquad \frac{1}{\sqrt{2\pi}} \int_{-\infty}^{+\infty} y e^{-y^2/2}\, dy = 0,$$

$$\frac{1}{\sqrt{2\pi}} \int_{-\infty}^{+\infty} y^2 e^{-y^2/2}\, dy = 1.$$

Introducing (131) into constraint (127), and using the transformation

$$\frac{y}{\sqrt{2}} = \sqrt{\gamma}x + \frac{\beta - 2\gamma\mu}{2\sqrt{\gamma}},$$

we get

$$1 = \int_{-\infty}^{+\infty} u(x)\, dx = e^{-\alpha} \int_{-\infty}^{+\infty} e^{-\beta x - \gamma(x-\mu)^2}\, dx =$$

$$= e^{-\alpha} e^{-\beta\mu + \beta^2/(4\gamma)} \int_{-\infty}^{+\infty} e^{-[\sqrt{\gamma}x + (\beta - 2\gamma\mu)/(2\sqrt{\gamma})]^2}\, dx =$$

$$= \frac{1}{\sqrt{2\gamma}} e^{-\alpha} e^{-\beta\mu + \beta^2/(4\gamma)} \int_{-\infty}^{+\infty} e^{-y^2/2} dy = \sqrt{\frac{\pi}{\gamma}} e^{-\alpha} e^{-\beta\mu + \beta^2/(4\gamma)},$$

from which, we get

$$e^{-\alpha} = \sqrt{\frac{\gamma}{\pi}}\, e^{\beta\mu - \beta^2/(4\gamma)}, \qquad (132)$$

which, introduced into (131), gives

$$u(x) = \sqrt{\frac{\gamma}{\pi}}\, e^{-[\sqrt{\gamma}(x-\mu) + \beta/(2\sqrt{\gamma})]^2}, \quad (-\infty < x < +\infty). \qquad (133)$$

Introducing (133) into constraint (128), and using the transformation

$$\frac{y}{\sqrt{2}} = \sqrt{\gamma}(x - \mu) + \frac{\beta}{2\sqrt{\gamma}},$$

we get

$$\mu = \int_{-\infty}^{+\infty} x\, u(x)\, dx = \sqrt{\frac{\gamma}{\pi}} \int_{-\infty}^{+\infty} x\, e^{-[\sqrt{\gamma}(x-\mu) + \beta/(2\sqrt{\gamma})]^2}\, dx =$$

$$= \frac{1}{\sqrt{2\pi}} \int_{-\infty}^{+\infty} \left[\mu - \frac{\beta}{2\gamma} + \frac{y}{\sqrt{2\gamma}}\right] e^{-y^2/2} dy = \mu - \frac{\beta}{2\gamma},$$

which implies $\beta = 0$, and (133) becomes

$$u(x) = \sqrt{\frac{\gamma}{\pi}} e^{-\gamma(x-\mu)^2}, \quad (-\infty < x < +\infty). \qquad (134)$$

Introducing (134) into constraint (129), and using the transformation

$$\frac{y}{\sqrt{2}} = \sqrt{\gamma}(x - \mu),$$

we get

$$\sigma^2 = \int_{-\infty}^{+\infty} (x - \mu)^2 u(x)\, dx = \sqrt{\frac{\gamma}{\pi}} \int_{-\infty}^{+\infty} (x - \mu)^2 e^{-\gamma(x-\mu)^2}\, dx =$$

$$= \frac{1}{2\gamma} \frac{1}{\sqrt{2\pi}} \int_{-\infty}^{+\infty} y^2 e^{y^2/2}\, dy = \frac{1}{2\gamma},$$

which implies

$$\gamma = \frac{1}{2\sigma^2}. \tag{135}$$

Introducing (135) into (134), we get (130). Also, using (132), (135) and the fact that $\beta = 0$, we obtain the expression of the entropy of the normal probability distribution $\mathbf{N}(\mu, \sigma^2)$, namely

$$\mathbf{N}(u) = \alpha + \beta\mu + \gamma\sigma^2 = \log(\sigma\sqrt{2\pi}) + \frac{1}{2}.$$

Therefore, if we know only the mean μ and the variance σ^2 of a continuous random variable, then the continuous entropy is maximized just by the normal distribution $\mathbf{N}(\mu, \sigma^2)$. Now we see why this probability distribution has been frequently used with good results in the applications of statistical inference and why it deserves the adjective 'normal'; in the infinite set of square integrable probability density functions defined on the real line with mean μ and variance σ^2, the normal distribution (or de Moivre-Laplace-Gauss distribution) is the distribution that maximizes entropy, being the most uncertain, or unbiased probability distribution, ignoring no possibilities, subject to the two given constraints: the mean μ, showing the central tendency, and the variance σ^2, showing the spread around the central tendency. Entropy would have had to be invented if only to demonstrate this variational property of the normal distribution!

Let u be a probability density function and v a nonnegative function, both defined on $D \subseteq \mathbf{R}^1$, such that u is absolutely continuous with respect to v, which means that $u(x) = 0$ if $v(x) = 0$. The relative entropy of u with respect to the reference measure of density v (also called the Kullback-Leibler (1951) indicator, or the divergence of u with respect to v) is defined by

$$\mathbf{H}(u \mid v) = <u \log \frac{u}{v}> = - <u \log v> - \mathbf{H}(u),$$

provided that the integrals exist. We have

$$\mathbf{H}(u \mid v) \geq 0,$$

with equality if and only if $u = v$, u-almost everywhere. Obviously, if the reference function v is constant on D, i.e., $v(x) = c$, then the relative entropy is just the negative absolute entropy up to an additive constant, namely

$$\mathbf{H}(u \mid v) = -\log c - \mathbf{H}(u).$$

In particular, if $D = [a, b]$ and v is the probability density of the uniform distribution on $[a, b]$, then u is absolutely continuous with respect to v and

$$\mathbf{H}(u \mid v) = \log(b - a) - \mathbf{H}(u),$$

which shows that $\mathbf{H}(u)$ measures how much the probability density u differs from the uniform distribution with respect to the logarithmic mean. According to the principle of minimum relative entropy, we determine the probability density u which is the closest one to the reference measure of density v subject to given mean values of some random variables. Thus, the principle of maximum (absolute) entropy may be viewed as a special case of the principle of minimum relative entropy when the reference measure is just the uniform distribution.

Proposition A13: If $D = [0, +\infty)$, and

$$v(x) = x^\beta, \quad (\beta > -1),$$

then the solution of the nonlinear program

$$\min_u \mathbf{H}(u \mid v)$$

subject to the constraints

$$< u > = 1, \qquad (136)$$

$$< x\, u(x) > = \mu, \qquad (137)$$

is the Gamma distribution $\mathbf{G}(\alpha, \beta + 1)$, *with parameters* α *and* $\beta + 1$, *whose density is*

$$u(x) = \frac{1}{\alpha^{\beta+1}\Gamma(\beta+1)} x^{\beta} e^{-x/\alpha}, \qquad (0 \le x < +\infty), \qquad (138)$$

where $\mu = \alpha(\beta+1)$.

Proof: Let γ and λ be the two Lagrange multipliers corresponding to the constraints (136) and (137), respectively. Using (105), we get

$$\mathbf{H}(u \mid v) + \gamma.1 + \lambda\mu =$$

$$= \int_{0}^{+\infty} \left[u(x) \log \frac{u(x)}{x^{\beta}} + \gamma u(x) + \lambda x u(x) \right] dx =$$

$$= \int_{0}^{+\infty} u(x) \log \left[\frac{u(x)}{x^{\beta}} e^{\gamma + \lambda x} \right] dx =$$

$$= \int_{0}^{+\infty} x^{\beta} e^{-\gamma - \lambda x} \left[\frac{u(x)}{x^{\beta}} e^{\gamma + \lambda x} \right] \log \left[\frac{u(x)}{x^{\beta}} e^{\gamma + \lambda x} \right] dx \ge$$

$$\ge \int_{0}^{+\infty} x^{\beta} e^{-\gamma - \lambda x} \left[\frac{u(x)}{x^{\beta}} e^{\gamma + \lambda x} - 1 \right] dx =$$

$$= 1 - \int_{0}^{+\infty} x^{\beta} e^{-\gamma - \lambda x} dx,$$

where the bound does not depend on $u(x)$, with equality if and only if

$$\frac{u(x)}{x^{\beta}} e^{\gamma + \lambda x} = 1,$$

133

which gives

$$u(x) = x^\beta e^{-\gamma - \lambda x}, \quad (0 \le x < +\infty), \tag{139}$$

in which case we have

$$\mathbf{H}(u(x) \mid x^\beta) + \gamma + \lambda\mu = 0. \tag{140}$$

Introducing (139) into constraint (136), and using the transformation $y = \lambda x$, we get

$$1 = \int_0^{+\infty} u(x)\, dx = \int_0^{+\infty} x^\beta e^{-\gamma - \lambda x} =$$

$$= e^{-\gamma} \frac{1}{\lambda^{\beta+1}} \int_0^{+\infty} y^\beta e^{-y} dy = e^{-\gamma} \frac{1}{\lambda^{\beta+1}} \Gamma(\beta + 1),$$

which implies

$$e^{-\gamma} = \frac{\lambda^{\beta+1}}{\Gamma(\beta + 1)}. \tag{141}$$

Introducing (141) into (139), we have

$$u(x) = \frac{\lambda^{\beta+1}}{\Gamma(\beta + 1)} x^\beta e^{-\lambda x}, \quad (0 \le x < +\infty). \tag{142}$$

Introducing (142) into constraint (137), and using again the transformation $y = \lambda x$, we get

$$\mu = \int_0^{+\infty} x\, u(x)\, dx = \frac{\lambda^{\beta+1}}{\Gamma(\beta + 1)} \int_0^{+\infty} x^{\beta+1} e^{-\lambda x} dx =$$

$$= \frac{\lambda^{\beta+1}}{\Gamma(\beta + 1)} \frac{1}{\lambda^{\beta+2}} \int_0^{+\infty} y^{\beta+1} e^{-y} dy = \frac{1}{\Gamma(\beta + 1)} \frac{1}{\lambda} \Gamma(\beta + 2) =$$

$$= \frac{(\beta + 1)\Gamma(\beta + 1)}{\Gamma(\beta + 1)\, \lambda} = \frac{\beta + 1}{\lambda},$$

which shows that

$$\lambda = \frac{\beta + 1}{\mu}. \tag{143}$$

Taking now

$$\alpha = \frac{1}{\lambda}, \tag{144}$$

in (142), we obtain

$$u(x) = \frac{1}{\alpha^{\beta+1}\Gamma(\beta+1)}x^{\beta}e^{-x/\alpha}, \quad (0 \leq x < +\infty),$$

i.e., the formula (138). According to (143) and (144), the mean value

$$\mu = <x\,u(x)>$$

is equal to

$$\mu = \alpha(\beta+1).$$

The probability distributions mentioned in this appendix (i.e., the uniform, canonical, exponential, normal, and Gamma distributions) had been introduced in mathematical modelling, in general, and in statistical inference, in particular, with different motivations, in a large variety of applications. It is remarkable that these basic probability distributions may be recovered in a unitary way, as being solutions of a unique variational principle involving the new measures of uncertainty. This fact makes us think to the old William of Ockham's 'razor', according to which it is futile to employ many principles when it is possible to employ fewer and the number of hypotheses and principles should not exceed the minimum necessary for explanation of the facts. The Greek thinkers held that Nature possesses an immanent tendency to simplicity, whereas Ockham demanded that in describing Nature one should avoid unnecessary complications. Variational principles seem to be the only effective tool we have for doing this. Fermat's principle of least time, Maupertuis' principle of least action and its generalizations by Euler, Lagrange, and Hamilton, have been used for making strictly deterministic predictions about the evolution of classic mechanical systems. Here, the variational tools are used in order to predict the random evolution of quantum systems.

APPENDIX 3. The Uniqueness of Entropy

We have seen that the properties of entropy proved in Appendix 1 have an interpretation in agreement with our intuition about what we expect from a measure of the mean amount of uncertainty contained by a probabilistic experiment. The inverse problem, however, remains open: are there other formulas for the mean amount of uncertainty contained by a probabilistic experiment compatible with the properties A1-A7? As mentioned at the end of Appendix 1, it is enough to take Properties A1-A5 as axioms for the measure of the mean amount of uncertainty contained by a probabilistic experiment and we obtain the formula (93). The following uniqueness theorem was proved by Khinchin (1957, p.9). It may be also found in Guiasu (1977, p.9).

Theorem A1: Let us consider the sequence of functions

$$\mathbf{H}_1(1), \mathbf{H}_2(p_1, p_2), \ldots, \mathbf{H}_n(p_1, \ldots, p_n), \ldots$$

where, for every n, $\mathbf{H}_n(p_1, \ldots, p_n)$ is defined on the set

$$\left\{ (p_1, \ldots, p_n); p_i \geq 0, \sum_{i=1}^{n} p_i = 1 \right\}.$$

Suppose that the following axioms hold:

(A1) *For every n,*
$$\mathbf{H}_n(p_1, \ldots, p_n) \geq 0.$$

(A2) *For every n, the function $\mathbf{H}_n(p_1, \ldots, p_n)$ is a continuous and symmetric function with respect to all its arguments.*

(A3) *For every n, we have*
$$\mathbf{H}_{n+1}(p_1, \ldots, p_n, 0) = \mathbf{H}_n(p_1, \ldots, p_n).$$

(A4) *For every n, we have the inequality*
$$\mathbf{H}_n(p_1, \ldots, p_n) \leq \mathbf{H}_n\left(\frac{1}{n}, \ldots, \frac{1}{n}\right).$$

(A5) *If*
$$\pi_{k\ell} \geq 0, \quad \sum_{k=1}^{n}\sum_{\ell=1}^{m} \pi_{k\ell} = 1, \quad p_k = \sum_{\ell=1}^{m} \pi_{k\ell},$$

136

we have

$$\mathbf{H}_{nm}(\pi_{11}, \ldots, \pi_{nm}) =$$

$$= \mathbf{H}_n(p_1, \ldots, p_n) + \sum_{k=1}^{n} p_k \mathbf{H}_m\left(\frac{\pi_{k1}}{p_k}, \ldots, \frac{\pi_{km}}{p_k}\right).$$

Then

$$\mathbf{H}_n(p_1, \ldots, p_n) = -\lambda \sum_{k=1}^{n} p_k \log p_k, \qquad (145)$$

where λ is a positive constant.

Proof: (a) First, let us prove the theorem for the particular case when

$$p_k = \frac{1}{n}, \quad (k = 1, \ldots, n).$$

Denoting by

$$L(n) = \mathbf{H}_n\left(\frac{1}{n}, \ldots, \frac{1}{n}\right),$$

we shall prove that

$$L(n) = \lambda \log n,$$

where λ is a positive constant. Applying (A3) and (A4), we obtain

$$L(n) = \mathbf{H}_n\left(\frac{1}{n}, \ldots, \frac{1}{n}\right) = \mathbf{H}_{n+1}\left(\frac{1}{n}, \ldots, \frac{1}{n}, 0\right) \leq$$

$$\leq \mathbf{H}_{n+1}\left(\frac{1}{n+1}, \ldots, \frac{1}{n+1}\right) = L(n+1),$$

so that $L(n)$ is a monotonic nondecreasing function of n.

Let r and n be positive integer numbers. We have

$$L(n^r) = rL(n). \qquad (146)$$

We prove this equality by induction. Let us put in (A5),

$$m = n, \quad \pi_{k\ell} = \frac{1}{n^2}, \quad p_k = \frac{1}{n}.$$

We obtain

$$\mathbf{H}_{n^2}\left(\frac{1}{n^2}, \ldots, \frac{1}{n^2}\right) =$$

$$= \mathbf{H}_n\left(\frac{1}{n}, \ldots, \frac{1}{n}\right) + \sum_{k=1}^{n} \frac{1}{n} \mathbf{H}_n\left(\frac{1}{n}, \ldots, \frac{1}{n}\right),$$

which means

$$L(n^2) = L(n) + L(n) = 2L(n).$$

Suppose now the equality (146) to be true for $r - 1$, namely,

$$L(n^{r-1}) = (r - 1)L(n),$$

and we want to prove it for r. Let us introduce, in the same condition (A5), the values

$$m = n^{r-1}, \quad \pi_{k\ell} = \frac{1}{n^r}, \quad p_k = \frac{1}{n}.$$

We obtain

$$\mathbf{H}_{n^r}\left(\frac{1}{n^r}, \ldots, \frac{1}{n^r}\right) =$$

$$= \mathbf{H}_n\left(\frac{1}{n}, \ldots, \frac{1}{n}\right) + \sum_{k=1}^{n} \frac{1}{n} \mathbf{H}_{n^{r-1}}\left(\frac{1}{n^{r-1}}, \ldots, \frac{1}{n^{r-1}}\right),$$

which may be written as

$$L(n^r) = L(n) + L(n^{r-1}) = L(n) + (r - 1)L(n) = rL(n).$$

Now let the numbers v, r, and t be given arbitrarily, but let the number s be determined by the inequalities

$$r^s \le v^t \le r^{s+1}, \tag{147}$$

whence

$$s \log r \le t \log v \le (s + 1) \log r,$$

which implies

$$\frac{s}{t} \le \frac{\log v}{\log r} \le \frac{s + 1}{t}. \tag{148}$$

It follows from the inequalities (147), by the monotonicity of the function $L(n)$, that

$$L(r^s) \le L(v^t) \le L(r^{s+1}),$$

and, consequently, by the equality (146),

$$sL(r) \leq tL(v) \leq (s+1)L(r),$$

so that

$$\frac{s}{t} \leq \frac{L(v)}{L(r)} \leq \frac{s+1}{t}. \tag{149}$$

Finally, it follows from (148)and (149) that

$$\left| \frac{L(v)}{L(r)} - \frac{\log v}{\log r} \right| \leq \frac{1}{t}.$$

Since the left side is independent of t and since t can be chosen arbitrarily large on the right side, we have

$$\frac{L(v)}{\log v} = \frac{L(r)}{\log r},$$

which, in view of the arbitrariness of v and r, means that

$$L(n) = \lambda \log n, \tag{150}$$

where λ is a constant. By the condition (A1) and monotonicity of the function $L(n)$, we have $\lambda > 0$, and the assertion is proved.

(b) The equality (150) represents the special case

$$p_k = \frac{1}{n}, \quad (k = 1, \ldots, n),$$

of the theorem to be proved. We now consider the more general case where the probabilities p_k, $(k = 1, \ldots, n)$ are any rational numbers, i.e.,

$$p_k = \frac{m_k}{m}, \quad (k = 1, \ldots, n),$$

where

$$m_k > 0, \quad (k = 1, \ldots, n), \quad \sum_{k=1}^{n} m_k = m.$$

Let us put

$$q_{\ell|k} = \frac{\pi_{k\ell}}{p_k}, \quad (p_k > 0).$$

139

If we consider now the particular values

$$
q_{\ell|k} = \begin{cases} 0, & \text{if} \quad 1 \le \ell \le \sum_{s=1}^{k-1} m_s, \\ 1/m_k, & \text{if} \quad \sum_{s=1}^{k-1} m_s + 1 \le \ell \le \sum_{s=1}^{k} m_s, \\ 0, & \text{if} \quad \sum_{s=1}^{k} m_s + 1 \le \ell \le m, \end{cases}
$$

then we obtain the values

$$
\pi_{k\ell} = \begin{cases} 0, & \text{if} \quad 1 \le \ell \le \sum_{s=1}^{k-1} m_s, \\ 1/m, & \text{if} \quad \sum_{s=1}^{k-1} m_s + 1 \le \ell \le \sum_{s=1}^{k} m_s, \\ 0, & \text{if} \quad \sum_{s=1}^{k} m_s + 1 \le \ell \le m. \end{cases}
$$

For these values, let us apply again the equality (A5); we obtain

$$
\mathbf{H}_m \left(\frac{1}{m}, \ldots, \frac{1}{m} \right) =
$$

$$
= \mathbf{H}_n(p_1, \ldots, p_n) + \sum_{k=1}^{n} p_k \mathbf{H}_{m_k} \left(\frac{1}{m_k}, \ldots, \frac{1}{m_k} \right), \tag{151}
$$

because, taking into account the condition (A3), we have

$$
\mathbf{H}_{nm}(\pi_{11}, \ldots, \pi_{nm}) =
$$

$$
= \mathbf{H}_{m_1 + \ldots + m_n} \left(\frac{1}{m}, \ldots, \frac{1}{m} \right) = \mathbf{H}_m \left(\frac{1}{m}, \ldots, \frac{1}{m} \right),
$$

and

$$
\mathbf{H}_m(q_{1|k}, \ldots, q_{m|k}) = \mathbf{H}_{m_k} \left(\frac{1}{m_k}, \ldots, \frac{1}{m_k} \right).
$$

Therefore, we can write the equality (151) as

$$
L(m) = \mathbf{H}_n(p_1, \ldots, p_n) + \sum_{k=1}^{n} p_k L(m_k),
$$

or

$$
\lambda \log m = \mathbf{H}_n(p_1, \ldots, p_n) + \lambda \sum_{k=1}^{n} p_k \log m_k,
$$

140

or

$$\lambda \log m =$$

$$= \mathbf{H}_n(p_1, \ldots, p_n) + \lambda \sum_{k=1}^{n} p_k \log m + \lambda \sum_{k=1}^{n} p_k \log p_k.$$

This means

$$\mathbf{H}_n(p_1, \ldots, p_n) = -\lambda \sum_{k=1}^{n} p_k \log p_k. \qquad (152)$$

(c) As the set of rational numbers is dense in $\mathbf{R}^1$, the formula (152), proved for rational probabilities, remains valid for any real values of $p_1, \ldots, p_n$, due to axiom (A2) of continuity of the function $\mathbf{H}$.

Remark 1: There are several uniqueness theorems for the probabilistic entropy, depending on which of its properties are taken as axioms. Shannon (1948) gave the first such uniqueness theorem in his seminal paper. This appendix contains the uniqueness theorem proved by Khinchin (1957), which is perhaps the most elegant one and uses very intuitive axioms. Other uniqueness theorems may be found in Guiasu (1977).

Remark 2: Shannon's entropy (145) is a measure of uncertainty that depends only on the probability distribution of the possible outcomes of a probabilistic experiment. It measures how much uncertainty we have *before* performing the corresponding experiment, or how much prior uncertainty has been removed when we know the outcome *after* performing the probabilistic experiment. Taking the amount of information supplied by a probabilistic experiment to be equal to the amount of prior uncertainty removed by knowing, a posteriori, the outcome of the experiment, Shannon's entropy is the main tool in the probabilistic information theory. Sometimes, however, we assign some qualitative weights $\{w_k; \ k = 1, \ldots, n\}$ to the possible outcomes, representing their utility with respect to a certain goal. In such a case, Shannon's entropy is generalized by the weighted entropy

$$-\sum_{k=1}^{n} w_k p_k \log p_k.$$

The properties, uniqueness theorem, and details about the weighted entropy may be found in Guiasu (1971, 1977).

APPENDIX 4. The Variational Theorem and the Least Weighted Deviation

1. The Variational Theorem. Let $\{E_n; n = 0, 1, \ldots\}$ be the possible values of the energy of a quantum system corresponding to the complete system of orthonormal wave functions $\{\psi_n; n = 0, 1, \ldots\}$,

$$\hat{H}\psi_n = E_n\psi_n,$$

where E_0 is the ground state energy of the system. Any suitable wave function ψ may be expressed in terms of $\{\psi_n; n = 0, 1, \ldots\}$, namely,

$$\psi = \sum_{n=0}^{+\infty} c_n\psi_n.$$

The energy corresponding to ψ becomes

$$E_\psi = <\hat{E}> = \frac{<\psi^*\hat{H}\psi>}{<\psi^*\psi>} = \frac{\sum_{n=0}^{+\infty} c_n^*c_n E_n}{\sum_{n=0}^{+\infty} c_n^*c_n}.$$

Subtracting E_0 from the left-hand side and

$$\frac{E_0\sum_{n=0}^{+\infty} c_n^*c_n}{\sum_{n=0}^{+\infty} c_n^*c_n}$$

from the right-hand side, we get

$$E_\psi - E_0 = \frac{\sum_{n=0}^{+\infty} c_n^*c_n(E_n - E_0)}{\sum_{n=0}^{+\infty} c_n^*c_n}.$$

But, by definition, E_0 being the ground state energy, we have $E_n - E_0 \geq 0$ for all values of n, and because $c_n^*c_n \geq 0$, we obtain $E_\psi - E_0 \geq 0$, or

$$E_\psi \geq E_0,$$

which is just the variational theorem.

2. The Least Weighted Deviation. Let H be a Hilbert space of functions $x : S \to \mathbf{R}^r$ with respect to the inner product

$$< \cdot \cdot >: \mathrm{H} \times \mathrm{H} \to \mathbf{R}^1.$$

In what follows, we assume that the set S is one of the following three special cases:

(a) $S = \{i; i = 1, \ldots, n\}$, in which case the inner product of

$$x = (x_1, \ldots, x_n) \in \mathrm{H} \quad \text{and} \quad y = (y_1, \ldots, y_n) \in \mathrm{H}$$

is defined by

$$< xy > = \sum_{i=1}^{n} x_i y_i$$

(b) $S = \{(i, j); i = 1, \ldots, n; j = 1, \ldots, m\}$, in which case the inner product of

$$x = (x_{11}, \ldots, x_{nm}) \in \mathrm{H}, \quad \text{and} \quad y = (y_{11}, \ldots, y_{nm}) \in \mathrm{H}$$

is defined by

$$< xy > = \sum_{i=1}^{n} \sum_{j=1}^{m} x_{ij} y_{ij}$$

(c) $S = [a_1, b_1] \times \ldots \times [a_s, b_s] \subset \mathbf{R}^s$, where each $[a_i, b_i]$ is a finite interval of $\mathbf{R}^1$, in which case the inner product of the square integrable functions

$$x = x(t_1, \ldots, t_s) \in \mathrm{H}, \quad \text{and} \quad y = y(t_1, \ldots, t_s) \in \mathrm{H}$$

is defined by

$$< xy > = \int_{a_1}^{b_1} \ldots \int_{a_s}^{b_s} x(t_1, \ldots, t_s) y(t_1, \ldots, t_s) \, dt_1 \ldots dt_s.$$

The norm is defined, as usual, by

$$\|x\| = +\sqrt{< x^2 >}$$

If a is a constant real number, we denote by the same letter the vector whose components are all equal to a, or the constant function that assigns the same value a to each element of the domain of definition S. Let y, z, and w be three elements of H such that $w > 0$. The *first order mean deviation* of y from z weighted by w is defined by

$$\mathbf{D}_1(y : z \mid w) = < w(y - z) > .$$

The *second order mean deviation* of y from z weighted by w is defined by

$$\mathbf{D}_2(y : z \mid w) = < w(y - x)^2 > .$$

Obviously, the following equalities hold:

$$\mathbf{D}_1(y : z \mid w) + \mathbf{D}_1(z : y \mid w) = 0,$$

$$\mathbf{D}_2(y : z \mid w) - \mathbf{D}_2(z : y \mid w) = 0.$$

If p and q are probability distributions or probability density functions from H such that $q > 0$, we have

$$\mathbf{D}_1\left(\log \frac{p}{q} : \log 1 \mid p\right) = <p \log \frac{p}{q}> = \mathbf{I}(p : q),$$

which is the Kullback-Leibler divergence of p from q. Also,

$$\mathbf{D}_1\left(\log \frac{q}{p} : \log 1 \mid q\right) = <q \log \frac{q}{p}> = \mathbf{I}(q : p),$$

and, if $p > 0$,

$$\mathbf{D}_1\left(\log \frac{p}{q} : \log 1 \mid p\right) + \mathbf{D}_1\left(\log \frac{q}{p} : \log 1 \mid q\right) =$$

$$= <p \log \frac{p}{q} + q \log \frac{q}{p}> = \mathbf{I}(p : q) + \mathbf{I}(q : p) = \mathbf{J}(p, q),$$

which is the Kullback-Leibler divergence between p and q. Also,

$$\mathbf{D}_2\left(\frac{p}{q} : 1 \mid q\right) = <q \left(\frac{p}{q} - 1\right)^2> = \tilde{\chi}^2(p : q),$$

which is Pearson's (reduced) chi-square indicator of p relative to q. Obviously,

$$\mathbf{D}_2\left(\frac{q}{p} : 1 \mid p\right) = <p \left(\frac{q}{p} - 1\right)^2> = \tilde{\chi}^2(q : p),$$

which implies

$$\tilde{\chi}^2(p : q) + \tilde{\chi}^2(q : p) =$$

$$= <\frac{p + q}{pq}(p - q)^2> = \mathbf{D}_2\left(p : q \mid \frac{p + q}{pq}\right).$$

If $S = \{1, \ldots, n\}$, q is the expected probability distribution, p is the observed probability distribution in a random sample of size K, then

$$\mathbf{D}_2\left(\frac{p}{q} : 1 \mid Kq\right) = K \sum_{i=1}^{n} q_i \left(\frac{p_i}{q_i} - 1\right)^2 =$$

144

$$= \sum_{i=1}^{n} \frac{(Kp_i - Kq_i)^2}{Kq_i} = \sum_{i=1}^{n} \frac{(O_i - E_i)^2}{E_i} = \tilde{\chi}^2(O : E),$$

which is Pearson's chi-square indicator showing the deviation of the observed absolute frequencies O from the expected absolute frequencies E in a random sample of size K.

Let f and u be probability density functions on an interval $[a, b] \subseteq \mathbf{R}^1$ and let

$$\{U_n;\ n = 0, 1, \ldots\}, \quad U_0 \equiv 1,$$

be a sequence of orthonormal polynomials with the weight u. We have

$$< U_n\, u >= 0, \quad (n = 1, 2, \ldots), \qquad < U_0\, u >=< u >= 1,$$

$$< U_n^2 u >= 1, \quad (n = 0, 1, \ldots), \qquad < U_n U_k\, u >= 0, \quad (n \neq k).$$

The chi-square mean deviation of f from u is

$$\tilde{\chi}^2 = \tilde{\chi}^2(f : u) =< \frac{(f - u)^2}{u} > .$$

Proposition A14: *The solution of the quadratic program:*

$$\min_f \tilde{\chi}^2(f : u) =< \frac{(f - u)^2}{u} >$$

subject to the constraints

$$< U_n\, f >= c_n, \qquad (n = 1, 2, \ldots, N)$$

is

$$f = u \left(1 + \sum_{n=1}^{N} c_n U_n \right).$$

Proof: In an isoperimetric problem (see Elsgolts (1977), for instance), we want to determine the extremum of a functional

$$\int_a^b F(t, f, f')dt, \qquad f = f(t),$$

given the isoperimetric conditions

$$\int_a^b F_n(t, f, f')dt = \ell_n, \quad (n = 1, \ldots, N).$$

where $\ell_n, (n = 1, \ldots, N)$, are constants. Then, it is necessary to form the auxiliary functional

$$\int_a^b \left(F(t, f, f') + \sum_{n=1}^N \lambda_n F_n(t, f, f') \right) dt,$$

where the Lagrange multipliers $\lambda_n, (n = 1, \ldots, N)$, are constants, and write the Euler equation for it, namely

$$\frac{\partial \tilde{F}}{\partial f} - \frac{d}{dt}\frac{\partial \tilde{F}}{\partial f'} = 0,$$

where

$$\tilde{F} = \tilde{F}(t, f, f', \lambda_1, \ldots, \lambda_N) =$$

$$= F(t, f, f') + \sum_{n=1}^N \lambda_n F_n(t, f, f').$$

In our case,

$$\tilde{F} = \frac{(f - u)^2}{u} + \sum_{n=1}^N \lambda_n U_n f,$$

and the Euler equation becomes

$$\frac{\partial \tilde{F}}{\partial f} = 0,$$

which implies

$$f = u \left(1 - \sum_{n=1}^N \frac{1}{2}\lambda_n U_n \right).$$

Introducing this expression into the constraints and using the orthonormality of the system $\{U_n; n = 1, \ldots, N\}$ with respect to the weight u, we get

$$c_n = < U_n f >=$$

$$=< U_n u > - \sum_{k=1}^N \frac{1}{2}\lambda_k < U_n U_k u >= -\frac{1}{2}\lambda_n,$$

from which we get

$$f = u \left(1 + \sum_{n=1}^N c_n U_n \right),$$

where c_n may be interpreted as being the generalized U_n-moment of f. Obviously, $< f > = < u > = 1$.

Let u and v be two probability density functions on $[a_1, b_1] \subseteq \mathbf{R}^1$, and $[a_2, b_2] \subseteq \mathbf{R}^1$, respectively, and let

$$\{U_n; \, n = 0, 1, \ldots\}, \quad \{V_m; \, m = 0, 1, \ldots\}, \quad (U_0 \equiv 1, V_0 \equiv 1)$$

be two complete systems of orthonormal functions on $[a_1, b_1]$ and $[a_2, b_2]$ with the weights u and v, respectively. If there is independence between marginals, then the joint probability density on $[a_1, b_1] \times [a_2, b_2]$ is simply the direct product uv. But what does it happen when there is interdependence between the two components? Consider the system of functions

$$\{U_n V_m; \, n, m = 0, 1, \ldots\}$$

on $[a_1, b_1] \times [a_2, b_2]$ with the weight uv. Let f be a joint probability density function on $[a_1, b_1] \times [a_2, b_2]$. The chi-square mean deviation of f from the independent direct product uv is

$$\tilde{\chi}^2 = \tilde{\chi}^2(f : uv) = < \frac{(f - uv)^2}{uv} > .$$

Proposition A15: The solution of the quadratic program:

$$\min_f \tilde{\chi}^2(f : uv) = < \frac{(f - uv)^2}{uv} >$$

subject to the constraints

$$< U_n V_m f > = c_{nm}, \quad (n = 0, 1, \ldots, N; m = 0, 1, \ldots, M; (n, m) \neq (0, 0)),$$

is the density

$$f = u\,v\,(1 + \sum_{\substack{n=0 \\ (n,m) \neq (0,0)}}^{N} \sum_{m=0}^{M} c_{nm}\, U_n V_m).$$

Proof: In this isoperimetric problem we want to determine the extremum of a functional

$$\int_{a_1}^{b_1} \int_{a_2}^{b_2} F(t_1, t_2, f, \frac{\partial f}{\partial t_1}, \frac{\partial f}{\partial t_2}) dt_1 dt_2, \quad f = f(t_1, t_2),$$

147

given the isoperimetric conditions

$$\int_{a_1}^{b_1} \int_{a_2}^{b_2} F_{nm}(t_1, t_2, f, \frac{\partial f}{\partial t_1}, \frac{\partial f}{\partial t_2}) dt_1 dt_2 = \ell_{nm},$$

$$(n = 0, 1, \ldots, N; m = 0, 1, \ldots, M; (n, m) \neq (0, 0)).$$

where

$$\ell_{nm}, (n = 0, 1, \ldots, N; m = 0, 1, \ldots, M; (n, m) \neq (0, 0)),$$

are constants. Then, it is necessary to form the auxiliary functional

$$\int_{a_1}^{b_1} \int_{a_2}^{b_2} \tilde{F}(t_1, t_2, f, \frac{\partial f}{\partial t_1}, \frac{\partial f}{\partial t_2}, \lambda_{10}, \lambda_{01}, \ldots, \lambda_{nm}) dt_1 dt_2,$$

where the new function is

$$\tilde{F}(t_1, t_2, f, \frac{\partial f}{\partial t_1}, \frac{\partial f}{\partial t_2}, \lambda_{10}, \lambda_{01}, \ldots, \lambda_{nm}) =$$

$$= F(t_1, t_2, f, \frac{\partial f}{\partial t_1}, \frac{\partial f}{\partial t_2}) + \sum_{\substack{n=0 \\ (n,m) \neq (0,0)}}^{N} \sum_{m=0}^{M} \lambda_{nm} F_{nm}(t_1, t_2, f, \frac{\partial f}{\partial t_1}, \frac{\partial f}{\partial t_2}),$$

where the Lagrange multipliers

$$\lambda_{nm}, \quad (n = 0, 1, \ldots, N; m = 0, 1, \ldots, M, (n, m) \neq (0, 0)),$$

are constants, and write the Euler equation for it, namely

$$\frac{\partial \tilde{F}}{\partial f} - \frac{d}{dt_1} \frac{\partial \tilde{F}}{\partial f_{t_1}} - \frac{d}{dt_2} \frac{\partial \tilde{F}}{\partial f_{t_2}} = 0,$$

where

$$f_{t_1} = \frac{\partial f}{\partial t_1}, \quad f_{t_2} = \frac{\partial f}{\partial t_2}.$$

In our case,

$$\tilde{F} = \frac{(f - uv)^2}{uv} + \sum_{\substack{n=0 \\ (n,m) \neq (0,0)}}^{N} \sum_{m=0}^{M} \lambda_{nm} U_n V_m f,$$

and the Euler equation becomes

$$\frac{\partial \tilde{F}}{\partial f} = 0,$$

which implies

$$f = uv(1 - \sum_{\substack{n=0 \\ (n,m) \neq (0,0)}}^{N} \sum_{m=0}^{M} \frac{1}{2} \lambda_{nm} U_n V_m).$$

Introducing this expression into the constraints and using the orthonormality of the systems

$$\{U_n; \ n = 0, 1, \ldots, N\}, \quad \{V_m; \ m = 0, 1, \ldots, M\}$$

with respect to the weights u and v, respectively, we get

$$c_{nm} = < U_n V_m f > =$$

$$= < U_n V_m uv > - \sum_{\substack{k=0 \\ (k,r) \neq (0,0)}}^{N} \sum_{r=0}^{M} \frac{1}{2} \lambda_{kr} < U_n U_k V_m V_r uv > = -\frac{1}{2} \lambda_{nm},$$

because, as the variables are separable,

$$< U_n U_k V_m V_r uv > = < U_n U_k u > < V_m V_r v > = \delta_{nk} \delta_{mr},$$

where δ_{ij} is the Kronecker's symbol, namely

$$\delta_{ij} = 1, \text{ if } i = j, \text{ and } \delta_{ij} = 0, \text{ if } i \neq j.$$

Thus, we get

$$f = u\,v\,(1 + \sum_{\substack{n=0 \\ (n,m) \neq (0,0)}}^{N} \sum_{m=0}^{M} c_{nm} U_n V_m).$$

Obviously, $< f > = < uv > = 1$. Such a joint probability density function is the closest one, in the $\tilde{\chi}^2$ sense, to the direct independent product uv subject to the generalized mixed moments (or correlations)

$$c_{nm} = < U_n V_m f >, \quad (n = 0, 1, \ldots, N; m = 0, 1, \ldots, M; (n, m) \neq (0, 0)).$$

Example: Suppose that we have two standard normal $\mathbf{N}(0, 1)$ random variables X and Y, the corresponding probability density functions u and v being

$$u(x) = \frac{1}{\sqrt{2\pi}} e^{-x^2/2}, \quad x \in (-\infty, +\infty),$$

$$v(y) = \frac{1}{\sqrt{2\pi}} e^{-y^2/2}, \quad y \in (-\infty, +\infty),$$

respectively. Taking $N = M = 2$, the corresponding orthonormal functions are the Hermite polynomials:

$$U_0(x) \equiv 1, \quad U_1(x) = x, \quad U_2(x) = \frac{1}{\sqrt{2}}(x^2 - 1),$$

$$V_0(y) \equiv 1, \quad V_1(y) = y, \quad V_2(y) = \frac{1}{\sqrt{2}}(y^2 - 1).$$

If X and Y are independent, then the joint probability distribution of the random vector (X, Y) has the probability density function

$$u(x)v(y) = \frac{1}{2\pi} e^{-(x^2 + y^2)/2}.$$

If, however, X and Y are dependent and the only information about their interdependence is given by the generalized correlations

$$c_{11} = <U_1 V_1 f>, \qquad c_{12} = <U_1 V_2 f>,$$

$$c_{21} = <U_2 V_1 f>, \qquad c_{22} = <U_2 V_2 f>,$$

then, the closest joint density function f to the independent direct product uv of the marginals is

$$f(x, y) = u(x)v(y)[1 + c_{11}U_1 V_1 + c_{12}U_1 V_2 + c_{21}U_2 V_1 + c_{22}U_2 V_2] =$$

$$= \frac{1}{2\pi} e^{-(x^2 + y^2)/2}[1 + c_{11}xy + c_{12}\frac{1}{\sqrt{2}}x(y^2 - 1) +$$

$$+ c_{21}\frac{1}{\sqrt{2}}(x^2 - 1)y + c_{22}\frac{1}{2}(x^2 - 1)(y^2 - 1)],$$

where closeness is meant with respect to the $\tilde{\chi}^2$ mean deviation. Its plot, for

$$c_{11} = 0.5, \quad c_{12} = 0.1, \quad c_{21} = 0.1, \quad c_{22} = 0.75,$$

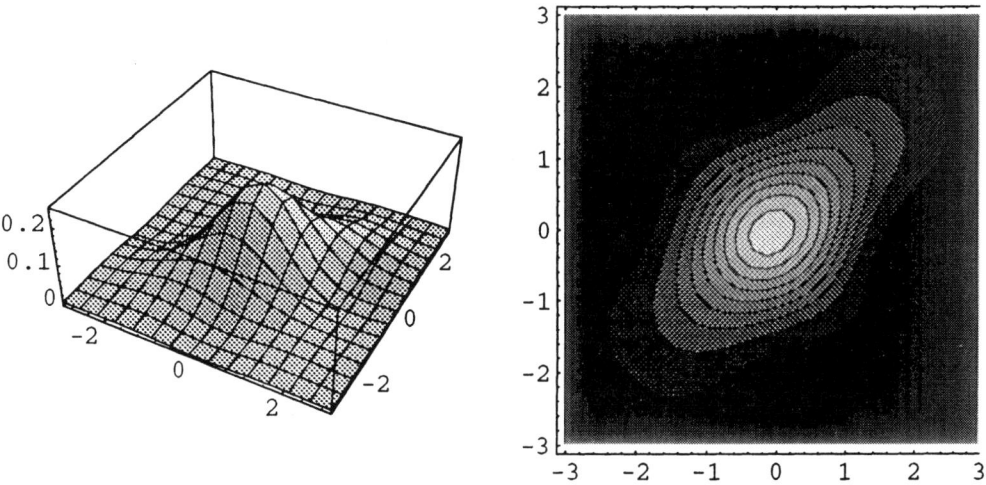

Figure 28: The plot of the joint density $f(x, y)$.

is shown in Figure 28. Such a function is the most independent joint density subject to the only information about the dependence between X and Y expressed by the four generalized correlations mentioned above. The corresponding probability wave function is

$$\chi(x,y) = \frac{f(x,y) - u(x)v(y)}{\sqrt{u(x)v(y)}} =$$

$$= \sqrt{u(x)v(y)}[c_{11}U_1V_1 + c_{12}U_1V_2 + c_{21}U_2V_1 + c_{22}U_2V_2] =$$

$$= \frac{1}{\sqrt{2\pi}}e^{-(x^2+y^2)/4}[c_{11}xy + c_{12}\frac{1}{\sqrt{2}}x(y^2 - 1)+$$

$$+c_{21}\frac{1}{\sqrt{2}}(x^2 - 1)y + c_{22}\frac{1}{2}(x^2 - 1)(y^2 - 1)],$$

and its plot, for the values of the generalized correlations mentioned above, is shown in Figure 29. The probability wave function describes the deviations from the state of independence of two components X and Y in statistical equilibrium, due to random fluctuations, when the only information available is represented by the generalized correlations c_{11}, c_{12}, c_{21}, and c_{22} between the generalized coordinates (directions) $\{U_1, V_1\}$, $\{U_1, V_2\}$, $\{U_2, V_1\}$, and $\{U_2, V_2\}$, respectively. The probability density induced by the probability wave function χ is

$$\frac{\chi^2(x,y)}{c_{11}^2 + c_{12}^2 + c_{21}^2 + c_{22}^2}.$$

It describes the probability distribution of the deviations from the state of independence of two components in statistical equilibrium, due to fluctuations, when the only information available about the change is provided by some generalized correlations, namely c_{11}, c_{12}, c_{21}, and c_{22}.

In general, if f is the closest joint density function to the independent direct product uv of the marginal probability density functions u and v which describe the statistical equilibrium, where closeness is measured by the mean chi-square deviation, the function $\chi = (f - uv)/\sqrt{uv}$ represents the weighted fluctuation of f from uv. Both f, the wave of minimum deviation from statistical equilibrium, and χ, the fluctuation from statistical equilibrium, may be taken as representatives of the probability wave functions of the joint system whose statistical equilibrium was perturbed by random disturbances.

152

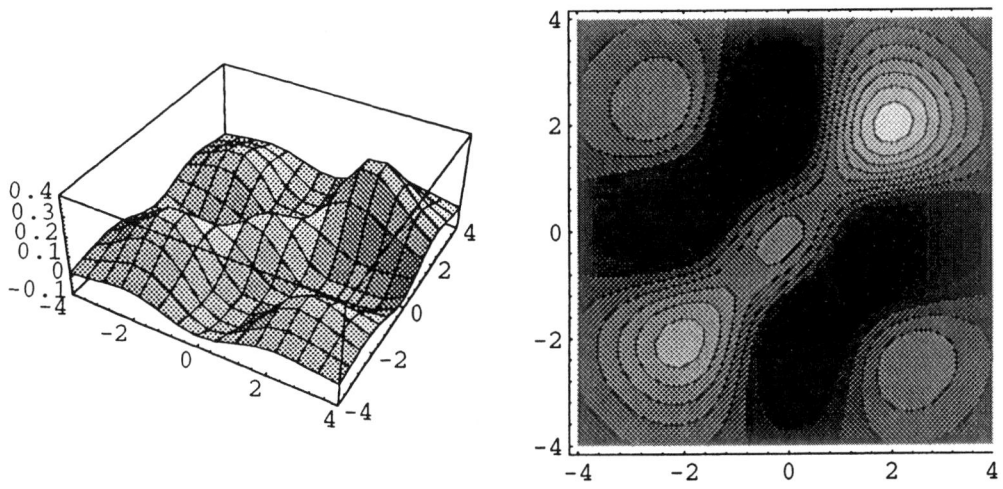

Figure 29: The plot of the probability wave function $\chi(x, y)$.

153

APPENDIX 5. A Series Representation of $\frac{1}{r_{12}}$

A definition of Legendre polynomial $P_n(x)$ is

$$P_n(x) = \frac{1}{n!2^n} \frac{d^n}{dx^n}[(x^2 - 1)^n],$$

from which, taking into account that

$$(x^2 - 1)^n = (x + 1)^n(x - 1)^n,$$

we get

$$P_n(x) = \frac{1}{n!2^n} \left[(x + 1)^n \left(\frac{d^n}{dx^n}(x - 1)^n\right) + \right. \tag{153}$$

$$\left. + \left(\frac{d}{dx}(x + 1)^n\right) \left(\frac{d^{n-1}}{dx^{n-1}}(x - 1)^n\right) + \ldots + \left(\frac{d^n}{dx^n}(x + 1)^n\right)(x - 1)^n\right].$$

Taking into account that

$$\frac{d^n}{dx^n}(x - 1)^n = n!,$$

$$\frac{d^k}{dx^k}(x - 1)^n \bigg|_{x=1} = 0, \quad \text{for } k < n,$$

we get from (153)

$$P_n(1) = 1. \tag{154}$$

In general, orthogonal polynomials $\{f_n(x); n = 0, 1, \ldots\}$ may be defined as being the terms of a series representation of a generating function $g(x, z)$, namely,

$$g(x, z) = \sum_{n=0}^{+\infty} a_n f_n(x) z^n.$$

Legendre polynomials $\{P_n(x); n = 0, 1, \ldots\}$ are obtained by taking $a_n = 1, (n = 0, 1, \ldots)$ and the generating function given by (Abramowitz and Stegun, 1972, p.784)

$$g(x, z) = \frac{1}{\sqrt{1 - 2xz + z^2}}. \tag{155}$$

Thus, we have

$$\frac{1}{\sqrt{1 - 2xz + z^2}} = \sum_{n=0}^{+\infty} P_n(x) z^n. \tag{156}$$

Let us take two electrons and let r_1 and r_2 be the distances from the nucleus, located at the origin of the coordinate system, and let θ be the angle between r_1 and r_2. We choose one of the radius vectors, say r_1, to be the z-axis. Let us take $r_1 = 1$, and denote $r_2 = r$. Using the law of cosines,

$$r_{12} = \sqrt{r_1^2 + r_2^2 - 2r_1 r_2 \cos\theta} = \sqrt{1 - 2r\cos\theta + r^2},$$

and, taking $x = \cos\theta$ and $z = r$, (156) becomes

$$\frac{1}{r_{12}} = \frac{1}{\sqrt{1 - 2r\cos\theta + r^2}} = \sum_{n=0}^{+\infty} P_n(\cos\theta) r^n. \tag{157}$$

Solving the equation

$$1 - 2xz + z^2 = 0,$$

we obtain the roots

$$z = x \pm \sqrt{x^2 - 1} = x \pm i\sqrt{1 - x^2}, \quad (i = \sqrt{-1}). \tag{158}$$

As $x = \cos\theta$, it must be a real number belonging to the interval $-1 < x < +1$. Then, the roots (158) are complex conjugate and the square of the module of each such root is equal to 1, namely

$$x^2 + (\sqrt{1 - x^2})^2 = 1.$$

For $x = \pm 1$, the roots (158) coincide and are both equal to ± 1. Thus, under the condition $-1 \le x \le +1$, the singular points af the function (155) are located at a distance equal to 1 from the origin of the coordinate system and, consequently, the series (156) is convergent for $|z| < 1$. In particular, the equality (157) holds for $r < 1$, i.e., for all the points that are inside the unit sphere. For the points outside the unit sphere we get another expression. Indeed, for $r > 1$, we can write

$$\frac{1}{\sqrt{1 - 2r\cos\theta + r^2}} = \frac{1}{r} \frac{1}{\sqrt{1 - 2(1/r)\cos\theta + (1/r)^2}}.$$

In this case we have $(1/r) < 1$, and using (157), we obtain the following representation outside the unit sphere

$$\frac{1}{\sqrt{1 - 2r\cos\theta + r^2}} = \sum_{n=0}^{+\infty} \frac{P_n(\cos\theta)}{r^{n+1}}. \tag{159}$$

Every term of this series has no singularity outside the unit sphere and tends to zero when r tends to $+\infty$.

Switching from $r_1 = 1$ and $r_2 = r$ to arbitrary r_1 and r_2, which means to replace the unit sphere in the above considerations by an arbitrary sphere of radius r_1, the formulas (157) and (159) become

$$\frac{1}{\sqrt{r_1^2 - 2r_1 r_2 \cos\theta + r_2^2}} = \sum_{n=0}^{+\infty} P_n(\cos\theta) \frac{r_2^n}{r_1^{n+1}}, \quad (r_2 < r_1);$$

$$\frac{1}{\sqrt{r_1^2 - 2r_1 r_2 \cos\theta + r_2^2}} = \sum_{n=0}^{+\infty} P_n(\cos\theta) \frac{r_1^n}{r_2^{n+1}}, \quad (r_1 < r_2);$$

which may be written in a condensed form

$$\frac{1}{r_{12}} = \sum_{n=0}^{+\infty} \frac{r_<^n}{r_>^{n+1}} P_n(\cos\theta), \tag{160}$$

where

$$r_< = \min\{r_1, r_2\} \quad \text{and} \quad r_> = \max\{r_1, r_2\},$$

and θ is the angle between r_1 and r_2.

Using spherical coordinates, if the two electrons are located at $(r_1, \theta_1, \omega_1)$ and $(r_2, \theta_2, \omega_2)$, respectively, let r_{12} be the distance between them. If we have the series

$$\frac{1}{r_{12}} = \sum_{n=0}^{+\infty} c_n P_n(\cos\theta_1) P_n(\cos\theta_2), \tag{161}$$

and we perform the rotation which makes the radius vectors r_1 to be the z-axis, then $\theta_1 = 0$, θ_2 becomes the angle θ between r_1 and r_2, the distance r_{12} between r_1 and r_2 remains invariant, and (161) becomes

$$\frac{1}{r_{12}} = \sum_{n=0}^{+\infty} c_n P_n(\cos\theta), \tag{162}$$

156

because, according to (154),

$$P_n(\cos 0) = P_n(1) = 1,$$

for each value of n, $(n = 0, 1, \ldots)$. But, from (160) and (162) we get

$$c_n = \frac{r_<^n}{r_>^{n+1}},$$

which, introduced into (161), gives

$$\frac{1}{r_{12}} = \sum_{n=0}^{+\infty} \frac{r_<^n}{r_>^{n+1}} P_n(\cos\theta_1) P_n(\cos\theta_2). \qquad (163)$$

Using $s = \cos\theta$, the series (162) may be written as

$$\frac{1}{r_{12}} = \sum_{n=0}^{+\infty} \frac{r_<^n}{r_>^{n+1}} P_n(s),$$

whereas using $s_1 = \cos\theta_1$ and $s_2 = \cos\theta_2$, the series (163) may be written as

$$\frac{1}{r_{12}} = \sum_{n=0}^{+\infty} \frac{r_<^n}{r_>^{n+1}} P_n(s_1) P_n(s_2).$$

The formulas (160) and (163) are frequently used in papers and textbooks dealing with the ground state of the helium atom, like Bethe and Salpeter (1957, p.221), Hylleraas (1963, p.427), Anderson (1971, p.381), Flügge (1974, vol.2, p.60), and McQuarrie (1983, p.340), for instance, but these formulas are given without a proof which, as seen above, is not trivial at all. In Part I of this monograph, (160) and (163) were used on pages 70-72. As it was discussed in Section 8, Hylleraas (1929) made a surprising breakthrough in the study of the ground state of helium atom by using the interelectron distance r_{12}, along with the radial distances r_1 and r_2 of the two electrons from the nucleus, in the trial function proposed by him. On the other hand, we have seen in Section 2 that the Legendre (spherical) polynomials $\{P_n(x); n = 0, 1, \ldots\}$ form a system of orthonormal polynomials with the weight given by the uniform probability distribution which describes the distribution of the spatial variables ν and ω. The formulas (160) and (163) just show that r_{12} may be expressed as a series of Legendre polynomials.

APPENDIX 6. Computer Program 1.

Using MATHEMATICA, Version 2.1 (Wolfram, (1991)), including the subroutine 'integExp' for calculating exponential integrals of the form

$$\int_0^{+\infty} r^n e^{-ar} dr,$$

from the package "Quantum" (Feagin, (1994)) which may be added to the packages of MATHEMATICA. In the program below, in order to simplify the writing, 'm', 'phi', and 'psi' are the symbols replacing μ, φ, and ψ, respectively. At the end of each line press the Enter key. The commands are numbered by the computer. In the programs, %23, for instance, is the output of command [23], and % refers to the output of the previous command. MATHEMATICA is case sensitive; its commands, except 'integExp', start with a capital letter. If a command ends with a semicolon ; then its output is not shown on the screen.

```
math
[1] Needs["Quantum'integExp'"]
[2] phi[r_,m_]:=((8*m^3*Pi)^(-1/2)) *Exp[-r/(2*m)]
[3] psi[r1_,r2_]:=phi[r1,m]*phi[r2,m]*(1 +a*(LaguerreL[1,2,r1/m]+
    LaguerreL[1,2,r2/m])+ b*LaguerreL[1,2,r1/m]*LaguerreL[1,2,r2/m]+
    c*(LaguerreL[2,2,r1/m]+LaguerreL[2,2,r2/m])+
    q*(LaguerreL[1,2,r1/m]*LaguerreL[2,2,r2/m]+
    LaguerreL[2,2,r1/m]*LaguerreL[1,2,r2/m])+
    g*LaguerreL[2,2,r1/m]*LaguerreL[2,2,r2/m])
[4] integExp[4*Pi*r1^2*psi[r1,r2]^2, {r1,0,Infinity}];
[5] integExp[4*Pi*r2^2*%,{r2,0,Infinity}];
[6] norm=Together[%]
[7] integExp[4*Pi*r1*psi[r1,r2]^2, {r1,0,Infinity}];
[8] integExp[4*Pi*r2^2*%,{r2,0,Infinity}];
[9] invr1=Together[%]
[10] laplace[r1_,r2_]:=D[psi[r1,r2],{r1,2}]+2* D[psi[r1,r2],r1]/r1
[11] Expand[4*Pi*r1^2*psi[r1,r2]* laplace[r1,r2]];
[12] integExp[%,{r1,0,Infinity}];
[13] integExp[4*Pi*r2^2*%,{r2,0,Infinity}];
[14] laplace1=Together[%]
[15] integExp[4*Pi*r2*psi[r1,r2]^2, {r2,0,Infinity}];
```

[16] integExp[4⋆Pi⋆r1^2⋆%,{r1,0,Infinity}];
[17] Integrate[4⋆Pi⋆r2⋆(r2−r1)⋆ psi[r1,r2]^2,{r2,0,r1}];
[18] integExp[4⋆Pi⋆r1⋆%,{r1,0,Infinity}];
[19] %+%%%;
[20] invr12=Together[%]
[21] energy=(−2⋆laplace1/2−2⋆2⋆invr1 +invr12)/norm
[22] FindMinimum[energy,{m,1},{a,0},{b,0},{c,0},{q,0}, {g,0}]
[23] Quit

Command [1] calls the subroutine 'integExp'. Commands [2] and [3] define the functions φ and ψ, respectively. Commands [4]–[6] calculate the integral of ψ^2, denoted by 'norm'. Commands [7]–[9] calculate the integral of $1/r_1$ with respect to the density ψ^2, denoted by 'invr1'. Because of symmetry, 'invr2' is not needed, beeing equal to 'invr1'. Commands [10]–[14] calculate the integral of $\nabla_1^2\psi$ with respect to the density ψ, denoted by 'laplace1'. Because of symmetry, 'laplace2' is not needed, beeing equal to 'laplace1'. Commands [15]–[20] calculate the integral of $1/r_{12}$ with respect to the density ψ^2, denoted by 'invr12'. Finally, command [21] gives the mean energy of the atom, denoted by 'energy', as a function of all variational parameters. Command [22] calls the fast subroutine 'FindMinimum' to locate the minimum of the mean energy starting from a search point given by the mean radial distance 1 a.u. and the variational parameters a, b, c, q, g equal to zero, corresponding to the statistical equilibrium of the atom. The outcome of command [22] is a message: "FindMinimum failed to converge to the prescribed accuracy within 30 iterations," followed by the current numerical results of the search, namely,

$$\{-2.87665, \{m- > 0.306391, a- > -0.0081968, b- > -0.0212219,$$

$$c- > 0.0231974, q- > -0.00126046, g- > -0.00128905\}\},$$

where the first number refers to the value of the mean energy. We repet command [22], but starting the search for minimum of the mean energy from the values of variational parameters given in the output of [22] mentioned above. This time, the minimum of the mean energy of the atom is attained.

APPENDIX 7. Computer Program 2.

Using MATHEMATICA, Version 2.1 (Wolfram, (1991)), including the sub-routine 'integExp' for calculating exponential integrals of the form

$$\int_0^{+\infty} r^n e^{-ar} dr,$$

from the package "Quantum" (Feagin, (1994)) which may be added to the software MATHEMATICA. In the program below, in order to simplify the writing, x, y, m, u, v and psi[x,y,u,v] are the symbols replacing r_1, r_2, μ, s_1, s_2 and $\psi(r_1, r_2, s_1, s_2)$, respectively. At the end of each line press the Enter key. The commands are numbered by the computer. In the programs, %23, for instance, is the output of command [23], and % refers to the output of the previous command. MATHEMATICA is case sensitive; its commands, except 'integExp', start with a capital letter. If a command ends with a semicolon ; then its output is not shown on the screen.

math

[1] Needs["Quantum'integExp'"]

[2] phi[x_]:=((8⋆(m^3)⋆Pi)^(−1/2))⋆ Exp[−x/(2⋆m)]

[3] psi[x_,y_,u_,v_]:=phi[x]⋆phi[y]⋆(1+a⋆ (LaguerreL[1,2,x/m]+
    LaguerreL[1,2,y/m])+b⋆LaguerreL[1,2,x/m]⋆LaguerreL[1,2,y/m]+
    c⋆(LaguerreL[2,2,x/m]+LaguerreL[2,2,y/m])+
    g⋆LaguerreL[2,2,x/m]⋆LaguerreL[2,2,y/m])+
    f⋆LegendreP[1,u]⋆LegendreP[1,v]);

[4] Expand[4⋆Pi^2⋆x^2⋆psi[x,y,u,v]^2];

[5] integExp[%,{x,0,Infinity}];

[6] integExp[y^2⋆%,{y,0,Infinity}];

[7] Integrate[%,{u,−1,1},{v,−1,1}];

[8] laplace[x_,y_,u_,v_]:=D[psi[x,y,u,v],{x,2}]+
    2⋆D[psi[x,y,u,v],x]/x−2⋆u⋆D[psi[x,y,u,v],u]/(x^2)

[9] Expand[4⋆Pi^2⋆x^2⋆psi[x,y,u,v]⋆ laplace[x,y,u,v]];

[10] integExp[%,{x,0,Infinity}];

[11] integExp[y^2⋆%,{y,0,Infinity}];

[12] Integrate[%,{u,−1,1},{v,−1,1}];

[13] integExp[4⋆Pi^2⋆x⋆psi[x,y,u,v]^2, {x,0,Infinity}];

[14] integExp[y^2⋆%,{y,0,Infinity}];

[15] Integrate[%,{u,−1,1},{v,−1,1}];

[16] Expand[((1/x−1/y+(y/x^2−x/y^2)⋆u⋆v)⋆
      (psi[x,y,u,v]/(phi[x]⋆phi[y]))^2];
[17] Integrate[%,{u,−1,1},{v,−1,1}];
[18] Expand[%⋆y^2⋆phi[y]^2⋆(2⋆Pi)];
[19] Integrate[%,{y,0,x}];
[20] Expand[((1/y+(x/y^2)⋆u⋆v) ⋆(psi[x,y,u,v]/(phi[x]⋆phi[y]))^2];
[21] Integrate[%,{u,−1,1},{v,−1,1}];
[22] Expand[%⋆y^2⋆psi[y]^2⋆(2⋆Pi)];
[23] integExp[%,{y,0,Infinity}];
[24] %19+%;
[25] integExp[%⋆x^2⋆phi[x]^2⋆(2⋆Pi), {x,0,Infinity}];
[26] energy=(−%12−4⋆%15+%)/%7
[27] FindMinimum[energy,{m,1},{a,0},{b,0},{c,0},{g,0},{f,0}]
[28] Quit

Command [1] calls the subroutine 'integExp'. Commands [2] and [3] define the functions φ and ψ, respectively. Commands [4]–[7] calculate the integral of ψ^2. Commands [8]–[12] calculate the integral of $\nabla_1^2\psi$ with respect to the density ψ. Because of symmetry, the integral of $\nabla_2^2\psi$ with respect to ψ is the same. Commands [13]–[15] calculate the integral of $1/r_1$ with respect to the density ψ^2. Because of symmetry, the integral of $1/r_2$ with respect to the density ψ^2 is the same. Commands [16]–[25] calculate the integral of $1/r_{12}$ with respect to the density ψ^2. Command [26] gives the mean energy of the atom, denoted by 'energy', as a function of all variational parameters. Command [27] calls the fast subroutine 'FindMinimum' to locate the minimum of the mean energy starting from a search point given by the mean radial distance 1 a.u. and the variational parameters a, b, c, g, f equal to zero, corresponding to statistical equilibrium. The outcome of command [27] is a message: "FindMinimum failed to converge to the prescribed accuracy within 30 iterations," followed by the current numerical results of the search,

$$\{-2.88081, \{m- > 0.292157, a- > -0.0307317, b- > -0.0219266,$$
$$c- > 0.0223345, g- > -0.00132335, f- > -0.0530259\}\},$$

where the first number refers to the value of the mean energy. We repet command [27], but starting the search for minimum from the values of variational parameters given by the output of [27] mentioned above. This time, the minimum of the mean energy of the atom is attained.

APPENDIX 8. Computer Program 3.

Using MATHEMATICA, Version 2.1 (Wolfram, (1991)), including the subroutine 'integExp' for calculating exponential integrals of the form

$$\int_0^{+\infty} r^n e^{-ar} dr,$$

from the package "Quantum" (Feagin, (1994)) which may be added to the software MATHEMATICA. In the program below, in order to simplify the writing, x, y, z, psi[x,y,z], m, and hampsi are the symbols replacing r_1, r_2, r_{12}, $\psi(r_1, r_2, r_{12})$, μ, and $\hat{H}\psi$, respectively. At the end of each line press the Enter key. The commands are numbered by the computer. In the programs, %23, for instance, is the output of command [23], and % refers to the output of the previous command. MATHEMATICA is case sensitive; its commands, except 'integExp', start with a capital letter. If a command ends with a semicolon ; then its output is not shown on the screen.

math

[1] Needs["Quantum'integExp'"]
[2] psi[x_,y_,z_]:=Exp[−(x+y)/(2⋆m)]⋆(1+c⋆z+
    b⋆LaguerreL[1,1,x/m]⋆LaguerreL[1,1,y/m]+
    a⋆(LaguerreL[2,1,x/m]+LaguerreL[2,1,y/m]))
[3] hampsi[x_,y_,z_]:=−D[psi[x,y,z],{x,2}]/2−
    D[psi[x,y,z],x]/x−D[psi[x,y,z],{y,2}]/2−
    D[psi[x,y,z],y]/y−D[psi[x,y,z],{z,2}]−
    2⋆D[psi[x,y,z],z]/z−(x^2−y^2+z^2)⋆
    D[D[psi[x,y,z],x],z]/(2⋆x⋆z)− (y^2−x^2+z^2)⋆
    D[D[psi[x,y,z],y],z]/(2⋆y⋆z)− (2/x+2/y−1/z)⋆psi[x,y,z]
[4] Expand[psi[x,y,z]⋆hampsi[x,y,z]⋆x⋆y⋆z];
[5] Integrate[%,{z,x−y,x+y}];
[6] Integrate[%,{y,0,x}];
[7] integExp[%,{x,0,Infinity}];
[8] Integrate[%4,{z,y−x,x+y}];
[9] integExp[%,{y,0,Infinity}];
[10] Integrate[%%,{y,0,x}];
[11] %%−%;
[12] integExp[%,{x,0,Infinity}];
[13] %7+%;

162

[14] Expand[psi[x,y,z]^2*x*y*z];
[15] Integrate[%,{z,x−y,x+y}];
[16] Integrate[%,{y,0,x}];
[17] integExp[%,{x,0,Infinity}];
[18] Integrate[%14,{z,y−x,x+y}];
[19] integExp[%,{y,0,Infinity}];
[20] Integrate[%%,{y,0,x}];
[21] %%−%;
[22] integExp[%,{x,0,Infinity}];
[23] %17+%;
[24] energy=%13/%
[25] FindMinimum[energy,{m,1},{c,0},{b,0},{a,0}]
[26] Quit

The output of command [24] gives the mean energy of the atom, denoted by 'energy', as a function of all variational parameters. Its expression here is

$$energy = (8 + 144a^2 - 16b - 96ab + 128b^2 - 54m - 12am - 687a^2m +$$
$$+12bm + 42abm - 435b^2m + 50cm + 36acm - 86bcm -$$
$$-416cm^2 - 400acm^2 - 16bcm^2 + 128c^2m^2 - 1012c^2m^3)/$$
$$(8m^2(4 + 72a^2 + 8b + 48ab + 64b^2 + 35cm + 42acm + 63bcm + 96c^2m^2))$$

Command [25] calls the fast subroutine 'FindMinimum' to locate the minimum of the mean energy starting from a search point given by the mean radial distance 1 a.u. and the variational parameters a, b, c equal to zero, corresponding to statistical equilibrium. The output of [25] has the form:

$$\{-2.90252, \{m- > 0.281424, c- > 0.281515, b- > -0.0171302,$$
$$a- > 0.0204191\}\},$$

where the first number is the minimum value of the ground state mean energy of the helium atom, an excellent approximation using only four variational parameters m, a, b, c. The experimental value is -2.90336 a.u., as mentioned in Lide (2000, p.**10**-175). On an IBM ThinkPad 380 personal computer with Intel Pentium processor, 80MB RAM and 400MB available on the hard disk, the typing and running of the above MATHEMATICA program took about 30 minutes.

APPENDIX 9. Computer Program 4.

Using MATHEMATICA, Version 2.1 (Wolfram, (1991)), including the sub-routine 'integExp' for calculating exponential integrals of the form

$$\int_0^{+\infty} r^n e^{-ar} dr,$$

from the package "Quantum" (Feagin, (1994)) which may be added to the software MATHEMATICA. The program below starts from the probability wave function (88), with the Hamiltonian (87) for which the Laplacian is (89). In order to simplify the writing, m, psi[x,y], and hampsi are the symbols replacing μ, $\psi(r_1, r_2, r_3)$, and $\hat{H}\psi$, respectively. At the end of each line press the Enter key. The commands are numbered by the computer. In the programs, %23, for instance, is the output of command [23], and % refers to the output of the previous command. MATHEMATICA is case sensitive; its commands, except 'integExp', start with a capital letter. If a command ends with a semicolon ; then its output is not shown on the screen.

math

[1] Needs["Quantum'integExp'"]

[2] phi[r_,m_]:=((8⋆m^3⋆Pi)^(−1/2)) ⋆Exp[−r/(2⋆m)]

[3] psi[r1_,r2_,r3_]:=phi[r1,m]⋆phi[r2,m]⋆ phi[r3,2]⋆
    (1+c⋆ LaguerreL[1,2,r1/m]⋆LaguerreL[1,2,r2/m])

[4] integExp[4⋆Pi⋆r1^2⋆ psi[r1,r2,r3]^2, {r1,0,Infinity}];

[5] integExp[4⋆Pi⋆r2^2⋆%,{r2,0,Infinity}];

[6] integExp[4⋆Pi⋆r3^2⋆%,{r3,0,Infinity}];

[7] norm=Together[%]

[8] integExp[4⋆Pi⋆r1⋆psi[r1,r2,r3]^2, {r1,0,Infinity}];

[9] integExp[4⋆Pi⋆r2^2⋆%,{r2,0,Infinity}];

[10] integExp[4⋆Pi⋆r3^2⋆%,{r3,0,Infinity}];

[11] invr1=Together[%]

[12] integExp[4⋆Pi⋆r3⋆psi[r1,r2,r3]^2, {r3,0,Infinity}];

[13] integExp[4⋆Pi⋆r1^2⋆%,{r1,0,Infinity}];

[14] integExp[4⋆Pi⋆r2^2⋆%,{r2,0,Infinity}];

[15] invr3=Together[%]

[16] laplace[r1_,r2_,r3_]:=D[psi[r1,r2,r3],{r1,2}]+
    2⋆D[psi[r1,r2,r3],r1]/r1

[17] Expand[4⋆Pi⋆r1^2⋆psi[r1,r2,r3]⋆ laplace[r1,r2,r3]];

[18] integExp[%,{r1,0,Infinity}];
[19] integExp[4⋆Pi⋆r2^2⋆%,{r2,0,Infinity}];
[20] integExp[4⋆Pi⋆r3^2⋆%,{r3,0,Infinity}];
[21] laplace1=Together[%]
[22] laplace[r1_,r2_,r3_]:=D[psi[r1,r2,r3],{r3,2}]+
    2⋆D[psi[r1,r2,r3],r3]/r3
[23] Expand[4⋆Pi⋆r3^2⋆psi[r1,r2,r3]⋆ laplace[r1,r2,r3]];
[24] integExp[%,{r3,0,Infinity}];
[25] integExp[4⋆Pi⋆r1^2⋆%,{r1,0,Infinity}];
[26] integExp[4⋆Pi⋆r2^2⋆%,{r2,0,Infinity}];
[27] laplace3=Together[%]
[28] integExp[4⋆Pi⋆r3^2⋆psi[r1,r2,r3]^2, {r3,0,Infinity}];
[29] integExp[4⋆Pi⋆r2⋆%,{r2,0,Infinity}];
[30] integExp[4⋆Pi⋆r1^2⋆%,{r1,0,Infinity}];
[31] Integrate[4⋆Pi⋆r2⋆(r2−r1)⋆%%%, {r2,0,r1}];
[32] integExp[4⋆Pi⋆r1⋆%,{r1,0,Infinity}];
[33] %+%%%;
[34] invr12=Together[%]
[35] integExp[4⋆Pi⋆r2^2⋆psi[r1,r2,r3]^2, {r2,0,Infinity}];
[36] integExp[4⋆Pi⋆r3⋆%,{r3,0,Infinity}];
[37] integExp[4⋆Pi⋆r1^2⋆%,{r1,0,Infinity}];
[38] Integrate[4⋆Pi⋆r3⋆(r3−r1)⋆%%%, {r3,0,r1}];
[39] integExp[4⋆Pi⋆r1⋆%,{r1,0,Infinity}];
[40] %+%%%;
[41] invr13=Together[%]
[42] energy=(−(2⋆laplace1+laplace3)/2− 3⋆(2⋆invr1+invr3)+
    (invr12+2⋆invr13))/norm
[43] FindMinimum[energy,{m,1},{c,0}]
[44] Quit

Command [1] calls the subroutine 'integExp'. Commands [2] and [3] define the functions φ and ψ, respectively. Commands [4]–[7] calculate the integral of ψ^2, denoted by 'norm'. Commands [8]–[15] calculate the integrals of $1/r_1$ and $1/r_3$ with respect to the density ψ^2, denoted by 'invr1' and 'invr3', respectively. Because of symmetry, 'invr2' is not needed, beeing equal to 'invr1'. Commands [16]–[27] calculate the integrals of $\nabla_1^2\psi$ and $\nabla_3^2\psi$ with respect to the density ψ, denoted by 'laplace1' and 'laplace3', respectively.

Because of symmetry, 'laplace2' is not needed, beeing equal to 'laplace1'. Commands [28]-[41] calculate the integrals of $1/r_{12}$ and $1/r_{13}$ with respect to the density ψ^2, denoted by 'invr12' and 'invr13', respectively. Because of symmetry, 'invr23' is equal to 'invr13'. Finally, command [42] gives the mean energy of the atom, denoted by 'energy', as a function of all variational parameters. Its expression is

$$\text{energy} = (256 + 5376c^2 - 2112m + 384cm - 12192c^2m - 6464m^2 +$$
$$+960cm^2 - 52656c^2m^2 - 7120m^3 + 960cm^3 - 62400c^2m^3 - 4176m^4 +$$
$$+480cm^4 - 42312c^2m^4 - 1452m^5 + 120cm^5 - 13242c^2m^5 - 284m^6 +$$
$$+12cm^6 - 2583c^2m^6 - 23m^7 - 207c^2m^5)/(32(1 + 9c^2)m^2(2 + m)^5)$$

Command [43] calls the fast subroutine 'FindMinimum' to locate the minimum of the mean energy starting from a search point given by the mean radial distance $m = 1$ a.u. and the variational parameter c equal to zero, corresponding to the statistical equilibrium of the lithium atom. The output is

$$\{-7.46071, \{m- > 0.18622, c- > -0.0122754\}\},$$

where the first number is the minimum value of the ground state mean energy, namely

$$-7.46071 \text{ a.u.} = -7.46071 \times 27.2116 \text{ eV} = -203.018 \text{ eV},$$

whereas, the experimental value (Lide, (2000, p.**10**-175)) is -203.486 eV. We have to add, however, that two debatable assumptions have been made here. First, we have taken here a simplified model of the lithium atom, where the three electrons keep all their four quantum numbers unchanged during their random motion around the nucleus. Second, we have placed the third electron at a fixed mean distance from the nucleus, equal to 2 a.u. in order to separate it from the first two electrons making up the first subshell of the atom. There was no such problem in dealing with the helium atom. The results show, however, that the shell structure of the lithium atom has to be essentially taken into account, but the question is how? Part II of the monograph will address this dilemma and the structure of the lithium atom will be analyzed again.

PART II.
THE LINEAR APPROXIMATION.

1. INTRODUCTION.

The maximum entropy probability distributions are generally used as probabilistic models for describing statistical equilibrium subject to given mean values of some random variables. The paper uses probability wave functions obtained by minimizing the mean deviation from statistical equilibrium subject to generalized moments and correlations, whose values are determined looking for stationary points of the mean energy of the quantum system. In Part I of this monograph, the mathematical formalism was applied to the study of some quantum systems (the free particle in a one- or three-dimensional box, the two noninteracting free particles in a one-dimensional box, the harmonic oscillator, and the hydrogen atom) without using the standard approach based on solving the corresponding Schrödinger equations, and to the study of the ground state of helium and lithium atoms. Part II of the monograph shows how the mathematical tool provided by the minimum mean deviation from statistical equilibrium may be used for building up four probabilistic models for the ground state of more complex atoms and uses the probability wave functions induced by this mathematical formalism both for calculating global, standard, or generalized correlations between interacting electrons, and for measuring the degree of stability of different systems of electrons or the amount of interdependence among subsystems of electrons of the atom. Dealing with models for more complex atoms, the main objective is to reinstate the orbits of the electrons, so familiar and useful to chemists, replacing, however, the rigid ellipses of the old quantum mechanics with a probabilistic description of the subshells of the atom, generated by the random motion of the electrons around the nucleus, whose mean orbits are given by weighted Bohr's quantified distances from the nucleus. The second part of Part II will deal with applications of the general results obtained to several atoms, like hydrogen, lithium, beryllium, boron, carbon, nitrogen, and argon atoms, together with the corresponding computer programs using the symbolic mathematics software package MATHEMATICA.

As stated above, the main objective of this Part II is to show how the variational principle of minimum mean deviation from statistical equilibrium may be used for building new probabilistic models for the structure of atoms which allow us to calculate both different correlations between electrons and the interdependence among different subsystems of electrons. Building a new model requires, however, to look to past similar models.

The atoms of the chemical elements are complex systems. The path towards a model of atomic structure has been a long one and was thoroughly analyzed in Mehra and Rechenberg (1982). Selecting briefly some relevant contributions, we start with André Marie Ampère (1814) who pictured the atoms of chemical elements as being composed of subatomic particles. Gustav Theodor Fechner (1826, 1828) assumed that the fundamental force in atoms, as in astronomy, was gravitation. Wilhelm Weber (1871,1875) replaced Fechner's gravitational attraction by electric forces. About 30 years later, Joseph John Thomson (1904) proposed "the view that the atoms of elements consist of a number of negatively charged electrified corpuscles enclosed in a sphere of uniform positive electrification." He stated that when the number of electrons exceeded three they tended to form rings or spherical shells and the number of electrons in these shells was somewhat reminiscent of the periods in the periodic system of chemical elements. John William Nicholson (1911) imagined the primary atoms to consist of small spheres of negative electricity, rotating in a ring about an even smaller sphere of positive electricity, the atomic nucleus. Ernest Rutherford (1911) suggested a model in which the atom had to be pictured as consisting of a point-like nucleus having a positive charge $Z|e|$, surrounded at some distance by Z electrons. He referred to the 'Saturnian' atom proposed by Hantaro Nagaoka (1904) who had assumed that the atom resembled the system of the planet Saturn with its rings, taking this idea from an essay of James Clerk Maxwell on the stability of Saturn's rings; in particular, the negative electrons should move in rings around the attractive centre of a positively charged mass. Niels Bohr (1913) published three papers about a theory of the constitution of atoms on the basis of Rutherford's atomic model and Planck's elementary quantum of action. He pictured these atoms as systems consisting of a positively charged nucleus surrounded by electrons moving in circular orbits and mentioned that the classical mechanics could be used for describing the dynamical equilibrium of the atomic systems in the stationary states, whereas the transition of the systems between stationary states had to be explained by the emission of a homogeneous radiation where the relation between the frequency and the amount of energy emitted obeyed Max Planck's quantum theory. Alfred Landé (1919) replaced Bohr's one-electron-ring helium atom with a new model where the two electrons moved in opposite directions on different orbits. Gilbert Newton Lewis (1916) emphasized the importance of octets of electrons in the structure of atoms and visualized these octets to occupy the corners of a cube. Lewis' cubical model of atom was generalized by Irving Langmuir (1919) who assumed that the electrons in any given atom are either stationary or rotate, revolve or oscillate, being distributed in equal cells on a series of concentric almost spherical shells of equal thickness, whose mean radii form an arithmetic series 1,2,3,.... John William Nicholson (1914a, 1914b) claimed that a concentric ring-structure in the atom was not stable and argued that either the electrons had to move in different planes or each electron described a properly oriented elliptical orbit. On the same line of thought, Arnold Sommerfeld (1918) wrote: "One has to assume that the n electrons are not distributed on the same ellipse; rather, each electron describes its own ellipse, congruent to the others, around the nucleus as the focus." He also dealt with the occupation of the levels in atoms of elements in the periodic system of elements, assuming that the light atoms contained 2,8,8,18 electrons on the K-, L-, M-, and N-rings, respectively, in agreement with the chemical properties. For the purpose of calculating the energy states of the atoms he returned to the arrangement of electrons in concentric rings. Walther Kossel (1916) summarized the simplest representation of the structure of the atoms corresponding to the elements of the periodic system of elements in the old quantum theory: "The organization of the inner electrons should always remain similar for successive elements; it should only be changed continuously in its dimensions, due to the continuous increase of the nuclear charge.... The next electron, which appears in the heavier element, should always be added at the periphery. ...The electrons, which are added further, should be put into concentric rings or shells, on each of which ... only a certain number of electrons ...should be arranged. As soon as one ring or shell is completed, a new one has to be started for the next element; the number of electrons, which are most easily accessible and lie at the outermost periphery, increases again from element to element and, therefore, in the formation of each new shell the chemical periodicity is repeated."

Summarizing, in the old quantum mechanics, the electrons moved on rigid orbits under the attractive force of nuclei and the repulsion of neighbouring electrons. After the birth of the new quantum mechanics, the Schrödinger equation (Schrödinger (1926)) became one of the basic postulates of the non-relativistic quantum mechanics. Its solution is the so called wave function ψ of the corresponding quantum system. It was Max Born (1926) who interpreted $|\psi|^2$ as being a probability density that may be used for making probabilistic predictions about the behavior of the quantum systems. An atom is now a quantum system consisting of electrons moving randomly around a nucleus. Unfortunately, the Schrödinger equation may be solved exactly only for a very limited number of quantum systems, namely the free particle in a box, the harmonic oscillator, and the hydrogen atom. For other quantum systems, like the atoms of the chemical elements involving two or more electrons, approximate methods had to be used, namely the perturbation theory and the variational method. Perturbation theory expresses the solution to one problem in terms of another problem that has been solved previously. Specifically, it starts from the unrealistic case of completely non-interacting particles and goes through a set of successive corrections to the starting unperturbed problem. The variational method, commonly used in quantum chemistry, provides an upper bound to the ground state energy of a system and the wave function is approximated starting from an arbitrary trial function whose parameters are determened by minimizing the mean energy of the respective quantum system.

Both perturbation theory and the variational method give good results for helium atom, but both methods become cumbersome as the number of electrons in the atom increases. Generally, the Z-electron atomic wave function is expressed in terms of Z single-electron wave functions called orbitals. Electron spin is taken into account by using products of spatial orbitals and spin eigenfunctions. Such products are called spin orbitals and are described by the four quantum numbers n, ℓ, m, and s. According to Pauli's exclusion principle, the electronic wave function has to be antisymmetric under the interchange of the coordinates of any two electrons. A way of getting an antisymmetric wave function, called the Hartree-Fock wave function, is to take it as being the determinant of spin orbitals, called the Slater determinant. Details may be found in McQuarrie (1983). For the helium atom, Hylleraas (1929) suggested an ingenious trial function and, applying the variational method, obtained a surprisingly accurate approximation of the ground state energy. His line of thought has been followed until these days, but its applications to more complex atoms are difficult computationally and have abandoned the orbital concept altogether. Or, as underlined by McQuarrie (1983, p.294), "the orbital concept has been of great use to chemists and so the trend nowadays is to find the Hartree-Fock orbitals ... and to correct these by perturbation theory." Nevertheless, the probabilistic justification and interpretation of the corrected wave functions obtained in this way are far from being clear.

Proposing a new statistical model for quantum mechanics, our objectives may be grouped in the following categories:

(a) Erwin Schrödinger's aim in 1926 was to construct the mathematical tool for an undulatory theory of matter, initiated by Louis de Broglie in 1924. There was no probability involved in his work. As, due to Max Born, the ultimate use of the Schrödinger equation nowadays is to provide us with a probabilistic model for the behavior of quantum systems, it is more natural to start by building up a probabilistic model of the quantum systems without using the Schrödinger equation as a postulate of the nonrelativistic quantum mechanics. The maximum entropy probability distributions are generally used as probabilistic models for describing statistical equilibrium subject to given mean values of some random variables. Random fluctuations due to external perturbations or/and internal interdependence among the components of the quantum system alter this ideal statistical equilibrium. The paper uses probability wave functions obtained by minimizing the mean deviation from statistical equilibrium subject to generalized moments and correlations, whose values are determined looking for stationary points of the mean energy of the quantum system. In Part I of this monograph, this mathematical formalism was applied to the study of some quantum systems (the free particle in a one- or three-dimensional box, the two noninteracting free particles in a one-dimensional box, the harmonic oscillator, and the hydrogen atom), without using the standard approach based on solving the corresponding Schrödinger equations, and to the study of the ground state of helium and lithium atoms. Part II shows how the linear approximation of the mathematical tool provided by the minimum mean deviation from statistical equilibrium may be used for building up four models for the ground state of more complex atoms, namely the calm, nervous, turbulent, and nebulous models, depending on how the probability wave function is made antisym-

171

metric, as required by Pauli's principle, and uses the wave functions induced by this mathematical formalism both for calculating global, standard, or generalized correlations between interacting electrons and for measuring the degree of stability of different systems of electrons or the amount of interdependence among subsystems of electrons of the atom. Dealing with models for more complex atoms, the main objective is to reinstate the orbits of the electrons, so familiar and useful to chemists, but replacing the rigid ellipses of the old quantum mechanics with a probabilistic description of the subshells of the atom, generated by the possible trajectories of the random motion of the electrons around the nucleus, whose mean orbits are given by weighted Bohr's quantified distances from the nucleus. Subsequently, the general results will be applied to the structure of several atoms, like hydrogen, helium, lithium, beryllium, boron, carbon, nitrogen, and argon atoms, in the ground state. The corresponding computer programs using the symbolic mathematics software package MATHEMATICA will be also included.

(b) Dealing with atoms of chemical elements, the Schrödinger equation may be solved exactly only for the hydrogen atom. For the helium atom, if the variational method is used, a trial function is chosen, depending on some variational parameters whose values are determined by minimizing the mean energy of the quantum system. Such an approach is based on the variational theorem which states that the mean energy of a quantum system reaches its minimum value for the wave function of the ground state of the respective system, offering a lower bound for the mean energy when arbitrary trial functions are taken as candidates for the wave function of the ground state of the quantum system. Helium atom, as the first atom for which the corresponding Schrödinger equation cannot be solved exactly, has raised many problems because there is no unique way of choosing a suitable trial function and no clear methodology for how to handle this problem. If the two electrons of the helium atom are supposed to be independent, the corresponding trial functions do not give values of the mean energy of the ground state of this atom close enough to the experimental value. The interdependence between the two atoms has to be taken into account, but how? The ingenious trial function proposed by Hylleraas (1929) gave an astonishingly good approximation but, unfortunately, it cannot be generalized to more complex atoms. The atom of the next chemical element, the lithium atom, has raised new difficulties. Thus, unlike what happened with the helium atom, if the three electrons of the lithium atom are supposed to be independent in their motion around the nucleus, then the corresponding mean energy of the ground state of the atom proves to be smaller than the experimental value, contradicting the variational theorem! This strange result is a consequence of ignoring the complex structure of the lithium atom which, unlike the helium atom, has two subshells. But how to take this shell-structure of complex atoms into account when we construct trial functions for the variational method? The mathematical formalism based on minimum mean deviation from statistical equilibrium we are dealing with here, provides a methodology for choosing suitable trail functions which works not only for helium atom but for more complex atoms as well. In Bohr's planetary model of atom the orbits of the electrons were rigid trajectories. In the probabilistic model presented here, the electrons move randomly, but each subshell of the atom is characterized by a mean orbit which applies to the electrons belonging to this subshell; the mean orbits obey a weighted form of Bohr's rule for the succession of the orbits. The numerical values are surprisingly good and show that somehow Bohr was right afterall, but only on average!

(c) Obviously, the electrons are not independent inside the atom. The inner structure of different atoms is customarily described using mainly experimental values of the ionization potentials (Lide (2000, p.10-175)) and the electron binding energies of the elements (Lide (2000, p.10-200)). As McQuarrie (1983, p.297) has noticed, however, "the inclusion of electron correlations in atomic and molecular wave functions is a problem of current interest." The mathematical formalism presented here allows the numerical computation of the global, generalized, and standard correlations between the electrons of an atom. The formalism may also be applied for calculating the degree of stability of a system of electrons and the interdependence among different subsystems of electrons of the atom.

(d) If the electrons of an atom are independent and in statistical equilibrium, then the joint radial probability distribution which describes the behavior of all the electrons of the atom, taken together, is a direct product of marginal Gamma radial probability distributions of the electrons. But the electrons are not independent and the statistical equilibrium is affected by random fluctuations induced by external perturbations and by internal interdependence both between electrons and nucleus and between electrons themselves. Building up a model of the atom based on minimum mean deviation from statistical equilibrium subject to generalized correlations induced by random fluctuations, we obtain a joint radial probability distribution which

173

contains not only the marginal Gamma radial probability distributions of the electrons but also a multivariate series involving second-order Laguerre polynomials. The more such polynomials are taken into account, the better the approximation of the wave function of the atom is. But there is a "curse of complexity" showing up as well. The more such polynomials are taken into account, the more computation and time-machine are needed. For simpler atoms, like helium, lithium, beryllium, boron, and carbon atoms, we can use second order Laguerre polynomials of degree three, four, or even greater, but we need powerful computers due to Pauli's exclusion principle that forces us to make the wave function antisymmetric, increasing the amount of computation needed when we minimize the mean ground state energy of the corresponding atom. In our numerical applications we prefered to take linear second order Laguerre polynomials and first order products in the approximation series of the wave function. Such an approach allowed very good approximations, which will be presented in the second half of Part II, for the helium, lithium, beryllium, boron, carbon, nitrogen, and argon atoms, in a reasonable amount of computing time, using Wolfram's symbolic mathematics software package MATHEMATICA, version 2.1, on an IBM ThinkPad 380 personal computer with Intel Pentium processor, 80MB RAM, and 400MB available on the hard disk. This approximation of the wave function based on minimum mean deviation from statistical equilibrium induced by linear mean random fluctuations may be applied in a straightforward way to more complex atoms but, of course, we need more computing power.

Following this introduction, Sections 2 and 3 contain the mathematical formalism based on the linear approximation of the solution of the variational principle based on the minimum mean deviation from statistical equilibrium induced by generalized mean random fluctuations. Section 4 deals with four probabilistic models for the structure of more complex atoms, namely the calm, nervous, turbulent, and nebulous models, depending on how the probability wave function of the atom is made antisymmetric, as required by Pauli's exclusion principle. It also contains the formulas for the global, generalized and standard correlations between electrons, the degree of stability of a system of electrons, and the amount of interdependence among different subsystems of electrons of the atom.

2. THE MATHEMATICAL FORMALISM.

We start this section with a reminder from calculus of variations. Let D be a measurable set from the n-dimensional Euclidean space $\mathbf{R}^n$, $x = (x_1, \ldots, x_n) \in D$, g an arbitrary continuous function on D, and $F, G_i, (i = 1, \ldots, m)$, continuous functions of (x, g). A particular case of an isoperimetric problem (Elsgolts, (1977, pp.399-402)), when the functional which has to be optimized does not depend on the derivatives of the unknown function, tells us that a necessary condition for the differentiable functional

$$\int_D G(x, g)\, dx \tag{1}$$

to have an extremum for $g = g^*$, subject to the constraints

$$\int_D G_i(x, g)\, dx = b_i, \quad (i = 1, \ldots, m), \tag{2}$$

is that g^* be a solution of the equation

$$\frac{\partial F}{\partial g} = 0, \quad \text{with} \quad F = G + \sum_{i=1}^{m} \lambda_i G_i, \tag{3}$$

where the Lagrange multipliers are to be determined by introducing the solution of equation (3) into the constraints (2).

Let now g be an arbitrary probability density function defined and integrable on the measurable set D on the real line. The amount of uncertainty contained by the probability density g is measured by the Shannon entropy

$$\mathbf{H}(g) = -\int_D g(x) \log g(x) dx.$$

If the probability density g is known, then we can find all the moments of it, in particular its mean value, provided that the corresponding integrals may be calculated. The converse is not true. Generally, if the mean value μ is given, there are infinitely many probability densities g having the mean μ. We want to find the most uncertain probability density g on D when we know the mean value μ of the probability distribution. Such a probability distribution is the most unbiased one, ignoring no possibility, subject to the given mean value μ. It descibes statistical equilibrium corresponding to the

175

mean value μ. According to the maximum entropy principle, mentioned implicitly in von Neumann (1932) and explicitly in Jaynes (1957, 1979) and Ingarden (1963), we are looking for the probability distribution of maximum entropy subject to the given constraints.

Proposition 1: *The solution of the maximization problem*

$$\max_{g} \mathbf{H}(g)$$

subject to

$$\int_{D} g(x)\, dx = 1, \tag{4}$$

$$\int_{D} xg(x)\, dx = \mu, \tag{5}$$

has the form

$$g(x) = \frac{1}{\Phi(\beta)} e^{-\beta x}, \qquad x \in D. \tag{6}$$

where

$$\Phi(\beta) = \int_{D} e^{-\beta x} dx, \tag{7}$$

and β is the solution of the equation

$$\frac{d \log \Phi(\beta)}{d\beta} = -\mu. \tag{8}$$

Proof: This is an isoperimetric variational problem with two constraints, and the functional which is optimized does not depend on the derivatives of the unknown function $g(x)$. Taking

$$F(g, x) = g(x)[-\log g(x) - \alpha - \beta x],$$

equation (3) becomes

$$\frac{\partial F}{\partial g} = -\log g(x) - \alpha - \beta x - 1 = 0,$$

which gives

$$g(x) = e^{-(\alpha+1)-\beta x}, \qquad x \in D, \tag{9}$$

where the Lagrange multipliers α and β are determined from the constraints (4) and (5), i.e.

$$\int_D e^{-(\alpha+1)-\beta x} dx = 1, \tag{10}$$

$$\int_D x e^{-(\alpha+1)-\beta x} dx = \mu. \tag{11}$$

We can see that the expression (9), satisfying (10) and (11), maximizes the entropy $\mathbf{H}(g)$. Indeed, if $g(x)$ is an arbitrary probability density on D, and we take into account the elementary inequality

$$-t \log t \le e^{-1}, \quad \text{with equality if and only if} \quad t = e^{-1},$$

and the constraints (4) and (5), we get

$$\mathbf{H}(g) - \alpha - \beta\mu =$$

$$= -\int_D e^{-\alpha-\beta x} g(x) e^{\alpha+\beta x} \log[g(x) e^{\alpha+\beta x}] \, dx \le$$

$$\le \int_D e^{-(\alpha+1)-\beta x} dx = 1.$$

Therefore, for any probability density g on D, we have

$$\mathbf{H}(g) \le \alpha + 1 + \beta\mu,$$

where the bound in the right hand side of the inequality does not depend on g, with equality if and only

$$g(x) e^{\alpha+\beta x} = e^{-1}, \quad (x \in D),$$

which is equivalent to (9). Taking into account (10), the expression (9) becomes

$$g(x) = \frac{1}{\Phi(\beta)} e^{-\beta x}, \quad (x \in D),$$

where $\Phi(\beta)$ is defined by (7). Introducing the last expression into the second constraint (5), we get

$$\frac{\Phi'(\beta)}{\Phi(\beta)} = -\mu,$$

which is equivalent to (8).

Proposition 2: *If $D = [a, b]$, the solution of the maximization problem*

$$\max_g \mathbf{H}(g)$$

subject to the constraint

$$\int_a^b g(x)\, dx = 1, \tag{12}$$

is the uniform probability distribution $\mathbf{U}(a, b)$ with parameters a, b, having the density

$$g(x) = \frac{1}{b - a}, \qquad (a \leq x \leq b), \tag{13}$$

Proof: Using Proposition 1 with the only constraint (12), we have

$$\beta = 0, \quad \Phi(0) = \int_a^b dx = b - a, \quad g(x) = \frac{1}{b - a}, \quad a \leq x \leq b.$$

Proposition 3: *If $D = [0, \infty)$, the solution of the maximization problem*

$$\max_g \mathbf{H}(g)$$

subject to the constraints

$$\int_0^{+\infty} g(x)\, dx = 1, \tag{14}$$

$$\int_0^{+\infty} x g(x)\, dx = \mu, \tag{15}$$

is the exponential probability distribution $\mathbf{E}(\mu)$ with parameter μ, having the density

$$g(x) = \frac{1}{\mu} e^{-x/\mu}, \qquad (x \geq 0). \tag{16}$$

Proof: Using Proposition 1 with the constraints (14) and (15), we have

$$\Phi(\beta) = \int_0^{+\infty} e^{-\beta x} dx = \frac{1}{\beta}, \quad \Phi'(\beta) = -\frac{1}{\beta^2},$$

in which case equation (8) gives $\beta = 1/\mu$, and (6) becomes (16).

178

Let g be a strictly positive probability density function on the measurable set D from the real line $\mathbf{R}^1$. A sequence of orthonormal polynomials $\{G_n; n = 0, 1, \ldots\}$, $G_0 \equiv 1$ defined on D with the weight g satisfies the equalities

$$\int_D G_n(x) G_\ell(x) g(x)\, dx = 0,$$

if $\ell \neq n$, and

$$\int_D G_n^2(x) g(x)\, dx = 1,$$

for all $n, \ell = 0, 1, \ldots$. We look at $\{G_n; n = 0, 1, \ldots\}$ as being a system of generalized coordinates with respect to the probability distribution with density g.

A system of orthogonal polynomials with the weight (13) on $D = [a, b]$ is

$$G_n(x) = (2n + 1)^{1/2} P_n \left(\frac{2}{b - a} x - \frac{a + b}{b - a} \right), \quad (n = 0, 1, \ldots),$$

where $P_n(x)$ is the Legendre (spherical) polynomial of degree n. The first ones are

$$P_0(x) = 1, \quad P_1(x) = x, \quad P_2(x) = \tfrac{1}{2}(3x^2 - 1),$$

$$P_3(x) = \tfrac{1}{2}(5x^3 - 3x), \quad P_4(x) = \tfrac{1}{8}(35x^4 - 30x^2 + 3),$$

$$P_5(x) = \tfrac{1}{8}(63x^5 - 70x^3 + 15x).$$

A system of orthonormal polynomials with the weight (16) on $D = [0, +\infty)$ is

$$G_n(x) = L_n \left(\frac{x}{\mu} \right), \quad (n = 0, 1, \ldots),$$

where $L_n(x)$ is the Laguerre polynomial of degree n. The first ones are:

$$L_0(x) = 1, \quad L_1(x) = -x + 1, \quad L_2(x) = \tfrac{1}{2}(x^2 - 4x + 2),$$

$$L_3(x) = \tfrac{1}{6}(-x^3 + 9x^2 - 18x + 6),$$

$$L_4(x) = \tfrac{1}{24}(x^4 - 16x^3 + 72x^2 - 96x + 24),$$

$$L_5(x) = \tfrac{1}{120}(-x^5 + 25x^4 - 200x^3 + 600x^2 - 600x + 120).$$

Let now f be another strictly positive probability density on D such that the following equalities exist, namely,

$$\int_D G_n(x)f(x)\,dx = \tilde{c}_n, \quad (n = 1, 2, \ldots, N). \tag{17}$$

Notice that $\tilde{c}_n = 0, n = 1, 2, \ldots, N$, if $f = g$. There are infinitely many probability densities compatible with the equalities (17). We want to find the closest probability density f to g subject to the constraints (17). As a measure of closeness between probability densities we prefer to take the Pearson's chi-square indicator, i.e.,

$$\tilde{\chi}^2(f : g) = \int_D \frac{[f(x) - g(x)]^2}{g(x)}\,dx = \int_D \left(\frac{f(x)}{g(x)} - 1\right)^2 g(x)\,dx.$$

Proposition 4: *A strictly positive probability density f on D which minimizes the quadratic functional $\tilde{\chi}^2(f : g)$ subject to the constraints (17) is given by the expression*

$$f(x) = g(x)\left[1 + \sum_{n=1}^N \tilde{c}_n G_n(x)\right] =$$

$$= g(x) \sum_{n=0}^N \tilde{c}_n G_n(x), \quad (x \in D, \tilde{c}_0 = 1). \tag{18}$$

Proof: This is another isoperimetric variational problem, which involves the minimization of a quadratic functional which does not contain derivatives of the unknown function f. Let us take

$$F(x, \lambda, f) = \chi^2(x) + \sum_{n=1}^N \lambda_n G_n(x)f(x),$$

where

$$\chi^2(x) = \frac{[f(x) - g(x)]^2}{g(x)},$$

and

$$\lambda = (\lambda_1, \ldots \lambda_N).$$

A necessary condition for the existence of an extremum of the integral of F on D is that its first variation be equal to zero, i.e.,

$$\int_D \left(\frac{\partial F}{\partial f} \delta f + \frac{\partial F}{\partial f'} \delta f' \right) dx = 0,$$

which, taking into account that

$$\frac{\partial F}{\partial f'} = 0,$$

implies

$$\frac{\partial F}{\partial f} = 0,$$

which is equivalent to

$$f(x) = g(x) \left[1 - \frac{1}{2} \sum_{n=1}^{N} \lambda_n G_n(x) \right], \quad (x \in D). \tag{19}$$

The Lagrange multipliers $\lambda_1, \ldots, \lambda_N$ are determined by introducing (19) into the constraints (17). Taking into account the orthonormality of the system $\{G_n; n = 1, \ldots, N\}$ with respect to g, and the fact that

$$\int_D G_n(x) g(x) dx = 0, \quad (n = 1, \ldots, N),$$

we get

$$\tilde{c}_n = -\frac{1}{2} \sum_{i=1}^{N} \lambda_i \delta_{in} = -\frac{1}{2} \lambda_n, \quad (n = 1, \ldots, N), \tag{20}$$

where δ_{in} is the Kronecker symbol, equal to 0 if $i \neq n$ and equal to 1 if $i = n$. From (19) and (20), we get (18).

A sufficient condition for a minimum of a functional is for the second variation

$$\int_D \left(\frac{\partial^2 F}{\partial f \partial f} \delta f^2 + \frac{\partial^2 F}{\partial f \partial f'} \delta f \delta f' + \frac{\partial^2 F}{\partial f' \partial f'} \delta f'^2 \right) dx = \int_D \frac{\partial^2 F}{\partial f \partial f} \delta f^2 dx$$

to keep a positive sign, which is so indeed here because

$$\frac{\partial^2 F}{\partial f \partial f} = \frac{2}{g} > 0.$$

Remark: The linear approximation of (18) is obtained by taking $N = 1$, i.e.,

$$f(x) = g(x)[1 + \tilde{c}_1 G_1(x)], \quad (x \in D). \tag{21}$$

In our context, g is a probability density describing statistical equilibrium on the one-dimensional set D of possible values of a random variable assigned to a system (like the exponential distribution, the most unbiased probability distribution of the values taken on by a positive random variable when only its mean value is known, or the uniform distribution, the most unbiased probability distribution of the possible values of a random variable when only its finite range is known). A system of orthonormal polynomials $\{G_n; n = 0, 1, \ldots\}$, $G_0 \equiv 1$, with weight g is taken to be a system of corresponding generalized coordinates attached to g. The integrals of $G_n, (n = 1, 2, \ldots)$, with respect to g, called generalized G_n-moments associated to g, or the generalized moments of g in the directions $G_n, (n = 1, 2, \ldots)$, are all equal to zero. Random fluctuations, due to external or/and internal interdependences, could alter the statistical equilibrium described by the probability density g. In such a case, the initial probability density g no longer describes accurately what happens with our system and it has to be replaced by a new probability density f on D, knowing that some generalized G_n-moments associated to f are no longer equal to zero, due to random fluctuations induced by internal or/and external interactions. The probability density f given by (18) represents the closest probability density to the initial probability density g, subject to the values $\tilde{c}_1, \ldots, \tilde{c}_N$ of the first N generalized G_n-moments, where the closeness is measured by Pearson's chi-square mean deviation. The numbers $\tilde{c}_1, \ldots, \tilde{c}_N$ represent the mean fluctuations along the directions $G_1, \ldots, G_N$, and f is called the probability density of minimum mean deviation from statistical equilibrium subject to the mean fluctuations $\tilde{c}_1, \ldots, \tilde{c}_N$ along the directions $G_1, \ldots, G_N$.

This approach may be generalized to the construction of the joint probability density describing the joint probabilistic behavior of interdependent components. Let $\{g_i; i = 1, \ldots, M\}$ be a set of marginal probability densities, where g_i is defined and positive on the measurable set D_i from the real line. In what follows, D_i is either the positive real line or a finite interval of it. Denote by x_i an arbitrary element of D_i. The independent joint probability density on the product set $D_1 \times D_2 \times \ldots \times D_M$ is the direct product of the marginal probability densities, namely $g_1 g_2 \ldots g_M$. If

$$\{G_{n(i)i}; \, n(i) = 0, 1, \ldots\}, \quad G_{0i} \equiv 1,$$

is an orthonormal sequence of polynomials on D_i with the weight g_i, then

$$\{G_{n(1)1} \ldots G_{n(M)M}; \, n(1) = 0, 1, \ldots; \ldots; n(M) = 0, 1, \ldots\}$$

is an orthonormal set of polynomials on $D_1 \times \ldots \times D_M$ with the weight $g_1 \ldots g_M$. Let now f be a joint probability density on the product set $D_1 \times \ldots \times D_M$ such that

$$\int_{D_1} \ldots \int_{D_M} G_{n(1)1}(x_1) \ldots G_{n(M)M}(x_M) f(x_1, \ldots, x_M) dx_1 \ldots dx_M =$$

$$= \tilde{c}_{n(1)\ldots n(M)}, \tag{22}$$

$$(n(1) = 0, 1, \ldots, N(1); \ldots; n(M) = 0, 1, \ldots, N(M)),$$
$$(n(1), \ldots, n(M)) \neq (0, \ldots, 0).$$

Let us notice that if f is replaced by $g_1 \ldots g_M$ in (22), then we get $\tilde{c}_{n(1)\ldots n(M)} = 0$ for all $(n(1), \ldots, n(M)) \neq (0, \ldots, 0)$.

We are looking for the closest joint probability density f to the independent direct product $g_1 \ldots g_M$ on the product set $D_1 \times \ldots \times D_M$ subject to the generalized covariances (22). With these notations, using again Pearson's chi-square mean deviation, from Proposition 4, we obtain:

Proposition 5: *The joint probability density f on the product set $D_1 \times \ldots \times D_M$ which minimizes Pearson's chi-square mean deviation $\tilde{\chi}^2(f : g_1 \ldots g_M)$ of f from the direct independent product $g_1 \ldots g_M$ of the marginal probability densities subject to the generalized mixed moments (22) is given by*

$$f(x_1, \ldots, x_M) = g_1(x_1) \ldots g_M(x_M) \times$$

$$\times [1 + \sum_{\substack{n(1)=0 \\ (n(1),\ldots,n(M)) \neq (0,\ldots,0)}}^{N(1)} \ldots \sum_{n(M)=0}^{N(M)} \tilde{c}_{n(1)\ldots n(M)} G_{n(1)1}(x_1) \ldots G_{n(M)M}(x_M). \tag{23}$$

Remark: The linear approximation is obtained by taking only M linear terms in (23) and $N(1) = 1, \ldots, N(M) = 1$, i.e.,

$$f(x_1, \ldots, x_M) = g_1(x_1) \ldots g_M(x_M) \times$$
$$\times \left[1 + \tilde{c}_{10\ldots0} G_{11}(x_1) + \ldots + \tilde{c}_{00\ldots1} G_{1M}(x_M)\right]. \tag{24}$$

3. PROBABILITY WAVE FUNCTIONS.

Summarizing what was done in the previous section, we start from statistical equilibrium described by a maximum entropy probability distribution subject to constraints induced by given mean values of some random variables. This variational problem, known as the maximum entropy principle (MEP), is very well studied in literature nowadays. Various applications of it in quantum chemistry may be found in Bernstein and Levine (1972a,1972b), Alhassid and Levine (1977), Sears, Parr, and Dinur (1980), Ghosh, Berkowitz, and Parr (1984), Morrison and Parr (1991), Nagy and Parr (1996). Thus, on the positive real axis $[0, +\infty)$, the solution of MEP subject to the mean value μ is the exponential distribution $\mathbf{E}(\mu)$. On an arbitrary interval $[a, b]$ of the real axis, if we have no constraints attached, then the probability distribution of maximum entropy is the uniform distribution $\mathbf{U}(a, b)$. Once the probability distribution that describes statistical equilibrium is obtained, and g is its density, we choose a sequence of orthogonal functions with the weight g as a system of generalized coordinates. Such systems of orthogonal functions are the Laguerre polynomials for the exponential distribution $\mathbf{E}(\mu)$, and the Legendre (spherical) polynomials for the uniform distribution $\mathbf{U}(a, b)$. As long as the statistical equilibrium is not perturbed, the mean values of the generalized coordinates remain equal to zero. If, however, random perturbations induced by internal or/and external interactions alter the statistical equilibrium, then some of these mean values cease to be equal to zero and we have to find a new probability distribution which describes what happens under the new circumstances. We do this by minimizing the mean deviation from statistical equilibrium subject to given generalized mean values. The mean deviation is measured using Pearson's chi-square indicator from statistical inference. When we are dealing with more complicated systems consisting of several components with random behavior, the system is described by a joint probability distribution which is eaqual to the product of the marginal probability distributions if the components making up the system are independent. If the components are interdependent, however, then we construct the closest joint probability distribution to the independent product of the marginals subject to the given values of some generalized covariances. Again, closeness is obtained by minimizing Pearson's chi-square mean deviation from independence subject to the given generalized mixed moments called generalized covariances.

A central concept in quantum mechanics is the wave function of a quantum system. The standard approach is to introduce the wave function ψ first, as a solution of the corresponding Schrödinger equation, and to interpret the square of its absolute value, i.e., $\mid \psi \mid^2$, to be the probability density of the quantum system. In the approach discussed in this paper there are two ways of getting a wave function without solving the Schrödinger equation of the quantum system:

(a) Minimizing the mean deviation $\tilde{\chi}^2(f : g)$ from statistical equilibrium described by the probability density g, i.e.

$$\min_{f} \tilde{\chi}^2(f : g)$$

subject to

$$\int_D G_n(x)f(x)dx = \tilde{c}_n, \quad (n = 1, \ldots, N),$$

we get the probability density

$$f(x) = g(x)\left[1 + \sum_{n=1}^{N} \tilde{c}_n G_n(x)\right], \quad (x \in D),$$

where $\tilde{c}_n$ is the generalized moment in the direction G_n. Using the approximation

$$\sqrt{1+t} \approx 1 + \frac{1}{2}t,$$

for small t, the wave function associated to the probability density f is

$$\psi(x) = \sqrt{g(x)}\left[1 + \sum_{n=1}^{N} c_n G_n(x)\right], \quad (x \in D), \tag{25}$$

where $c_n = \frac{1}{2}\tilde{c}_n$.

(b) Let g be a probability density on $D \subseteq \mathbf{R}^1$. If $\{G_n;\ n = 0, 1, \ldots\}, G_0 \equiv 1$, is a complete orthonormal system on D with respect to the weight g, then any square integrable wave function may be approximated by a linear combination

$$\sum_{n=0}^{N} d_n G_n \sqrt{g}.$$

Dividing by d_0, we get the wave function

$$\psi(x) = \sqrt{g(x)} \left[1 + \sum_{n=1}^{N} c_n G_n(x) \right], \tag{26}$$

with $c_n = d_n/d_0$. Using the approximation $(1 + t)^2 \approx 1 + 2t$ for small t, we get the density

$$f(x) = \psi^2(x) \approx g(x) \left[1 + \sum_{n=1}^{N} \tilde{c}_n G_n(x) \right], \tag{27}$$

with $\tilde{c}_n = 2c_n$.

Let us notice that from (27) we get the deviation of the density f from the statistical equilibrium probability density g,

$$\chi(x) = \frac{f(x) - g(x)}{\sqrt{g(x)}} \approx \sqrt{g(x)} \sum_{n=1}^{N} \tilde{c}_n G_n(x). \tag{28}$$

We have from (28) and (26),

$$\chi(x) \approx 2(\psi(x) - \sqrt{g(x)}),$$

and $\chi(x)$ may be interpreted as being the deviation of the wave function $\psi(x)$ from the statistical equilibrium wave function $\sqrt{g(x)}$. The Pearson's chi-square indicator is just the integral

$$\tilde{\chi}^2(f : g) = \int_D \chi^2(x) dx.$$

The linear approximation of (26) is the wave function

$$\psi(x) = \sqrt{g(x)}[1 + c_1 G_1(x)]. \tag{29}$$

186

As we have

$$\int_D \psi^2(x)dx = 1 + c_1^2,$$

the corresponding probability density is

$$f(x) = g(x)[1 + c_1 G_1(x)]^2/(1 + c_1^2).$$

In the two-dimensional case, let g and h be two probability densities on the sets D_1 and D_2 from $\mathbf{R}^1$, respectively, both describing statistical equilibrium. Let $\{G_n; n = 0, 1, \ldots\}, G_0 \equiv 1$, and $\{H_m; m = 0, 1, \ldots\}, H_0 \equiv 1$ be two sequences of orthonormal polynomials with the weight g and h, respectively. We have

$$\int_{D_1} G_n(x)G_\ell(x)f(x)dx = \delta_{n\ell},$$

$$\int_{D_2} H_m(y)H_k(y)g(y)dy = \delta_{mk},$$

where $\delta_{n\ell}$ is equal to 1, if $n = \ell$, and to 0, if $n \neq \ell$. Let us follow the two ways of getting a wave function in the two-dimensional case. The generalization to more dimensions is straightforward:

(a) If the two marginal probability densities g and h, describing statistical equilibrium of two components of a system, are independent, then the statistical equilibrium of the system itself is described by the joint probability density gh. Minimizing the mean deviation $\tilde{\chi}^2(f : gh)$ of f from the joint statistical equilibrium described by gh, i.e.,

$$\min_f \tilde{\chi}^2(f : gh)$$

subject to the mixed moments

$$\int_{D_1} \int_{D_2} G_n(x)H_m(y)f(x, y)dxdy = \tilde{c}_{nm},$$

$$(n = 0, 1, \ldots, N; m = 0, 1, \ldots, M; (n, m) \neq (0, 0)),$$

we get the joint probability density

$$f(x, y) = g(x)h(y)[1 + \sum_{\substack{n=0 \\ (n,m)\neq(0,0)}}^{N} \sum_{m=0}^{M} \tilde{c}_{nm}G_n(x)H_m(y)], \qquad (30)$$

for $x \in D_1, y \in D_2$, where $\tilde{c}_{nm}$ is the generalized mixed moment in the "directions" G_n and H_m. Using the approximation

$$\sqrt{1+t} \approx 1 + \frac{1}{2}t,$$

for small t, the wave function associated to the joint density f is

$$\psi(x,y) = \sqrt{g(x)h(y)}[1+ \sum_{\substack{n=0 \\ (n,m)\neq(0,0)}}^{N} \sum_{m=0}^{M} c_{nm}G_n(x)H_m(y)], \qquad (31)$$

for $x \in D_1, y \in D_2$, where $c_{nm} = \frac{1}{2}\tilde{c}_{nm}$.

(b) If $\{G_n; n = 0, 1, \ldots\}, G_0 \equiv 1$ is a complete orthonormal system on D_1, with respect to the weight g, and $\{H_m; m = 0, 1, \ldots\}, H_0 \equiv 1$ is a complete orthonormal system on D_2, with respect to the weight h, then any square integrable wave function may be approximated by a linear combination

$$\sum_{n=0}^{N} \sum_{m=0}^{M} d_{nm}G_nH_m\sqrt{gh}.$$

Dividing by d_{00}, we get the wave function

$$\psi(x,y) = \sqrt{g(x)h(y)}[1+ \sum_{\substack{n=0 \\ (n,m)\neq(0,0)}}^{N} \sum_{m=0}^{M} c_{nm}G_n(x)H_m(y)], \qquad (32)$$

for $x \in D_1, y \in D_2$, where $c_{nm} = d_{nm}/d_{00}$. Using the elementary approximation $(1 + t)^2 \approx 1 + 2t$, for small t, we get the joint density

$$f(x,y) = \psi^2(x,y) \approx g(x)h(y) [1+ \sum_{\substack{n=0 \\ (n,m)\neq(0,0)}}^{N} \sum_{m=0}^{M} \tilde{c}_{nm}G_n(x)H_m(y)], \quad (33)$$

for $x \in D_1, y \in D_2$, where $\tilde{c}_{nm} = 2c_{nm}$.

Let us notice that from (33) we get the deviation of the joint density f from the statistical equilibrium joint probability density gh, namely

$$\chi(x,y) = \frac{f(x,y) - g(x)h(y)}{\sqrt{g(x)h(y)}} \approx$$

$$\approx \sqrt{g(x)h(y)} \sum_{\substack{n=0 \\ (n,m)\neq(0,0)}}^{N} \sum_{m=0}^{M} \tilde{c}_{nm} G_n(x) H_m(y). \tag{34}$$

for $x \in D_1$, and $y \in D_2$. From (34) and (32) we have

$$\chi(x,y) \approx 2(\psi(x,y) - \sqrt{g(x)h(y)}),$$

and $\chi(x,y)$ may be interpreted as being the deviation of the wave function $\psi(x,y)$ from the statistical equilibrium wave function $\sqrt{g(x)h(y)}$. Pearson's chi-square indicator is just the integral

$$\tilde{\chi}^2(f:gh) = \int_{D_1} \int_{D_2} \chi^2(x,y)dxdy.$$

The linear approximation of (32) is the wave function

$$\psi(x,y) = \sqrt{g(x)h(y)}[1 + c_{10}G_1(x) + c_{01}H_1(y)]. \tag{35}$$

As we have

$$\int_{D_1} \int_{D_2} \psi^2(x,y)dxdy = 1 + c_{10}^2 + c_{01}^2,$$

the corresponding probability density is

$$f(x,y) = g(x)h(y)[1 + c_{10}G_1(x) + c_{01}H_1(y)]^2/(1 + c_{10}^2 + c_{01}^2). \tag{36}$$

The generalized covariance in the "directions" G_1 and H_1 is

$$\int_{D_1} \int_{D_2} G_1(x)H_1(y)f(x,y)dxdy = \frac{2c_{10}\,c_{01}}{1 + c_{10}^2 + c_{01}^2} \tag{37}$$

In general, if $\{G_{k\ell}; \ k = 0, 1, \ldots\}, G_{0\ell} \equiv 1$ is an orthonormal system on D_ℓ with the weight g_ℓ, a probability density on D_ℓ, for each ℓ, $(1 \leq \ell \leq n)$, then the linear approximation of the wave function induced by the minimum mean deviation from independent statistical equilibrium is

$$\psi(x_1, \ldots, x_n) = \sqrt{g_1(x_1)\ldots g_n(x_n)}[1 + c_1 G_{11}(x_1) + \ldots + c_n G_{1n}(x_n)] \tag{38}$$

and the corresponding probability density is

$$f(x_1, \ldots, x_n) =$$

189

$$= g_1(x_1) \ldots g_n(x_n)[1 + c_1 G_{11}(x_1) + \ldots + c_n G_{1n}(x_n)]^2 / (1 + c_1^2 + \ldots + c_n^2). \quad (39)$$

The generalized covariance in the directions G_{1i} and G_{1j} is

$$\int_{D_1} \int_{D_2} G_{1i}(x_i) G_{1j}(x_j) f(x_1, \ldots, x_n) dx_1 \ldots dx_n = \frac{2 c_i c_j}{1 + c_1^2 + \ldots + c_n^2}. \quad (40)$$

In order to determine the coefficients $\{c_1, \ldots, c_n\}$ of the linear approximation of the wave function, we are looking for stationary points of the mean energy of the quantum system. In particular, in dealing with the ground state of the atom, we determine the unknown coefficients by minimizing the mean energy of the corresponding quantum system (the variational theorem),

$$\min_{\psi} \frac{< \psi \hat{H} \psi >}{< \psi^2 >}, \quad (41)$$

where $\hat{H}$ is the Hamiltonian of the respective quantum system and

$$< \psi \hat{H} \psi > = \int_{D_1} \ldots \int_{D_n} \psi(x_1, \ldots, x_n) \hat{H} \psi(x_1, \ldots, x_n) dx_1 \ldots dx_n,$$

$$< \psi^2 > = \int_{D_1} \ldots \int_{D_n} \psi^2(x_1, \ldots, x_n) dx_1 \ldots dx_n.$$

If the value of the mean energy obtained by solving the variational problem (41) proves to be very close to the experimental value of the ground energy, then this is an indirect justification of the choice of the respective probability wave function. The literature is full with all sorts of trial functions proposed to be wave functions for different quantum systems of interest, often without any sound rationale for making such choices, in a blind race to get a mean ground energy as close as possible to the experimental value. Whereas using the variational method for getting a mean energy of the atom as close as possible to the experimental value of the ground energy remains an important objective, it is more important, however, to test different probabilistic models of the atom structure in order to estimate different correlations describing the dependence between electrons, to evaluate the degree of stability of a system of electrons, and to calculate the amount of global interdependence among different subsystems of electrons inside the atom. Anyway, the methodology used here for introducing a probability wave function is entirely based on a probabilistic approach but the variational parameters of such a probability wave function are determined by minimizing (more exactly by looking for stationary points of) the mean energy of the atom in the ground state. Here we have the connection between the general probabilistic modelling (i.e., build up the most unbiased model subject to constraints expressed by mean values and generalized mixed moments) and the specific quantum system we are dealing with (i.e., the atom and its electrons moving randomly around the nucleus).

4. PROBABILISTIC MODELS FOR THE ATOM STRUCTURE.

a) Probability wave function of the one-electron atom.

Assume that an electron is moving randomly around the nucleus taken as the origin in a three-dimensional Euclidean space of rectangular coordinates (x, y, z). As the Coulomb potential depends only on the radial distance

$$r(x, y, z) = \sqrt{x^2 + y^2 + z^2},$$

it is natural to use a spherical coordinate system (r, θ, ω), by using the transformation

$$x = r \sin\theta \cos\omega, \quad y = r \sin\theta \sin\omega, \quad z = r \cos\theta,$$

where $0 \leq r < +\infty$, $0 \leq \theta \leq \pi$, $0 \leq \omega \leq 2\pi$, and eventually taking

$$\nu = \cos\theta, \qquad (0 \leq \theta \leq \pi),$$

which means that $-1 \leq \nu \leq 1$, we get a transformation from the Cartesian system (x, y, z) to the system of coordinates (r, ν, ω) for which the volume element becomes

$$dxdydz = r^2 \sin\theta drd\theta d\omega = r^2 drd\nu d\omega. \tag{42}$$

Focusing on the three coordinates (r, ν, ω) on the product set

$$[0, +\infty) \times [-1, 1] \times [0, 2\pi],$$

generally we have no restrictions on the possible values of the coordinates r and ω. Therefore, statistical equilibrium will be described by the uniform distributions $\mathbf{U}(-1, 1)$ for ν and $\mathbf{U}(0, 2\pi)$ for ω, with the probability densities

$$g_2(\nu) = \frac{1}{2}, \qquad (-1 \leq \nu \leq 1), \tag{43}$$

and

$$g_3(\omega) = \frac{1}{2\pi}, \qquad (0 \leq \omega \leq 2\pi), \tag{44}$$

respectively. Concerning the random distance $r(x, y, z)$ in the space of the Cartesian variables (x, y, z), if the mean distance of the electron from the nucleus, fixed at the origin of the coordinate system, is $1/\mu$, then the radial statistical equilibrium in the space of Cartesian variables (x, y, z) will be described by the exponential distribution $\mathbf{E}(\mu)$ whose probability density is

$$g(r(x, y, z)) = \frac{1}{\mu} e^{-r(x,y,z)/\mu}, \qquad (0 \leq r(x, y, z) < +\infty). \tag{45}$$

As the coordinates (r, ν, ω) are supposed to be independent, taking into account (45),(43),(44), and (42), the statistical equilibrium for the electron, when we pass from the Cartesian system of coordinates (x, y, z) to the space of the variables (r, ν, ω), is characterized by the joint independent probability density

$$h(x, y, z)dxdydz = \tilde{C}g(r(x, y, z))dxdydz = \tilde{C}g(r)r^2drd\nu d\omega =$$

$$= Cg(r)g_2(\nu)g_3(\omega)r^2drd\nu d\omega = C\frac{1}{\mu}e^{-r/\mu}\frac{1}{2}\frac{1}{2\pi}r^2drd\nu d\omega,$$

where C is a constant to be determined from the equality

$$C\int_0^{+\infty}\frac{1}{\mu}r^2e^{-r/\mu}dr\int_{-1}^1\frac{1}{2}d\nu\int_0^{2\pi}\frac{1}{2\pi}d\omega = 1,$$

which gives

$$C = \frac{1}{2\mu^2}, \qquad \tilde{C} = \frac{1}{8\pi\mu^2}.$$

Therefore, the statistical equilibrium for the electron in the space of coordinates (r, ν, ω), $(0 \leq r < +\infty, -1 \leq \nu \leq 1, 0 \leq \omega \leq 2\pi)$, is given by the density

$$\varphi^2(r, \mu)r^2drd\nu d\omega = \frac{1}{2\mu^2}g(r)g_2(\nu)g_3(\omega)r^2drd\nu d\omega = g_1(r)g_2(\nu)g_3(\omega)drd\nu d\omega =$$

192

$$= \frac{1}{2!\mu^3} r^2 e^{-r/\mu} \frac{1}{2} \frac{1}{2\pi} dr d\nu d\omega = \frac{1}{8\pi\mu^3} r^2 e^{-r/\mu} dr d\nu d\omega, \qquad (46)$$

where the radial probability density in the space of variables (r, ν, ω) is

$$g_1(r) = Cg(r)r^2 = \frac{1}{2!\mu^3} r^2 e^{-r/\mu}, \qquad (0 \leq r < +\infty). \qquad (47)$$

Remarkably, $g_1(r)$ is the probability density function of the Gamma distribution $\mathbf{G}(\mu, 3)$ with parameters μ and 3. A system of orthonormal polynomials on the interval $[0, +\infty)$ with the weight (47) is

$$U_n(r) = \left[\frac{2}{(n+1)(n+2)} \right]^{1/2} L_n^{(2)} \left(\frac{r}{\mu} \right), \qquad (n = 0, 1, \ldots), \qquad (48)$$

where $L_n^{(2)}(r)$ is the Laguerre polynomial of degree n and order 2, i.e.,

$$L_n^{(2)}(r) = \sum_{\ell=0}^n (-1)^\ell \binom{n+2}{n-\ell} \frac{1}{\ell!} r^\ell.$$

The first five Laguerre polynomials of order 2 are:

$$L_0^{(2)}(r) \equiv 1, \quad L_1^{(2)}(r) = 3 - r, \quad L_2^{(2)}(r) = \frac{1}{2}(12 - 8r + r^2),$$

$$L_3^{(2)}(r) = \frac{1}{6}(60 - 60r + 15r^2 - r^3), L_4^{(2)}(r) = \frac{1}{24}(360 - 480r + 180r^2 - 24r^3 + r^4),$$

$$L_5^{(2)}(r) = \frac{1}{120}(2520 - 4200r + 2100r^2 - 420r^3 + 35r^4 - r^5).$$

A system of orthonormal functions on $[-1, 1]$ with the weight (43) is

$$V_\ell(\nu) = (2\ell + 1)^{1/2} P_\ell(\nu), \qquad (\ell = 0, 1, \ldots), \qquad (49)$$

where $P_\ell(\nu)$ is the Legendre (spherical)polynomial of degree ℓ. The first five ones are:

$$P_0(\nu) \equiv 1, \quad P_1(\nu) = \nu, \quad P_2(\nu) = \frac{1}{2}(3\nu^2 - 1),$$

$$P_3(\nu) = \frac{1}{2}(5\nu^3 - 3\nu), \quad P_4(\nu) = \frac{1}{8}(35\nu^4 - 30\nu^2 + 3),$$

$$P_5(\nu) = \frac{1}{8}(63\nu^5 - 70\nu^3 + 15\nu).$$

A system of orthonormal functions on $[0, 2\pi]$ with the weight (44) is the trigonometric system:

$$W_0(\omega) \equiv 1, \quad W_k(\omega) = \sqrt{2}\sin\frac{k}{2}\omega, \quad (k = 1, 2, \ldots). \tag{50}$$

Therefore, a system of orthonormal functions on the product set

$$[0, +\infty) \times [-1, 1] \times [0, 2\pi]$$

with the weight $g_1(r)g_2(\nu)g_3(\omega)$ is

$$\{U_nV_\ell W_k; \, n = 0, 1, \ldots; \ell = 0, 1, \ldots; k = 0, 1, \ldots\}.$$

The closest density f to statistical equilibrium described by the probability density (46) subject to the generalized moments

$$\int_0^{+\infty}\int_{-1}^1\int_0^{2\pi} U_n(r)V_\ell(\nu)W_k(\omega)f(r, \nu, \omega)r^2 dr d\nu d\omega = \tilde{c}_{n\ell k},$$

$$(n = 0, 1, \ldots, N; \ell = 0, 1, \ldots, L; k = 0, 1, \ldots, K), (n, \ell, k) \neq (0, 0, 0),$$

is the density

$$f(r, \nu, \omega) = \varphi^2(r, \mu)[1 + \sum_{\substack{n=0 \\ (n,\ell,k)\neq(0,0,0)}}^N \sum_{\ell=0}^L \sum_{k=0}^K \tilde{c}_{n\ell k}U_n(r)V_\ell(\nu)W_k(\omega)], \tag{51}$$

and the corresponding probability wave function is

$$\psi(r, \nu, \omega) = \varphi(r, \mu)[1 + \sum_{\substack{n=0 \\ (n,\ell,k)\neq(0,0,0)}}^N \sum_{\ell=0}^L \sum_{k=0}^K c_{n\ell k}U_n(r)V_\ell(\nu)W_k(\omega)] \tag{52}$$

where $c_{n\ell k} = \frac{1}{2}\tilde{c}_{n\ell k}$.

Taking $L = 0$ and $K = 0$ we obtain the radial approximation of the wave function

$$\psi(r, \nu, \omega) = \varphi(r, \mu)\left[1 + \sum_{n=1}^N c_{n00} U_n(r)\right]. \tag{53}$$

As we have

$$\int_0^{+\infty} \int_{-1}^{+1} \int_0^{2\pi} \psi^2(r,\nu,\omega) r^2 dr d\nu d\omega = 1 + \sum_{n=1}^N c_{n00}^2,$$

the probability density induced by the radial approximation (53) of the wave function is

$$f(r,\nu,\omega) = \varphi^2(r,\mu) \left[1 + \sum_{n=1}^N c_{n00} U_n(r) \right]^2 \Big/ \left[1 + \sum_{n=1}^N c_{n00}^2 \right].$$

The linear radial approximation of the wave function is obtained by taking $N = 1$, i.e.,

$$\psi(r,\nu,\omega) = \varphi(r,\mu)[1 + c_{100} U_1(r)], \tag{54}$$

which, taking into account (43)-(46) and (48), becomes

$$\psi(r,\nu,\omega) = \frac{1}{\sqrt{8\pi\mu^3}} e^{-r/(2\mu)} \left[1 + c \left(3 - \frac{r}{\mu} \right) \right], \tag{55}$$

where $c = c_{100}/\sqrt{3}$. As

$$\int_0^{+\infty} \int_{-1}^{+1} \int_0^{2\pi} \psi^2(r,\nu,\omega) r^2 dr d\nu d\omega = 1 + 3c^2,$$

the probability density with respect to the measure $r^2 dr d\nu d\omega$, induced by the linear radial approximation (55) of the wave function is

$$f(r,\nu,\omega) = \frac{1}{8\pi\mu^3} e^{-r/\mu} \left[1 + c \left(3 - \frac{r}{\mu} \right) \right]^2 \Big/ (1 + 3c^2).$$

b) Probability wave functions for a multi-electron atom.

Assume that an atom has Z electrons moving randomly around the nucleus which is supposed to be fixed at the origin of the system of coordinates. For each $i, (i = 1, \ldots, Z)$, denote by

$$g^{(i)}(r_i) = \frac{1}{\mu_i} e^{-r_i/\mu_i}, \quad g_1^{(i)}(r_i) = \frac{1}{2!\mu_i^3} r_i^2 e^{-r_i/\mu_i}, \quad (0 \le r_i < +\infty), \tag{56}$$

195

$$g_2^{(i)}(\nu_i) = \frac{1}{2}, \quad (-1 \le \nu_i \le 1), \quad g_3^{(i)}(\omega_i) = \frac{1}{2\pi}, \quad (0 \le \omega_i \le 2\pi), \qquad (57)$$

$$\varphi(r_i, \mu_i) = \left[\frac{1}{2\mu_i^2} g^{(i)}(r_i) g_2^{(i)}(\nu_i) g_3^{(i)}(\omega_i) \right]^{1/2} = \frac{1}{\sqrt{8\pi\mu_i^3}} e^{-r_i/(2\mu_i)}, (0 \le r_i < +\infty). \tag{58}$$

The equalities (56) and (57) show the probability distributions describing the statistical equilibrium of the individual electrons with respect to the system of coordinates (r_i, ν_i, ω_i), as described in the previous subsection.

We use the notation $< F >_1$ as an abbreviation for the multiple integral of an integrable function $F(r_1, \nu_1, \omega_1, \ldots, r_Z, \nu_Z, \omega_Z)$ with respect to the variables $r_1, \nu_1, \omega_1, \ldots r_Z, \nu_Z, \omega_Z$ on the product set ($3Z$-dimensional rectangle)

$$D = [0, +\infty) \times [-1, 1] \times [0, 2\pi] \times \ldots \times [0, +\infty) \times [-1, 1] \times [0, 2\pi], \qquad (59)$$

with respect to the measure $r_1^2 dr_1 d\nu_1 d\omega_1 \ldots r_Z^2 dr_Z d\nu_Z d\omega_Z$, i.e.,

$$< F >_1 = \int \ldots \int_D F(r_1, \nu_1, \omega_1, \ldots, r_Z, \nu_Z, \omega_Z) r_1^2 dr_1 d\nu_1 d\omega_1 \ldots r_Z^2 dr_Z d\nu_Z d\omega_Z.$$

We can extend what was done in the previous subsection for the one-electron atom to a Z-electron atom. In particular, generalizing (54) and (55), the linear radial wave function for the atom consisting of Z electrons is

$$\psi(r_1, \nu_1, \omega_1, \ldots, r_Z, \nu_Z, \omega_Z) = \varphi(r_1, \mu_1) \ldots \varphi(r_Z, \mu_Z) \left[1 + \sum_{i=1}^{Z} c_i \left(3 - \frac{r_i}{\mu_i} \right) \right],$$

or, taking into account (58),

$$\psi(r_1, \nu_1, \omega_1, \ldots, r_Z, \nu_Z, \omega_Z) =$$

$$= (8\pi)^{-Z/2} \left\{ \prod_{i=1}^{Z} \mu_i^{-3/2} e^{-r_i/(2\mu_i)} \right\} \left[1 + \sum_{i=1}^{Z} c_i \left(3 - \frac{r_i}{\mu_i} \right) \right]. \qquad (60)$$

As we have

$$< \psi^2 >_1 = \int \ldots \int_D \psi^2(r_1, \nu_1, \omega_1, \ldots, r_Z, \nu_Z, \omega_Z) r_1^2 dr_1 d\nu_1 d\omega_1 \ldots r_Z^2 dr_Z d\nu_Z d\omega_Z =$$

$$= 1 + 3c_1^2 + \ldots + 3c_Z^2,$$

the probability density induced by the linear radial wave function ψ is

$$f(r_1, \nu_1, \omega_1, \ldots, r_Z, \nu_Z, \omega_Z) = \psi^2(r_1, \nu_1, \omega_1, \ldots, r_Z, \nu_Z, \omega_Z)/ <\psi^2> =$$

$$= \varphi^2(r_1, \mu_1) \ldots \varphi^2(r_Z, \mu_Z) \left[1 + \sum_{i=1}^{Z} c_i \left(3 - \frac{r_i}{\mu_i} \right) \right]^2 / (1 + 3 \sum_{i=1}^{Z} c_i^2),$$

where $\mu_1, \ldots, \mu_Z$ are not necessarily distinct, or, taking (58) and (60) into account,

$$f(r_1, \nu_1, \omega_1, \ldots, r_Z, \nu_Z, \omega_Z) =$$

$$= (8\pi)^{-Z} \left\{ \prod_{i=1}^{Z} \mu_i^{-3} e^{-r_i/\mu_i} \right\} \left[1 + \sum_{i=1}^{Z} c_i \left(3 - \frac{r_i}{\mu_i} \right) \right]^2 / (1 + 3 \sum_{i=1}^{Z} c_i^2). \qquad (61)$$

The coefficients $c_1, \ldots, c_Z$ are determined by minimizing the mean energy of the system, i.e.,

$$\min_{\psi} \frac{<\psi \hat{H} \psi>_1}{<\psi^2>_1}, \qquad (62)$$

where $\hat{H}$ is the Hamiltonian of the atom, whose expression in atomic units is

$$\hat{H} = -\frac{1}{2} \sum_{i=1}^{Z} \nabla_i^2 - Z \sum_{i=1}^{Z} \frac{1}{r_i} + \sum_{i=1}^{Z-1} \sum_{\substack{j=2 \\ i<j}}^{Z} \frac{1}{r_{ij}}, \qquad (63)$$

where the Laplacian corresponding to the electron i is

$$\nabla_i^2 = \frac{\partial^2}{\partial x_i^2} + \frac{\partial^2}{\partial y_i^2} + \frac{\partial^2}{\partial z_i^2},$$

and r_{ij} is the distance between electron i and electron j.

c) Generalized linear correlation between subshells.

If the i-th electron of the atom has the radial distance r_i and belongs to a subshell having the mean radial distance μ_i from the nucleus, let us denote by

$$U_{1i}(r_i, \mu_i) = \frac{1}{\sqrt{3}} L_1^{(2)}(r_i/\mu_i) = \frac{1}{\sqrt{3}} \left(3 - \frac{r_i}{\mu_i} \right).$$

197

The generalized linear covariance between an arbitrary electron, say i, belonging to a subshell having the mean radial distance from the nucleus equal to μ_i and an arbitrary electron, say j, belonging to a subshell having the mean radial distance from the nucleus equal to μ_j, which is a linear measure of the dependence between two electrons along the directions U_{1i} and U_{1j} is the integral

$$< U_{1i} U_{1j} f >_1 = \int \ldots \int_D U_{1i}(r_i, \mu_i) U_{1j}(r_j, \mu_j) f(r_1, \nu_1, \omega_1, \ldots, r_Z, \nu_Z, \omega_Z) \times$$

$$\times r_1^2 dr_1 d\nu_1 d\omega_1 \ldots r_Z^2 dr_Z d\nu_Z d\omega_Z = \frac{6 c_i c_j}{1 + 3 \sum_{\ell=1}^{Z} c_\ell^2}. \qquad (64)$$

Hölder's inequality implies that

$$< U_{1i} U_{1j} f >_1 = < U_{1i} \sqrt{f} \, U_{1j} \sqrt{f} >_1 \leq < U_{1i}^2 f >_1^{1/2} < U_{1j}^2 f >_1^{1/2},$$

where

$$< U_{1i}^2 f >_1 =$$

$$= \int \ldots \int_D U_{1i}^2(r_i, \mu_i) f(r_1, \nu_1, \omega_1, \ldots, r_Z, \nu_Z, \omega_Z) r_1^2 dr_1 d\nu_1 d\omega_1 \ldots r_Z^2 dr_Z d\nu_Z d\omega_Z =$$

$$= \frac{1 + 3 \sum_{\ell=1}^{Z} c_\ell^2 - 4 c_i + 12 c_i^2}{1 + 3 \sum_{\ell=1}^{Z} c_\ell^2}.$$

The generalized linear radial correlation between an electron of radial mean μ_i and an electron of radial mean μ_j, which may be also viewed as a correlation between the subshell with the radial mean μ_i and the subshell with the radial mean μ_j is

$$\mathrm{gencorr}(\mu_i, \mu_j) = \frac{< U_{1i} U_{1j} f >_1}{< U_{1i}^2 f >_1^{1/2} < U_{1j}^2 f >_1^{1/2}} =$$

$$= \frac{6 c_i c_j}{\sqrt{(1 + 3 \sum_{\ell=1}^{Z} c_\ell^2 - 4 c_i + 12 c_i^2)(1 + 3 \sum_{\ell=1}^{Z} c_\ell^2 - 4 c_j + 12 c_j^2)}}. \qquad (65)$$

d) Antisymmetrization of the probability wave function.

In order to cope with the electron indistinguishability, each electron is characterized by a set of values of four quantum numbers (n, ℓ, m, s), where

198

n is the principal quantum number, ℓ is the angular momentum quantum number, m is the magnetic quantum number, and s is the spin. The exclusion principle, due to Wolfgang Pauli, says that no more than one electron is allowed to occupy a given quantum state specified by the four quantum numbers (n, ℓ, m, s), where

$$n = 1, 2, 3, \ldots; \ell = 0, 1, \ldots, n-1; m = -\ell, -\ell+1, \ldots, \ell-1, \ell; s = -\frac{1}{2}, +\frac{1}{2}.$$

Sometimes, the two possible values of s are abbreviated as $-$ and $+$, respectively. The smaller values of ℓ give larger probabilities for finding the electron near the nucleus. The larger values of ℓ refer to states in which there are larger probabilities for finding the electron farther from the nucleus.

The four quantum numbers have been used for explaining the periodic table of the elements and the energy shells, the innermost shell corresponding to the minimum energy in the ground state electron configuration. A shell picture of the atom has been in force since the time of the old quantum theory. In fact, the exclusion principle emerged along with the spin to explain the picture before the coming of quantum mechanics. To date, it has not been possible to determine whether electronic shells are also predicted by the Schrödinger equation.

In what follows, the quantum number n will identify the n-th *shell* of the atom, the pair of quantum numbers (n, ℓ) will identify the ℓ-th *subshell* of the n-th shell of the atom, the triple of quantum numbers (n, ℓ, m) will identify the m-th *stratum* of the ℓ-th subshell of the n-th shell of the atom, and all four quantum numbers (n, ℓ, m, s), taken together, will identify an *individual electron* of the atom. Electrons with the same first two quantum numbers (n, ℓ), i.e., the electrons belonging to the subshell corresponding to the quantum numbers (n, ℓ), have the same energy level $E_{n,\ell}$. Inside the same energy level $E_{n,\ell}$, i.e., inside the subshell corresponding to the quantum numbers (n, ℓ), there is a $2(2\ell + 1)$ degeneracy with respect to the last two quantum numbers (m, s).

Let $\alpha_s(\sigma)$ be the *spin* eigenfunctions, i.e.,

$$\int \alpha_-^*(\sigma)\alpha_-(\sigma)d\sigma = \int \alpha_+^*(\sigma)\alpha_+(\sigma)d\sigma = 1,$$

$$\int \alpha_-^*(\sigma)\alpha_+(\sigma)d\sigma = \int \alpha_+^*(\sigma)\alpha_-(\sigma)d\sigma = 0,$$

where $*$ means the complex conjugate. The variable σ will be called the spin variable. The spin eigenfunctions are used in a model of the atom in which the electrons keep their spins unchanged during their random motion around the nucleus of the atom.

Let $\beta_m(\tau)$ be the *magnetic* eigenfunctions, i.e.,

$$\int \beta_m^*(\tau)\beta_{m'}(\tau)d\tau = \begin{cases} 1 & \text{if } m = m', \\ 0 & \text{otherwise.} \end{cases}$$

The magnetic eigenfunctions are used in a model of the atom in which the electrons keep their magnetic quantum numbers unchanged during their random motion around the nucleus of the atom.

Let $\beta_{\ell,m}(\tau)$ be the *subshell-magnetic* eigenfunctions, i.e.,

$$\int \beta_{\ell,m}^*(\tau)\beta_{\ell',m'}(\tau)d\tau = \begin{cases} 1 & \text{if } \ell = \ell', m = m', \\ 0 & \text{otherwise.} \end{cases}$$

The subshell-magnetic eigenfunctions are used in a model of the atom in which the electrons keep their angular momentum and magnetic quantum numbers unchanged during their random motion around the nucleus of the atom.

Let $\beta_{n,\ell,m}(\tau)$ be the *shell-subshell-magnetic* eigenfunctions, i.e.,

$$\int \beta_{n,\ell,m}^*(\tau)\beta_{n',\ell',m'}(\tau)d\tau = \begin{cases} 1 & \text{if } n = n', \ell = \ell', m = m', \\ 0 & \text{otherwise.} \end{cases}$$

The shell-subshell-magnetic eigenfunctions are used in a model of the atom in which the electrons keep their principal, angular momentum, and magnetic quantum numbers unchanged during their random motion around the nucleus of the atom. In all three cases mentioned above, the variable τ will be called shell variable.

Let $\psi(r_1, \nu_1, \omega_1, \ldots, r_Z, \nu_Z, \omega_Z)$ be the spatial probability wave function. Denote by $n(i), \ell(i), m(i)$, and $s(i)$ the quantum numbers of the i-th electron of the atom. The space-shell-subshell-magnetic-spin probability wave function of the probabilistic model of a *laminar* or *calm* atom is

$$\tilde{\psi}(1,\ldots,Z) = \psi(r_1,\nu_1,\omega_1,\ldots,r_Z,\nu_Z,\omega_Z)\prod_{i=1}^{Z}\beta_{n(i),\ell(i),m(i)}(\tau_i)\alpha_{s(i)}(\sigma_i). \quad (66)$$

The space-subshell-magnetic-spin probability wave function of the probabilistic model of a *nervous* atom is

$$\tilde{\psi}(1,\ldots,Z) = \psi(r_1,\nu_1,\omega_1,\ldots,r_Z,\nu_Z,\omega_Z) \prod_{i=1}^{Z} \beta_{\ell(i),m(i)}(\tau_i)\alpha_{s(i)}(\sigma_i).$$

The space-magnetic-spin probability wave function of the probabilistic model of a *turbulent* atom is

$$\tilde{\psi}(1,\ldots,Z) = \psi(r_1,\nu_1,\omega_1,\ldots,r_Z,\nu_Z,\omega_Z) \prod_{i=1}^{Z} \beta_{m(i)}(\tau_i)\alpha_{s(i)}(\sigma_i).$$

The space-spin probability wave function of the probabilistic model of a *nebulous* atom is

$$\tilde{\psi}(1,\ldots,Z) = \psi(r_1,\nu_1,\omega_1,\ldots,r_Z,\nu_Z,\omega_Z) \prod_{i=1}^{Z} \alpha_{s(i)}(\sigma_i). \tag{67}$$

The spatial probability wave function of the probabilistic model of a *chaotic* atom is

$$\tilde{\psi}(1,\ldots,Z) = \psi(r_1,\nu_1,\omega_1,\ldots,r_Z,\nu_Z,\omega_Z).$$

In the model of a laminar or calm atom, the electrons move randomly inside the atom but without changing their shells, subshells, magnetic quantum numbers, and spins. In the model of a nervous atom, the electrons move randomly inside the atom, free to switch shells, but without changing their subshells, magnetic quantum numbers, and spins. In the model of a turbulent atom, the electrons move randomly inside the atom, free to switch shells and subshells, but without changing their magnetic quantum numbers and spins. Finally, in the model of a nebulous atom, the electrons move randomly inside the atom, free to switch shells, subshells, and magnetic quantum numbers, but without changing their spins. Let us notice that presently the quantum mechanics textbooks deal with the model of a nebulous atom.

As the electrons are fermions, the probability wave function of a system of electrons must be antisymmetric with respect to the exchange of any two sets of space and spin variables. The expression changes sign when any two variables are exchanged and vanishes when any two space-spin indices denote the same state. The totally antisymmetrical probability wave function of the atom with Z electrons is

$$\Psi = \frac{1}{\sqrt{Z!}} \sum_p \text{sgn}(p)\, \tilde{\psi}(p(1), \ldots, p(Z)), \tag{68}$$

where the sum is taken with respect to all $Z!$ permutations $p = (p(1), \ldots, p(Z))$ of the set of positive integers $\{1, \ldots, Z\}$, and $\text{sgn}(p)$ is the signature, or sign of the permutation p. An inversion is said to occur in a permutation $p = (p(1), \ldots, p(Z))$ of $\{1, \ldots, Z\}$ whenever a larger integer preceeds a smaller one. A permutation is called even if the total number of inversions is an even integer and is called odd if the total number of inversions is an odd integer. More explicitly, a permutation is even or odd according as to whether there is an even or odd number of pairs (i, k) for which $i > k$ but i precedes k in the permutation. The sign or signature, or parity of the permutation is equal to 1 if the permutation is even and equal to -1 if the permutation is odd. Antisymmetrization of the probability wave function of the atom is required but may not always be essential for a description of electrons.

In what follows, if a function F depends on the space variables

$$r_1, \nu_1, \omega_1, \ldots, r_Z, \nu_Z, \omega_Z,$$

its multiple integral with respect to the measure

$$r_1^2 dr_1 d\nu_1 d\omega_1 \ldots r_Z^2 dr_Z d\nu_Z d\omega_Z$$

will be denoted by $< F >_1$, if the integral does exist. If a function F depends on the space and spin variables

$$r_1, \nu_1, \omega_1, \sigma_1, \ldots, r_Z, \nu_Z, \omega_Z, \sigma_Z,$$

its multiple integral with respect to the measure

$$r_1^2 dr_1 d\nu_1 d\omega_1 d\sigma_1 \ldots r_Z^2 dr_Z d\nu_Z d\omega_Z d\sigma_Z$$

will be denoted by $< F >_2$, if the integral does exist. Finally, if a function F depends on the space, shell, and spin variables

$$r_1, \nu_1, \omega_1, \tau_1, \sigma_1, \ldots, r_Z, \nu_Z, \omega_Z, \tau_Z, \sigma_Z,$$

its multiple integral with respect to the measure

$$r_1^2 dr_1 d\nu_1 d\omega_1 d\tau_1 d\sigma_1 \ldots r_Z^2 dr_Z d\nu_Z d\omega_Z d\tau_Z d\sigma_Z$$

will be denoted by $< F >_3$, if the integral does exist.

In Bohr-Rutherford old electronic model of atom, the electrons moved on rigid orbits. In a probabilistic model of the atom, the electrons are supposed to move randomly. The pair of quantum numbers (n, ℓ) identifies the energy subshells $E_{n,\ell}$ of the atom. The subshells of an atom may be numbered in increasing order, namely $k = 1, 2, \ldots, Y$, taking $\ell = 0, 1, \ldots, n - 1$ for $n = 1, 2, \ldots$. Thus, the subshell $k = 1$ corresponds to $n = 1, \ell = 0$, the subshell $k = 2$ corresponds to $n = 2, \ell = 0$, the subshell $k = 3$ corresponds to $n = 2, \ell = 1$, the subshell $k = 4$ corresponds to $n = 3, \ell = 0$, the subshell $k = 5$ corresponds to $n = 3, \ell = 1$, the subshell $k = 6$ corresponds to $n = 3, \ell = 2$, etc. In the probabilistic model, the electrons belonging to the k-th subshell move randomly, having the same mean radial distance μ_k from the nucleus. If all the electrons of the atom are supposed to be independent and move randomly with the same mean radial distance μ from the nucleus, then, by minimizing the mean energy of the atom, the corresponding ground state energy would be much smaller then the experimental value, which is absurd. This shows that indeed, the electrons of the atom are structured in subshells and the mean radial distance from the nucleus must be different for different energy subshells. In the old quantum mechanics, Niels Bohr assumed the existence of stationary electron orbits, the so called Bohr's orbitals, and the radius of the k-th orbit in the Bohr-Rutherford model of atom was taken to be

$$r_k \sim \frac{k^2 a_0}{Z},$$

where a_0 is the Bohr radius which is equal to 1 if atomic units are used. In a probabilistic model of the atom, instead of moving along Bohr's fixed and rigid orbits, the electrons move randomly, according to the probability density function induced by the square of the respective wave function, but a reliable model of the atom is obtained by assuming that the mean distance (not the distance, as in the Bohr's rigid model) from the nucleus of the electrons belonging to the k-th energy subshell is given by the weighted Bohr number

$$\mu_k = r\frac{k^2}{Z}, \tag{69}$$

where r is a positive radial weight to be determined by looking for stationary values of the mean energy of the atom. The results obtained in such a probabilistic model are in concordance with the experimental values obtained by chemists. In some sense we can say that Bohr was right but on average, or more precisely, on a weighted average.

The model of the calm atom assumes that the electrons keep all their four quantum numbers (n, ℓ, m, s) unchanged during their random motion around the nucleus. The model of the nervous atom assumes that the electrons keep only three of their quantum numbers, namely (ℓ, m, s), unchanged. The model of the turbulent atom assumes that the electrons keep only two of their quantum numbers, namely (m, s), unchanged. Finally, the model of the nebulous atom assumes that the electrons keep only one of their quantum numbers, namely s, unchanged, and otherwise are free to move randomly anywhere inside the atom. Nowadays, the quantum chemistry deals with what has been called here the model of the nebulous atom, but without relying on the mean weighted Bohr's orbits (69) proposed in this paper. As the model of a nebulous atom covers the standard description of atoms generally accepted today, why do we need more probabilistic models for describing the structure of an atom and the random motion of its electrons? The answer has two parts. First, if a given energy is assigned to each subshell of the atom, then the degeneracy inside a subshell may be eliminated only by using both the magnetic quantum number m and the spin s as electron identifiers. In such a case we do not allow two electrons in the same subshell to have the same pair (m, s). Such a thing is not possible in the model of a nebulous atom. In a probabilistic model of a turbulent atom, however, two electrons belonging to the same subshell must have distinct pairs of quantum numbers (m, s), which means that if the corresponding pairs are (m_1, s_1) and (m_2, s_2), then either $m_1 \neq m_2$ or $s_1 \neq s_2$ or both. Second, in a probabilistic model of a nebulous atom, the antisymmetrization of the probability wave function induces a very large number of terms, a difficulty we cannot cope with in the case of complex atoms even when we use powerful computers for doing the computations. Things are more manageable in probabilistic models of turbulent atoms and relatively easy in probabilistic models of nervous or calm atoms. Obviously, the probabilistic model of a calm atom provides a rigid and rough approximation of what happens inside an atom but at least the model may be implemented to complex atoms where the standard model of a nebulous atom, generally accepted today, is practically inapplicable. In what follows, we will focus mainly on the probabilistic models of turbulent and calm atoms.

If the electrons are supposed to be independent and in statistical equilibrium, then the probability density function

$$f(r_1, \nu_1, \omega_1, \ldots, r_Z, \nu_Z, \omega_Z) = (8\pi)^{-Z} \prod_{i=1}^{Z} \mu_i^{-3} e^{-r_i/\mu_i},$$

with respect to the measure $r_1^2 dr_1 d\nu_1 d\omega_1 \ldots r_Z^2 dr_Z d\nu_Z d\omega_Z$, is symmetric and, using the spin, we get an antisymmetric probability wave function in the model of the nebulous atom given by (67), i.e.,

$$\tilde{\psi}(1, \ldots, Z) = \psi(r_1, \nu_1, \omega_1, \ldots, r_Z, \nu_Z, \omega_Z) \prod_{i=1}^{Z} \alpha_{s(i)}(\sigma_i),$$

as requested by Pauli's exclusion principle. The variational parameters are the mean radial distances $\mu_1, \ldots, \mu_Z$ of the electrons from the nucleus of the atom. But minimizing the mean energy

$$\frac{< \tilde{\psi}^* \hat{H} \tilde{\psi} >_2}{< \tilde{\psi}^* \tilde{\psi} >_2},$$

we obtain that $\mu_1 = \mu_2 = \ldots = \mu_Z$, which means that the electrons are indistinguishable, form an unstructured quantum system, and the corresponding minimum mean energy is much smaller than the experimental value of the ground state energy, contradicting the variational theorem. Therefore, the electrons cannot be independent and must form a structured configuration inside the atom, as it was pointed out by the old quantum mechanics. In agreement with quantum chemistry, the electrons are grouped in energy subshells $E_{n,\ell}$ depending on the first two quantum numbers. The electrons belonging to the k-th subshell have the same mean radial distance μ_k from the nucleus of the atom. But how to discriminate between the different mean values μ_k, $(k = 1, 2, \ldots)$, in order to prevent them from being equal when we minimize the mean energy of the atom which, as mentioned before, induces an unacceptable value of the ground state energy of the atom. On the other hand, we cannot impose some prior, strict numerical values for μ_k, $(k = 1, 2, \ldots)$. It is more reasonable to allow them to depend on a variational parameter whose optimal value is obtained from minimizing the mean energy of the atom. Good models for the atom have been obtained by taking μ_k to be equal to a weighted form of Bohr's orbit values, namely,

$$\mu_k = r\frac{k^2}{Z}, \quad (k = 1, 2, \ldots),$$

where r is a strictly positive weight to be determined from the variational method. But by doing this, the probability wave function is no longer symmetrical and, according to Pauli's exclusion principle, we have to antisymmetrize it, as mentioned before. Unfortunately, the number of permutations involved in the antisymmetrization of the probability wave function increases tremendously for complex atoms. This is why we need several types of probabilistic models for the structure of atoms. Thus, if $Z = 3$, for instance, there are 6 terms in the probability wave function Ψ, whereas, if a function F depends only on the space variables $r_1, \nu_1, \omega_1, \ldots, r_Z, \nu_Z, \omega_Z$, then $< \Psi^* F \Psi >_1$ contains 36 terms in the model of a chaotic atom, $< \Psi^* F \Psi >_2$ contains 12 terms in the model of a nebulous atom, and $< \Psi^* F \Psi >_3$ contains six terms in the model of a calm atom, due to the respective different eigenfunctions taken into account. If a function F, depending only on the spatial variable $r_1, \nu_1, \omega_1, \ldots, r_Z, \nu_Z, \omega_Z$, is symmetric with respect to the triplets $(r_i, \nu_i, \omega_i), (i = 1, \ldots, Z)$, then $< \Psi^* F \Psi >_1$ contains six terms in the model of a chaotic atom, $< \Psi^* F \Psi >_2$ contains two terms in the model of a nebulous atom, and $< \Psi^* F \Psi >_3$ contains one term in the model of a calm atom.

e) Multiple integrals involving probability wave functions.

Let Ψ be the antisymmetric probability wave function defined by (68). Denote by

$$F(1, \ldots, Z) = F(r_1, \nu_1, \omega_1, \ldots, r_Z, \nu_Z, \omega_Z),$$

a function, or a linear operator, depending on the spatial variables. For simplifying the writing, denote the spatial probability wave function by

$$\psi(1, \ldots, Z) = \psi(r_1, \nu_1, \omega_1, \ldots, r_Z, \nu_Z, \omega_Z).$$

Thus, for the probabilistic model of a chaotic atom,

$$< \Psi^* F(1, \ldots, Z)\Psi >_1=$$

$$= \frac{1}{Z!} \sum_p < \psi(p(1), \ldots, p(Z)) F(1, \ldots, Z) \rho_{cha}(p(1), \ldots, p(Z)) >_1 =$$

$$= \frac{1}{Z!} \sum_p < \psi(1, \ldots, Z) F(p(1), \ldots, p(Z)) \rho_{cha}(1, \ldots, Z) >_1 =$$

$$= < \psi(1, \ldots, Z) \left[\frac{1}{Z!} \sum_p F(p(1), \ldots, p(Z)) \right] \rho_{cha}(1, \ldots, Z) >_1,$$

where the sum is taken with respect to all $Z!$ permutations $p = (p(1), \ldots, p(Z))$ of the set of positive integers $\{1, \ldots, Z\}$, and $\rho_{cha}(1, \ldots, Z)$ is the sum of all the functions of the form $\mathrm{sgn}(q)\, \psi(q(1), \ldots, q(Z))$, for all permutations $q = (q(1), \ldots, q(Z))$ of the set $\{1, \ldots, Z\}$.

For the probabilistic model of a nebulous atom,

$$< \Psi^* F(1, \ldots, Z) \Psi >_2 =$$

$$= \frac{1}{Z!} \sum_p < \psi(p(1), \ldots, p(Z)) F(1, \ldots, Z) \rho_{neb}(p(1), \ldots, p(Z)) >_1 =$$

$$= \frac{1}{Z!} \sum_p < \psi(1, \ldots, Z) F(p(1), \ldots, p(Z)) \rho_{neb}(1, \ldots, Z) >_1 =$$

$$= < \psi(1, \ldots, Z) \left[\frac{1}{Z!} \sum_p F(p(1), \ldots, p(Z)) \right] \rho_{neb}(1, \ldots, Z) >_1,$$

where the sum is taken with respect to all $Z!$ permutations $p = (p(1), \ldots, p(Z))$ of the positive integers $\{1, \ldots, Z\}$, and $\rho_{neb}(1, \ldots, Z)$ is the sum of functions of the form $\mathrm{sgn}(q)\, \psi(q(1), \ldots, q(Z))$, for some permutations $q = (q(1), \ldots, q(Z))$ of the set $\{1, \ldots, Z\}$ allowed by the corresponding spin eigenfunctions.

For the probabilistic model of a turbulent atom,

$$< \Psi^* F(1, \ldots, Z) \Psi >_3 =$$

$$= \frac{1}{Z!} \sum_p < \psi(p(1), \ldots, p(Z)) F(1, \ldots, Z) \rho_{tur}(p(1), \ldots, p(Z)) >_1 =$$

$$= \frac{1}{Z!} \sum_p < \psi(1, \ldots, Z) F(p(1), \ldots, p(Z)) \rho_{tur}(1, \ldots, Z) >_1 =$$

$$= < \psi(1, \ldots, Z) \left[\frac{1}{Z!} \sum_p F(p(1), \ldots, p(Z)) \right] \rho_{tur}(1, \ldots, Z) >_1,$$

where the sum is taken with respect to all $Z!$ permutations $p = (p(1), \ldots, p(Z))$ of the positive integers $\{1, \ldots, Z\}$, and $\rho_{tur}(1, \ldots, Z)$ is the sum of functions of the form $\text{sgn}(q)\,\psi(q(1), \ldots, q(Z))$, for some permutations $q = (q(1), \ldots, q(Z))$ of the set $\{1, \ldots, Z\}$ allowed by the corresponding magnetic and spin eigenfunctions.

Similarly, for the probabilistic model of a nervous atom,

$$< \Psi^* F(1, \ldots, Z)\Psi >_3=$$

$$=< \psi(1, \ldots, Z) \left[\frac{1}{Z!} \sum_p F(p(1), \ldots, p(Z)) \right] \rho_{ner}(1, \ldots, Z) >_1,$$

where the sum is taken with respect to all $Z!$ permutations $p = (p(1), \ldots, p(Z))$ of the positive integers $\{1, \ldots, Z\}$, and $\rho_{ner}(1, \ldots, Z)$ is the sum of functions of the form $\text{sgn}(q)\,\psi(q(1), \ldots, q(Z))$, for some permutations $q = (q(1), \ldots, q(Z))$ of the set $\{1, \ldots, Z\}$ allowed by the corresponding subshell-magnetic and spin eigenfunctions.

Finally, for the probabilistic model of a calm atom,

$$< \Psi^* F(1, \ldots, Z)\Psi >_3=$$

$$=< \psi(1, \ldots, Z) \left[\frac{1}{Z!} \sum_p F(p(1), \ldots, p(Z)) \right] \rho_{cal}(1, \ldots, Z) >_1,$$

where the sum is taken with respect to all $Z!$ permutations $p = (p(1), \ldots, p(Z))$ of the positive integers $\{1, \ldots, Z\}$, and $\rho_{cal}(1, \ldots, Z)$ is the sum of functions of the form $\text{sgn}(q)\,\psi(q(1), \ldots, q(Z))$, for some permutations $q = (q(1), \ldots, q(Z))$ of the set $\{1, \ldots, Z\}$ allowed by the corresponding shell-subshell-magnetic and spin eigenfunctions.

Thus, for the lithium atom,

$$
\begin{aligned}
\rho_{cha}(1,2,3) &= \psi(1,2,3) - \psi(2,1,3) + \psi(2,3,1) - \\
&\quad -\psi(3,2,1) + \psi(3,1,2) - \psi(1,3,2); \\
\rho_{neb}(1,2,3) &= \psi(1,2,3) - \psi(3,2,1); \\
\rho_{tur}(1,2,3) &= \psi(1,2,3) - \psi(3,2,1); \\
\rho_{ner}(1,2,3) &= \psi(1,2,3) - \psi(3,2,1); \\
\rho_{cal}(1,2,3) &= \psi(1,2,3).
\end{aligned}
$$

If F is symmetric with respect to the Z groups of variables (r_i, ν_i, ω_i), $(i = 1, \ldots, Z)$, then

$$F(1, \ldots, Z) = F(p(1), \ldots, p(Z))$$

for all permutations $p = (p(1), \ldots, p(Z))$ of the set of integers $\{1, \ldots, Z\}$ and for the probabilistic models of chaotic, nebulous, turbulent, nervous, and calm atoms we get

$$< \Psi^* F(1, \ldots, Z)\Psi >_1 = < \psi(1, \ldots, Z)F(1, \ldots, Z)\rho_{cha}(1, \ldots, Z) >_1,$$
$$< \Psi^* F(1, \ldots, Z)\Psi >_2 = < \psi(1, \ldots, Z)F(1, \ldots, Z)\rho_{neb}(1, \ldots, Z) >_1,$$
$$< \Psi^* F(1, \ldots, Z)\Psi >_3 = < \psi(1, \ldots, Z)F(1, \ldots, Z)\rho_{tur}(1, \ldots, Z) >_1,$$
$$< \Psi^* F(1, \ldots, Z)\Psi >_3 = < \psi(1, \ldots, Z)F(1, \ldots, Z)\rho_{ner}(1, \ldots, Z) >_1,$$
$$< \Psi^* F(1, \ldots, Z)\Psi >_3 = < \psi(1, \ldots, Z)F(1, \ldots, Z)\rho_{cal}(1, \ldots, Z) >_1,$$

respectively.

Dealing with the probabilistic model of a turbulent atom, let us give the expression of $\rho_{tur}(1, \ldots, Z)$ for some atoms:

For the helium atom, $Z = 2$, and there are $2! = 2$ permutations of the two electrons. For the turbulent model of this atom we have

$$\rho_{tur}(1, 2) = \psi(1, 2), \quad < \Psi^* \Psi >_2 = < \psi^2(1, 2) >_1,$$

$$< \Psi^* r_i \Psi >_2 = < \frac{1}{2}(r_1 + r_2)\psi^2(1, 2) >_1 = < r_i \psi^2(1, 2) >_1, (i = 1, 2).$$

For the lithium atom, $Z = 3$, and there are $3! = 6$ permutations of the three electrons. For the turbulent model of this atom, however, we have

$$\rho_{tur}(1, 2, 3) = \psi(1, 2, 3) - \psi(3, 2, 1), \quad < \Psi^* \Psi >_2 = < \psi(1, 2, 3)\rho_{tur}(1, 2, 3) >_1,$$

$$< \Psi^* r_i \Psi >_2 = < \frac{1}{3}(r_1 + r_2 + r_3)\psi(1, 2, 3)\rho_{tur}(1, 2, 3) >_1, (i = 1, 2, 3).$$

For the beryllium atom, $Z = 4$, and there are $4! = 24$ permutations of the four electrons. For the turbulent model of this atom, however, we have

$$\rho_{tur}(1, 2, 3, 4) = \psi(1, 2, 3, 4) - \psi(1, 4, 3, 2) - \psi(3, 2, 1, 4) + \psi(3, 4, 1, 2).$$

For the boron atom, $Z = 5$, and there are $5! = 120$ permutations of the five electrons. For the turbulent model of this atom, however, we have

$$\rho_{tur}(1, 2, 3, 4, 5) = \psi(1, 2, 3, 4, 5) - \psi(3, 2, 1, 4, 5) -$$

$$-\psi(1,4,3,2,5) + \psi(3,4,1,2,5).$$

For the carbon atom, $Z = 6$, and there are $6! = 720$ permutations of the six electrons. For the turbulent model of this atom, however, we have

$$\rho_{tur}(1,2,3,4,5,6) = \psi(1,2,3,4,5,6)-$$

$$-\psi(3,2,1,4,5,6) - \psi(1,4,3,2,5,6) + \psi(3,4,1,2,5,6).$$

For the nitrogen atom, $Z = 7$, and there are $7! = 5040$ permutations of the seven electrons. For the turbulent model of this atom, however, we have

$$\begin{aligned}
\rho_{tur}(1,2,3,4,5,6,7) &= \psi(1,2,3,4,5,6,7) - \psi(1,2,7,4,5,6,3) - \\
&\quad -\psi(3,2,1,4,5,6,7) + \psi(3,2,7,4,5,6,1) + \\
&\quad +\psi(7,2,1,4,5,6,3) - \psi(7,2,3,4,5,6,1) - \\
&\quad -\psi(1,4,3,2,5,6,7) + \psi(1,4,7,2,5,6,3) + \\
&\quad +\psi(3,4,1,2,5,6,7) - \psi(3,4,7,2,5,6,1) - \\
&\quad -\psi(7,4,1,2,5,6,3) + \psi(7,4,3,2,5,6,1).
\end{aligned}$$

Let us notice that for the helium atom the calm, nervous, turbulent, and nebulous models are identical. For the lithium atom the nervous, turbulent, and nebulous models are identical. The same thing happens for the beryllium atom. For the boron atom the nervous and turbulent models are identical. For its nebulous model, however, we have

$$\begin{aligned}
\rho_{neb}(1,2,3,4,5) &= \psi(1,2,3,4,5) - \psi(3,2,1,4,5) - \\
&\quad -\psi(5,2,3,4,1) - \psi(1,2,5,4,3) + \\
&\quad +\psi(3,2,5,4,1) + \psi(5,2,1,4,3) - \\
&\quad -\psi(1,4,3,2,5) + \psi(3,4,1,2,5) + \\
&\quad +\psi(5,4,3,2,1) + \psi(1,4,5,2,3) - \\
&\quad -\psi(3,4,5,2,1) - \psi(5,4,1,2,3).
\end{aligned}$$

For the carbon atom the nervous and turbulent models are identical too. For its nebulous model, however, we have

$$\rho_{neb}(1,2,3,4,5,6) =$$

$$+\psi(1,2,3,4,5,6) - \psi(3,2,1,4,5,6) - \psi(5,2,3,4,1,6) -$$

210

$$-\psi(1,2,5,4,3,6) + \psi(3,2,5,4,1,6) + \psi(5,2,1,4,3,6) -$$
$$-\psi(1,4,3,2,5,6) - \psi(3,4,1,2,5,6) + \psi(5,4,3,2,1,6) +$$
$$+\psi(1,4,5,2,3,6) - \psi(3,4,5,2,1,6) - \psi(5,4,1,2,3,6) -$$
$$-\psi(1,6,3,4,5,2) + \psi(3,6,1,4,5,2) + \psi(5,6,3,4,1,2) +$$
$$+\psi(1,6,5,4,3,2) - \psi(3,6,5,4,1,2) - \psi(5,6,1,4,3,2) -$$
$$-\psi(1,2,3,6,5,4) + \psi(3,2,1,6,5,4) + \psi(5,2,3,6,1,4) +$$
$$+\psi(1,2,5,6,3,4) - \psi(3,2,5,6,1,4) - \psi(5,2,1,6,3,4) +$$
$$+\psi(1,4,3,6,5,2) - \psi(3,4,1,6,5,2) - \psi(5,4,3,6,1,2) -$$
$$-\psi(1,4,5,6,3,2) + \psi(3,4,5,6,1,2) + \psi(5,4,1,6,3,2) +$$
$$+\psi(1,6,3,2,5,4) - \psi(3,6,1,2,5,4) - \psi(5,6,3,2,1,4) -$$
$$-\psi(1,6,5,2,3,4) + \psi(3,6,5,2,1,4) + \psi(5,6,1,2,3,4).$$

In general, the function $\rho_{neb}(1,\ldots,Z)$ contains

$$\left(\frac{Z-1}{2}!\right) \times \left(\frac{Z+1}{2}!\right)$$

terms if Z is odd and

$$\left(\frac{Z}{2}!\right) \times \left(\frac{Z}{2}!\right)$$

terms if Z is even.

For the nitrogen atom the calm, nervous, turbulent, and nebulous models are all distinct. In its case, $\rho_{neb}(1,2,3,4,5,6,7)$ contains $(3!) \times (4!) = 144$ terms.

For the probabilistic model of a turbulent atom, for instance, the following two special cases are important:

1^0. If $F(1,\ldots,Z) \equiv 1$, then

$$< \Psi^*\Psi >_3 = < \psi(1,\ldots,Z)\rho_{tur}(1,\ldots,Z) >_1 .$$

2^0. If $F(1,\ldots,Z) = \hat{H}$, where $\hat{H}$ is the Hamiltonian (63) of the system, then

$$< \Psi^*\hat{H}\Psi >_3 = < \psi(1,\ldots,Z)\hat{H}\rho_{tur}(1,\ldots,Z) >_1,$$

and the mean energy of the atom is

$$E = \frac{< \Psi^*\hat{H}\Psi >_3}{< \Psi^*\Psi >_3} = \frac{< \psi(1,\ldots,Z)\hat{H}\rho_{tur}(1,\ldots,Z) >_1}{< \psi(1,\ldots,Z)\rho_{tur}(1,\ldots,Z) >_1}.$$

Similar formulas are obtained replacing $\rho_{tur}(1,\ldots,Z)$ by $\rho_{ner}(1,\ldots,Z)$ and by $\rho_{cal}(1,\ldots,Z)$, for the probabilistic models of a nervous atom and of a calm atom, respectively, and replacing $< \cdot >_3$ and $\rho_{tur}(1,\ldots,Z)$ by $< \cdot >_2$ and $\rho_{neb}(1,\ldots,Z)$, for the probabilistic model of a nebulous atom, and by $< \cdot >_1$ and $\rho_{cha}(1,\ldots,Z)$, for the probabilistic model of a chaotic atom.

f) Global correlation between electrons.

Using the notations given above, for the probabilistic model of a turbulent atom, for instance, the following special cases are useful for defining a global correlations between an arbitrary pair of electrons of the atom:

1^0. If $F(1,\ldots,Z) = r_i$, then

$$< \Psi^* r_i \Psi >_3 = < \psi(1,\ldots,Z)\frac{1}{Z!}[r_1(Z-1)!+\ldots+r_Z(Z-1)!]\rho_{tur}(1,\ldots,Z) >_1 =$$

$$= < \psi(1,\ldots,Z)\frac{1}{Z}(r_1 + \ldots + r_Z)\rho_{tur}(1,\ldots,Z) >_1 .$$

Thus,

$$\bar{r} = \frac{< \Psi^* r_i \Psi >_3}{< \Psi^* \Psi >_3} =$$

$$= \frac{1}{Z}\frac{< \psi(1,\ldots,Z)(r_1 + \ldots + r_Z)\rho_{tur}(1,\ldots,Z) >_1}{< \psi(1,\ldots,Z)\rho_{tur}(1,\ldots,Z) >_1}. \qquad (70)$$

2^0. If $F(1,\ldots,Z) = (r_i - \bar{r})^2$, then

$$< \Psi^*(r_i - \bar{r})^2 \Psi >_3 =$$

$$= < \psi(1,\ldots,Z)\frac{1}{Z}[(r_1 - \bar{r})^2 + \ldots + (r_Z - \bar{r})^2]\rho_{tur}(1,\ldots,Z) >_1 . \qquad (71)$$

3^0. If $F(1,\ldots,Z) = (r_i - \bar{r})(r_j - \bar{r})$, then

$$< \Psi^*(r_i - \bar{r})(r_j - \bar{r})\Psi >_3 =$$

$$= < \psi(1,\ldots,Z)[\frac{1}{Z!} \sum_{i=1}^{Z}\sum_{\substack{j=1 \\ i \neq j}}^{Z} (r_i - \bar{r})(r_j - \bar{r})(Z-2)!]\rho_{tur}(1,\ldots,Z) >_1 =$$

$$=< \psi(1,\ldots,Z)[\frac{2}{(Z-1)Z} \sum_{\substack{i=1 \\ i<j}}^{Z-1} \sum_{j=2}^{Z} (r_i - \bar{r})(r_j - \bar{r})]\rho_{tur}(1,\ldots,Z) >_1$$

$$(72)$$

Using (70)-(72), we can calculate the global correlation between two arbitrary electrons i and j of the atom according to the formula

$$\text{globcorr}_{tur} = \frac{< \Psi^*(r_i - \bar{r})(r_j - \bar{r})\Psi >_3}{\sqrt{< \Psi^*(r_i - \bar{r})^2\Psi >_3 < \Psi^*(r_j - \bar{r})^2\Psi >_3}}. \qquad (73)$$

Similar formulas are obtained replacing $\rho_{tur}(1,\ldots,Z)$ by $\rho_{ner}(1,\ldots,Z)$ and by $\rho_{cal}(1,\ldots,Z)$, for the probabilistic models of a nervous atom and of a calm atom, respectively, and replacing $< \cdot >_3$ and $\rho_{tur}(1,\ldots,Z)$ by $< \cdot >_2$ and $\rho_{neb}(1,\ldots,Z)$, for the probabilistic model of a nebulous atom, and by $< \cdot >_1$ and $\rho_{cha}(1,\ldots,Z)$, for the probabilistic model of a chaotic atom.

g) Standard correlation between the subshells of the atom.

The standard correlation from statistics may be used for calculating the amount of linear dependence between different levels of an atom. It may be also viewed as measuring the linear standard correlation between an electron belonging to the subshell with mean radial distance from nucleus equal to μ_k and an electron belonging to a subshell with mean radial distance from nucleus equal to μ_ℓ. When $\mu_\ell = \mu_k = \mu$, it will measure the autocorrelation of the subshell with the mean radial distance μ, or the linear standard correlation between two arbitrary electrons belonging to the same subshell having the mean radial distance from the nucleus equal to μ. For a turbulent model of the atom, for instance, the standard correlation between an arbitrary electron, say i, belonging to the subshell of mean radial distance from the nucleus equal to μ_k and an arbitrary electron, say j, belonging to the subshell of mean radial distance from the nucleus equal to μ_ℓ is defined by

$$\text{stcorr}_{tur}(\mu_k, \mu_\ell) = \frac{< \Psi^*(r_i - \mu_k)(r_j - \mu_\ell)\Psi >_3}{\sqrt{< \Psi^*(r_i - \mu_k)^2\Psi >_3 < \Psi^*(r_j - \mu_\ell)^2\Psi >_3}}, \qquad (74)$$

where the covariance is

$$< \Psi^*(r_i - \mu_k)(r_j - \mu_\ell)\Psi >_3=$$

213

$$=< \psi(1,\ldots,Z)[\frac{1}{(Z-1)Z} \sum_{i=1}^{Z} \sum_{j=1}^{Z} (r_i - \mu_k)(r_j - \mu_\ell)]\rho_{tur}(1,\ldots,Z) >_1,$$

$$(75)$$

and the variance is

$$< \Psi^*(r_i - \mu_k)^2 \Psi >_3=$$

$$=< \psi(1,\ldots,Z)\frac{1}{Z}[(r_1 - \mu_k)^2 + \ldots + (r_Z - \mu_k)^2]\rho_{tur}(1,\ldots,Z) >_1 . \quad (76)$$

Similar formulas may be written for the other probabilistic models of the atom, replacing ρ_{tur} by ρ_{cal}, ρ_{ner}, or ρ_{neb}, respectively.

h) Stability and interdependence inside the atom.

All three types of correlation introduced so far have the disadvantage of comparing only pairs of entities whereas, in some cases, the interconnection among more than two entities is relevant. On the other hand, when a system has a probabilistic or random behavior, we are interested to know the degree of stability of its different subsystems because randomness does not generally mean chaos. The entropic tools from information theory, like Shannon's entropy and Watanabe's entropic measure of cohesion, are customarily used for measuring the amount of uncertainty and the global interdependence among the subsystems of a system with random behaviour. Unfortunately, in dealing with the continuous probability densities assigned to the subsystems of electrons, these tools cannot be practically used simply because the corresponding multiple integrals are much too difficult even for the simplest atoms. We can use, however, a simpler indicator of the amount of randomness, namely, the norm of the probability densities in the Hilbert space of square integrable functions.

In the finite discrete case, if $p = \{p_1,\ldots,p_n\}$ is a probability distribution, the information energy (Onicescu (1966)) is the number

$$O(p_1,\ldots,p_n) = \sum_{i=1}^{n} p_i^2.$$

It is easy to show that

$$\frac{1}{n} \leq O(p_1,\ldots,p_n) \leq 1,$$

214

the smalest value corresponding to maximum uncertainty, i.e., the uniform distribution $p_i = 1/n$, whereas the largest value corresponds to certainty, i.e., $p_i = 1$ if $i = i_0$ and $p_i = 0$ if $i \neq i_0$. The positive square root of the information energy is just the Euclidean norm of the n-dimensional vector $(p_1, \ldots, p_n)$.

In our context, for three arbitrary electrons belonging to the subshells having the mean radial distances μ_i, μ_j, and μ_k, repectively, denote by $d(\mu_i, \mu_j, \mu_k)$ the marginal density with respect to the space variables

$$r_i, \nu_i, \omega_i, \quad r_j, \nu_j, \omega_j, \quad r_k, \nu_k, \omega_k,$$

of the joint density

$$\frac{\psi\rho}{<\psi\rho>_1}.$$

The degree of stability of the subsystem of these three electrons is given by the norm

$$< d^2(\mu_i, \mu_j, \mu_k) >_1^{1/2} .$$

A larger value of it means less randomness in locating the subsystem of three electrons and, therefore, more stability of its random behavior. The smaller its value the more random is the joint evolution of the three electrons making up this subsystem. The amount of interdependence among the three electrons inside the subsystem formed by them, taken together, is measured by the deviation from independence,

$$I(\{\mu_i, \mu_j, \mu_k\} : \{\mu_i\}, \{\mu_j\}, \{\mu_k\}) = \qquad (77)$$

$$= | < d^2(\mu_i, \mu_j, \mu_k) >_1^{1/2} - < d^2(\mu_i) >_1^{1/2} < d^2(\mu_j) >_1^{1/2} < d^2(\mu_k) >_1^{1/2} |.$$

Similarly, the interdependence between the subsubsystem of the two arbitrary electrons belonging to the subshells of mean radial distances μ_i and μ_j, taken together, and the subsubsystem consisting of an arbitrary electron belonging to the subshell of mean radial distance μ_k, inside the subsystem formed by these three electrons is measured by the deviation from independence

$$I(\{\mu_i, \mu_j, \mu_k\} : \{\mu_i, \mu_j\}, \{\mu_k\}) =$$

$$= | < d^2(\mu_i, \mu_j, \mu_k) >_1^{1/2} - < d^2(\mu_i, \mu_j) >_1^{1/2} < d^2(\mu_k) >_1^{1/2} |.$$

If, for instance, a subshell of mean radial distance μ_i consists of two electrons and another subshell of mean radial distance μ_j consists of three electrons, then the interdependence between the two subshells is measured by

$$I(\{\mu_i, \mu_i, \mu_j, \mu_j, \mu_j\} : \{\mu_i, \mu_i\}, \{\mu_j, \mu_j, \mu_j\}),$$

and the interdependence among the electrons making up the subshells is measured by

$$I(\{\mu_i, \mu_i\} : \{\mu_i\}, \{\mu_i\}) \quad \text{and} \quad I(\{\mu_j, \mu_j, \mu_j\} : \{\mu_j\}, \{\mu_j\}, \{\mu_j\}),$$

repectively. The generalization of the above concepts to any number of subsystems consisting of any number of electrons of the atom is obvious.

i) Estimations for the diameter of the atom.

Let $\bar{\mu}$ be the weighted mean orbit of the atom, μ_{ext} the mean orbit of the external subshell of the atom, and $\bar{r}$ the mean distance from the nucleus of an arbitrary electron of the atom. In the case of the nebulous model for the lithium atom, for instance,

$$\bar{\mu} = \frac{1}{3}(2\mu_1 + \mu_2), \quad \mu_{ext} = \mu_2, \quad \bar{r} = \frac{1}{3}\frac{<\psi\,(r_1 + r_2 + r_3)\,\rho_{neb} >_1}{<\psi\rho_{neb}>_1}.$$

The corresponding approximations for the diameter of the respective atom are

$$D_1 = 2\bar{\mu}, \quad D_2 = 2\mu_{ext}, \quad D_3 = 2\bar{r},$$

where D_1 is the mean diameter induced by statistical equilibrium, D_2 is the mean diameter induced by the external subshell of the atom, and D_3 is the mean diameter induced by the individual random fluctuations of the electrons of the atom. We can use the experimental value obtained for the diameter of an atom in order to compare it with the three approximations D_1, D_2, and D_3 mentioned above and see which one is closer. If D_1 is the closest to the experimental value, then this is an indirect indication that the atom is relatively stable. If D_2 is the closest to the experimental value, then this is an indirect indication that the external subshell of the atom plays an essential role in the structure of the atom. Finally, if D_3 is the closest to the experimental value, then this is an indirect indication that the random

216

motion of all individual atoms is determinant for the structure of that atom. We can also use $\bar{r} - \bar{\mu}$ as a mean index of instability of the respective atom.

j) Summary.

We can now summarize the strategy open to us. We start from the linear approximation of the probability wave function ψ induced by the minimum mean deviation from statistical equilibrium as given by the expression (60). Its normed square (61) is the probability density function. It depends on some variational parameters which are determined by minimizing the mean energy (62) of the atom, where $\hat{H}$ is the Hamiltonian given by (63) and where all the electrons belonging to the same k-th energy subshell have the mean radial value given by weighted Bohr's number (69), where r is a positive radial weight. Depending on different ways of making the wave function antisymmetric, as required by Pauli's exclusion principle, we introduce four models (i.e., calm, nervous, turbulent, and nebulous models) of the atom. Dealing with a probabilistic model of a turbulent atom, for instance, we find the radial weight r and the coefficients $c_1, \ldots, c_Z$ of ψ from the optimization problem

$$\min_{\psi} \frac{< \psi \hat{H} \rho_{tur} >_1}{< \psi \rho_{tur} >_1}.$$

The numerical values obtained for $r, c_1, \ldots, c_Z$ are introduced in the expression of ψ. Once the probability wave function ψ and its normed probability density f, given by (61), are determined, we can calculate the generalized linear correlation, gencorr$_{tur}$, between subshells, given by (65), the global correlation, globcorr$_{tur}$, between any two electrons of the atom, given by (73), the standard correlation between subshells, stcorr$_{tur}$, given by (74), the degree of instability of different subsystems of electrons, the interdependence between such subsystems, the three approximations of the diameter of the atom, and the mean index of stability of the atom. The same analysis may be performed for the other probabilistic models of the atom, replacing ρ_{tur} by ρ_{cal}, ρ_{ner}, or ρ_{neb}, respectively. The second half of Part II will contain the applications of the topics discussed here to hydrogen, helium, lithium, beryllium, boron, carbon, nitrogen, and argon atoms, together with the correponding computer programs, using the powerful computer software package MATHEMATICA which does both numerical and symbolic computations.

5. EXAMPLES.

Throughout this chapter we denote by

$$\varphi(r_i, \mu_k) = \frac{1}{\sqrt{8\pi\mu_k^3}} \, e^{-r_i/(2\mu_k)},$$

which is nothing but the positive square root of the radial probability density given by (46), with respect to the measure $r_i^2 dr_i d\nu_i d\omega_i$. This wave function is assigned to an electron in statistical equilibrium with the radial distance r_i from the nucleus and located in the k-th subshell of the atom characterized by the mean radial distance μ_k from the nucleus given by (69). The experimental values for the ground state energy of the atoms studied in this section are taken from Lide (2000, p.**10**-175). The experimental values of the diameters of some atoms are taken from Gray (1972, p.**7**-6).

a) Hydrogen Atom ($Z = 1$).

The exact solution for this atom, based on the minimum mean deviation from statistical equilibrium and without using the Schrödinger equation, was given in Guiasu (1998a, pp.984-992) and in Section 7 of Part I of this monograph. Let us see, however, what the linear approximation of the probability wave function induced by the minimum deviation from statistical equilibrium can tell us about the behavior of this atom. The only electron in this case has the quantum numbers $n = 1, \ell = 0, m = 0, s = -1/2$ and the mean orbit (i.e., the mean radial distance from the nucleus corresponding to statistical equilibrium) is $\mu_1 = r$. For the hydrogen atom all four models (calm, nervous, turbulent, and nebulous) give the same result as far as the linear radial approximation is concerned. The probability wave function is

$$\Psi_{cal} = \tilde{\psi}_{cal}(1) = \psi(1)\beta_{1,0,0}(\tau_1)\alpha_-(\sigma_1),$$

in the calm model, and

$$\Psi_{neb} = \tilde{\psi}_{neb}(1) = \psi(1)\alpha_-(\sigma_1),$$

in the nebulous model, where, using the linear radial approximation,

$$\psi(1) = \psi(r_1, \nu_1, \omega_1) = \varphi(r_1, \mu_1)\left[1 + c_1\left(3 - \frac{r_1}{r}\right)\right],$$

where $0 \leq r_1 < +\infty, -1 \leq \nu_1 \leq 1, 0 \leq \omega_1 \leq 2\pi$, and $\mu_1 = r$. The variational parameters are c_1 and r. We have,

$$< \Psi_{cal}^* \Psi_{cal} >_3 = < \Psi_{neb}^* \Psi_{neb} >_2 = < \psi^2(1) >_1 = 1 + 3c_1^2$$

$$< \Psi_{cal}^* \nabla_1^2 \Psi_{cal} >_3 = < \Psi_{neb}^* \nabla_1^2 \Psi_{neb} >_2 = < \psi(1) \nabla_1^2 \psi(1) >_1 = -\frac{1 + 4c_1 + 7c_1^2}{4r^2},$$

$$< \Psi_{cal}^* \frac{1}{r_1} \Psi_{cal} >_3 = < \Psi_{neb}^* \frac{1}{r_1} \Psi_{neb} >_2 = < \psi(1) \frac{1}{r_1} \psi(1) >_1 = \frac{1 + 2c_1 + 3c_1^2}{2r}.$$

As the Hamiltonian is

$$\hat{H} = -\frac{1}{2} \nabla_1^2 - \frac{1}{r_1},$$

we get for the mean energy the expression

$$E(r, c_1) = \frac{< \Psi_{cal}^* \hat{H} \Psi_{cal} >_3}{< \Psi_{cal}^* \Psi_{cal} >_3} = \frac{< \Psi_{neb}^* \hat{H} \Psi_{neb} >_2}{< \Psi_{neb}^* \Psi_{neb} >_2} =$$

$$= \frac{< \psi(1) \hat{H} \psi(1) >_1}{< \psi^2(1) >_1} = \frac{1 + 4c_1 + 7c_1^2 - 4r - 8c_1 r - 12c_1^2 r}{8(1 + 3c_1^2)r^2}.$$

Minimizing the mean energy, we get

$$\mu_1 = r = 0.514778 \text{ a.u.}, \quad c_1 = 0.0148998,$$

and the corresponding minimum value of the ground state energy is

$$E = -0.5 \text{ a.u.} = -0.5 \times 27.2116 \text{ eV} = -13.6058 \text{ eV},$$

whereas the experimental value of the ground state energy is -0.499729527 a.u. or -13.59844 eV. As the atom has only one subshell, the weighted mean orbit $\bar{\mu} = \mu_1$ and therefore, the mean diameter induced by $\bar{\mu}$ is $D_1 = 2\bar{\mu} = 1.029556$ a.u. The mean diameter induced by the external orbit is also $D_2 = 2\mu_1 = 1.029556$ a.u.

Introducing $r = 0.514778$ and $c_1 = 0.0148998$ in the expression of the probability wave function $\psi(1)$, abbreviated as ψ, we obtain

$$\psi = \frac{1}{\sqrt{8\pi(0.514778)^3}} e^{-r_1/(2(0.514778))} \left[1 + (0.0148998) \left(3 - \frac{r_1}{0.514778} \right) \right],$$

and we obtain the mean radial distance from the nucleus

$$\bar{r} = \frac{< r_1 \psi^2 >_1}{< \psi^2 >_1} = \frac{1.5003}{1.00067} = 1.49903 \text{ a.u.},$$

which gives the mean diameter of the hydrogen atom induced by $\bar{r}$ equal to $D_3 = 2\bar{r} = 2.99806$ a.u. As the experimental value of the diameter is $D_{exp} = 3.0$, this shows that D_3 is the closest approximation of the diameter, and the structure of the hydrogen atom depends more on the random fluctuations of its electron than on a regular evolution along an orbit of statistical equilibrium. The mean index of instability is $\bar{r} - \bar{\mu} = 0.984252$. The variance given by (71) is 0.747707 and the standard deviation of the electron is 0.864699. Denoting by

$$d(\mu_1) = \frac{\psi^2}{< \psi^2 >_1} = \frac{\psi^2}{1.00067},$$

the degree of stability of the only electron of the hydrogen atom is

$$< d^2(\mu_1) >_1^{1/2} = 0.17282.$$

b) Helium Atom ($Z = 2$).

The two electrons of this atom belong to the same subshell, whose mean orbit (69), i.e., the mean radial distance corresponding to statistical equilibrium, is $\mu_1 = r/2$. Their quantum numbers (n, ℓ, m, s) are $(1, 0, 0, -1/2)$ and $(1, 0, 0, +1/2)$, respectively. The Schrödinger equation of this atom cannot be solved exactly. Let us see what the linear approximation of the probability wave function induced by the minimum deviation from statistical equilibrium tells us about the behavior of this atom. For the helium atom, all four models (calm, nervous, turbulent, and nebulous) give the same result as far as the linear radial approximation is concerned. Denoting by

$$\psi(1, 2) = \psi(r_1, \nu_1, \omega_1; r_2, \nu_2, \omega_2) =$$

$$= \varphi(r_1, \mu_1)\varphi(r_2, \mu_1) \left[1 + c_1 \left(3 - \frac{r_1}{\mu_1} \right) + c_2 \left(3 - \frac{r_2}{\mu_1} \right) \right], \qquad (78)$$

the probability wave function is

$$\Psi_{cal} = \frac{1}{\sqrt{2!}} [\tilde{\psi}_{cal}(1, 2) - \tilde{\psi}_{cal}(2, 1)],$$

in the calm model of this atom, where

$$\tilde{\psi}_{cal}(1,2) = \psi(1,2)\,\beta_{1,0,0}(\tau_1)\,\alpha_-(\sigma_1)\,\beta_{1,0,0}(\tau_2)\,\alpha_+(\sigma_2),$$

and

$$\Psi_{neb} = \frac{1}{\sqrt{2!}}[\tilde{\psi}_{neb}(1,2) - \tilde{\psi}_{neb}(2,1)],$$

in the nebulous model of this atom, where

$$\tilde{\psi}_{neb}(1,2) = \psi(1,2)\alpha_-(\sigma_1)\alpha_+(\sigma_2).$$

The variational parameters are c_1, c_2 and r. We have:

$$<\Psi^*_{cal}\Psi_{cal}>_3 = <\Psi^*_{neb}\Psi_{neb}>_2 = \frac{1}{2}[<\psi^2(1,2)>_1 + <\psi^2(2,1)>_1] =$$

$$= <\psi^2(1,2)>_1 = 1 + 3c_1^2 + 3c_2^2,$$

$$<\Psi^*_{cal}\nabla_1^2\Psi_{cal}>_3 = <\Psi^*_{neb}\nabla_1^2\Psi_{neb}>_2 = <\psi(1,2)\nabla_1^2\psi(1,2)>_1 =$$

$$= -\frac{1 + 4c_1 + 7c_1^2 + 3c_2^2}{r^2},$$

$$<\Psi^*_{cal}\nabla_2^2\Psi_{cal}>_3 = <\Psi^*_{neb}\nabla_2^2\Psi_{neb}>_2 = <\psi(1,2)\nabla_2^2\psi(1,2)>_1 =$$

$$= -\frac{1 + 3c_1^2 + 4c_2 + 7c_2^2}{r^2},$$

$$<\Psi^*_{cal}\frac{1}{r_1}\Psi_{cal}>_3 = <\Psi^*_{neb}\frac{1}{r_1}\Psi_{neb}>_2 = <\psi(1,2)\frac{1}{r_1}\psi(1,2)>_1 =$$

$$= \frac{1 + 2c_1 + 3c_1^2 + 3c_2^2}{r},$$

$$<\Psi^*_{cal}\frac{1}{r_2}\Psi_{cal}>_3 = <\Psi^*_{neb}\frac{1}{r_2}\Psi_{neb}>_2 = <\psi(1,2)\frac{1}{r_2}\psi(1,2)>_1 =$$

$$= \frac{1 + 3c_1^2 + 2c_2 + 3c_2^2}{r},$$

$$<\Psi^*_{cal}\frac{1}{r_{12}}\Psi_{cal}>_3 = <\Psi^*_{neb}\frac{1}{r_{12}}\Psi_{neb}>_2 = <\psi(1,2)\frac{1}{r_{12}}\psi(1,2)>_1 =$$

$$= \frac{5 + 5c_1 + 12c_1^2 + 5c_2 + 6c_1c_2 + 12c_2^2}{8r}.$$

As the Hamiltonian is

$$\hat{H} = -\frac{1}{2}(\nabla_1^2 + \nabla_2^2) - 2(\frac{1}{r_1} + \frac{1}{r_2}) + \frac{1}{r_{12}},$$

and is symmetric with respect to the two electrons, we get for the mean energy the expression

$$E(r, c_1, c_2) = \frac{< \Psi_{cal}^* \hat{H} \Psi_{cal} >_3}{< \Psi_{cal}^* \Psi_{cal} >_3} = \frac{< \Psi_{neb}^* \hat{H} \Psi_{neb} >_2}{< \Psi_{neb}^* \Psi_{neb} >_2} =$$

$$= \frac{< \psi(1,2) \hat{H} \psi(1,2) >_1}{< \psi^2(1,2) >_1} =$$

$$= \frac{8 + 16c_1 + 40c_1^2 + 16c_2 + 40c_2^2 - 27r - 27c_1r - 84c_1^2r - 27c_2r + 6c_1c_2r - 84c_2^2r}{8(1 + 3c_1^2 + 3c_2^2)r^2}.$$

Minimizing the mean energy, we get

$$r = 0.601972 \text{ a.u.}, \qquad c_1 = c_2 = 0.00782582,$$

and the corresponding minimum value of the ground state energy is

$$E = -2.84766 \text{ a.u.} = -2.84766 \times 27.2116 \text{ eV} = -77.4894 \text{ eV},$$

representing 98.0814% of the experimental value of the ground state energy -2.90336 a.u. $= -79.0052$ eV. The weighted mean orbit of the atom is equal to the mean orbit μ_1, namely

$$\bar{\mu} = \frac{2\mu_1}{2} = \mu_1 = \frac{r}{2} = 0.300986 \text{ a.u.},$$

and the mean diameter induced by statistical equilibrium is $D_1 = 2\bar{\mu} = 0.601972$ a.u. The mean diameter induced by the external subshell is the same $D_2 = 2\mu_1 = 0.601972$ a.u.

We have $< \psi^2 >_1 = 1.00037$, and the mean radial distance of an electron from the nucleus is

$$\bar{r} = \frac{1}{2} \frac{< \psi (r_1 + r_2) \psi >_1}{< \psi^2 >_1} = 0.888941 \text{ a.u.},$$

and the mean diameter of the atom induced by the individual random fluctuations of its electrons is $D_3 = 2\bar{r} = 1.777882$ a.u. The mean index of instability of the atom is $\bar{r} - \bar{\mu} = 0.587955$.

Taking $Z = 2, \mu_1 = 0.300986$ a.u., $c_1 = c_2 = 0.00782582$ in (65), we obtain that the generalized linear correlation between the two electrons of the helium atom is

$$\text{gencorr}_{cal}(\mu_1, \mu_1) = \text{gencorr}_{ner}(\mu_1, \mu_1) =$$

$$= \text{gencorr}_{tur}(\mu_1, \mu_1) = \text{gencorr}_{neb}(\mu_1, \mu_1) = 0.000378904.$$

Taking $\mu_1 = 0.300986$ a.u. and $c_1 = c_2 = 0.00782582$ in (78), the standard correlation between the two electrons, as defined by (74)-(76), is

$$\text{stcorr}_{cal}(\mu_1, \mu_1) = \text{stcorr}_{ner}(\mu_1, \mu_1) =$$

$$= \text{stcorr}_{tur}(\mu_1, \mu_1) = \text{stcorr}_{neb}(\mu_1, \mu_1) = 0.567509.$$

In order to calculate the global correlation between two arbitrary electrons of the helium atom we introduce again

$$\mu_1 = 0.300986 \text{ a.u.}, \quad c_1 = c_2 = 0.00782582$$

into (78), and using (70)-(73) we get

$$\text{globcorr} = -0.000367089.$$

Taking $\mu_1 = 0.300986$ a.u. and $c_1 = c_2 = 0.00782582$ in (78) we denote by

$$d(\mu_1, \mu_1) = \frac{\psi^2}{<\psi^2>_1}$$

the joint probability density of the system of the two electrons and by $d(\mu_1)$ the marginal density, referring to the behavior of only one of the electrons, which is obtained by integrating $d(\mu_1, \mu_1)$ with respect to the space variables (r_2, ν_2, ω_2) or equivalently, due to symmetry, by integrating $d(\mu_1, \mu_1)$ with respect to the space variables (r_1, ν_1, ω_1). Then the degrees of stability of the helium atom and of an individual electron of the atom are

$$<d^2(\mu_1, \mu_1)>_1^{1/2} = 0.0503559, \quad <d^2(\mu_1)>_1^{1/2} = 0.224404,$$

223

whereas the interdependence between the two electrons of the helium atom, as defined by (77), is

$$I(\{\mu_1, \mu_1\} : \{\mu_1\}, \{\mu_1\}) = 1.43369 \times 10^{-6}.$$

Remark: If we choose a better approximation for the radial wave function than the linear approximation (78), namely

$$\psi(1,2) = \varphi(r_1, \mu_1)\varphi(r_2, \mu_1)\{1 + c\,[L_1^{(2)}(r_1/\mu_1) + L_1^{(2)}(r_2/\mu_1)] +$$

$$+ d\,[L_2^{(2)}(r_1/\mu_1) + L_2^{(2)}(r_2/\mu_1)] + e\,[L_3^{(2)}(r_1/\mu_1) + L_3^{(2)}(r_2/\mu_1)] +$$

$$+ f\,[L_4^{(2)}(r_1/\mu_1) + L_4^{(2)}(r_2/\mu_1)] + g\,[L_5^{(2)}(r_1/\mu_1) + L_5^{(2)}(r_2/\mu_1)]\},$$

which uses the first five Laguerre polynomials of order two, we obtain the mean radial distance of the two electrons from the nucleus $\mu_1 = 0.41906$ a.u., and the ground state mean energy of the helium atom to be

$$E = -2.87618 \text{ a.u.} = -78.2655 \text{ eV},$$

whose absolute value represents 99.0637% of the absolute value of the experimental value -2.90336 a.u. $= -79.0052$ eV. For other trial functions and better approximations of the ground state energy see Guiasu (1998b, pp.1024-1039) and Section 8 of Part I of this monograph.

c) Lithium Atom $(Z = 3)$.

This is the first atom with two subshells and more than two electrons orbiting around the nucleus. The ground state of this atom was studied in Guiasu (1998b, pp.1039-1043) and in Section 9 of Part I of this monograph. It was shown there that if the three electrons are free and independent, then the mean energy of the atom is considerably smaller than the experimental value. But if the mean radial distance from the nucleus to the most remote electron is kept equal to 2 a.u., then the mean energy becomes very close to this experimental value. But why fix the mean radial distance at the level 2 a.u. and what can we learn from this when we analyze more complex atoms? At this point we need a flexible rule for describing the mean radial distance corresponding to the subshells of the respective atom. Using the weighted Bohr approximation (69), the mean orbits of the two subshells of the lithium atom, i.e., the mean radial distances of the two subshells corresponding to statistical equilibrium, are

$$\mu_1 = \frac{r}{3}, \qquad \mu_2 = \frac{4r}{3}.$$

The three electrons have the quantum numbers (n, ℓ, m, s) given by $(1, 0, 0, -1/2)$, $(1, 0, 0, +1/2)$, and $(2, 0, 0, -1/2)$, respectively. The Hamiltonian is

$$\hat{H} = -\frac{1}{2}(\nabla_1^2 + \nabla_2^2 + \nabla_3^2) - 3\left(\frac{1}{r_1} + \frac{1}{r_2} + \frac{1}{r_3}\right) + \left(\frac{1}{r_{12}} + \frac{1}{r_{13}} + \frac{1}{r_{23}}\right). \tag{79}$$

As there are only components referring to individual electrons or to pairs of electrons but no component about the triplet of electrons taken together, we notice that from now on the Hamiltonian will not reflect all the interactions between the electrons of the atom. The wave function is

$$\Psi = \frac{1}{\sqrt{3!}} [\tilde{\psi}(1, 2, 3) - \tilde{\psi}(1, 3, 2) - \tilde{\psi}(2, 1, 3) + \tilde{\psi}(2, 3, 1) + \tilde{\psi}(3, 1, 2) - \tilde{\psi}(3, 2, 1)],$$

where

$$\tilde{\psi}(1, 2, 3) =$$

$$= \psi(1, 2, 3)\beta_{1,0,0}(\tau_1)\alpha_-(\sigma_1)\beta_{1,0,0}(\tau_2)\alpha_+(\sigma_2)\beta_{2,0,0}(\tau_3)\alpha_-(\sigma_3),$$

in the calm model of the atom, and

$$\tilde{\psi}(1, 2, 3) = \psi(1, 2, 3)\alpha_-(\sigma_1)\alpha_+(\sigma_2)\alpha_-(\sigma_3),$$

in the nebulous model, where, using the linear radial approximation, we have

$$\psi(1, 2, 3) = \psi(r_1, \nu_1, \omega_1; r_2, \nu_2, \omega_2; r_3, \nu_3, \omega_3) =$$

$$= \varphi\left(r_1, \frac{r}{3}\right)\varphi\left(r_2, \frac{r}{3}\right)\varphi\left(r_3, \frac{4r}{3}\right)\left[1 + c_1\left(3 - \frac{3r_1}{r}\right) + c_2\left(3 - \frac{3r_2}{r}\right) + c_3\left(3 - \frac{3r_3}{4r}\right)\right].$$

Let us notice that for the lithium atom three models, namely the nervous, turbulent, and nebulous models, coincide.

1^0. For the *calm model* of the atom,

$$< \Psi_{cal}^* \Psi_{cal} >_3 = \frac{1}{3!}[< \psi^2(1, 2, 3) >_1 + < \psi^2(2, 1, 3) >_1 + < \psi^2(1, 3, 2) >_1 +$$

$$+ < \psi^2(2,3,1) >_1 + < \psi^2(3,1,2) >_1 + \psi^2(3,2,1) >_1] =$$

$$= \frac{1}{3!} 6 < \psi^2(1,2,3) >_1 = < \psi^2(1,2,3) >_1,$$

and for the Hamiltonian we have, similarly,

$$< \Psi^*_{cal} \hat{H} \Psi_{cal} >_3 =$$

$$= \frac{1}{3!} \sum_p < \psi(p(1), p(2), p(3)) \hat{H} \psi(p(1), p(2), p(3)) >_1 = < \psi(1,2,3) \hat{H} \psi(1,2,3) >_1,$$

where the sum above is taken with respect to all 3! permutations of the set $\{1, 2, 3\}$, denoted by $p = (p(1), p(2), p(3))$. In such a case,

$$< \Psi^*_{cal} \Psi_{cal} >_3 = 1 + 3c_1^2 + 3c_2^2 + 3c_3^2,$$

and the mean energy is

$$E(r, c_1, c_2, c_3) = \frac{< \Psi^*_{cal} \hat{H} \Psi_{cal} >_3}{< \Psi^*_{cal} \Psi_{cal} >_3} = \frac{< \psi \hat{H} \psi >_1}{< \psi^2 >_1} =$$

$$= (3(309375 + 600000c_1 + 1528125c_1^2 + 600000c_2 + 1528125c_2^2 + 37500c_3 +$$

$$+965625c_3^2 - 1132200r - 1064120c_1 r - 3483888c_1^2 r - 1064120c_2 r +$$

$$+150000c_1 c_2 r - 3483888c_2^2 r - 136160c_3 r + 24576c_1 c_3 r +$$

$$+ 24576c_2 c_3 r - 3421176c_3^2 r)/(400000(1 + 3c_1^2 + 3c_2^2 + 3c_3^2)r^2). \qquad (80)$$

Minimizing $E(r, c_1, c_2, c_3)$ we get

$$r = 0.52248 \text{ a.u.,} \quad c_1 = -0.0394564, \quad c_2 = -0.0394564, \quad c_3 = 0.36027,$$

and the corresponding ground state mean energy is

$$E^* = -7.96682 \text{ a.u.} = -7.96682 \times 27.2116 \text{ eV} = -216.790 \text{ eV,}$$

which is smaller than the experimental value

$$-7.47791 \text{ a.u.} = -203.486 \text{ eV,}$$

contradicting the variational theorem and showing that the Hamiltonian (79), while taking into account the interactions between pairs of electrons, ignores

226

the interaction between all three electrons taken together. This insufficiency may be corrected, allowing the electrons of the lithium atom to repel each other and move, on average, closer to the nucleus, by taking r to be a correction factor in the expression of the mean energy $E(r, c_1, c_2, c_3)$. In fact, the real problem is not only to get the experimental ground state energy, but to use its known value in order to better approximate the model of the atom. Denoting by $a = c_1 = c_2$, and $b = c_3$, the mean energy (80) becomes

$$E(r, a, b) = [3(309375 + 1200000a + 3056250a^2 + 37500b + 965625b^2 -$$

$$-1132200r - 2128240ar - 6817776a^2r - 136160br + 49152abr -$$

$$-3421176b^2r)]/[400000(1 + 6a^2 + 3b^2)r^2].$$

Taking different values for r, and minimizing $E(r, a, b)$ with respect to a and b, we find that for $r = 0.84$, the mean energy $E(0.84, a, b)$ has the minimum value

$$E^* = -7.47582 \text{ a.u.} = -7.47582 \times 27.2116 \text{ eV} = -203.429 \text{ eV},$$

for the values

$$c_1 = c_2 = a = 0.187395, \quad c_3 = b = 0.171296, \tag{81}$$

and its absolute value represents 99.972% of the absolute value of the experimental value of the ground state mean energy. The mean orbits of the two subshells of the lithium atom are

$$\mu_1 = \frac{r}{3} = 0.28 \text{ a.u.}, \quad \mu_2 = \frac{4r}{3} = 1.12 \text{ a.u.}, \tag{82}$$

and the weighted mean orbit of the atom is equal to

$$\bar{\mu} = \frac{2\mu_1 + \mu_2}{3} = 0.56 \text{ a.u.}$$

Therefore, the mean diameters of the atom induced by the statistical equilibrium and by the external subshell are

$$D_1 = 2\bar{\mu} = 1.12 \text{ a.u.}, \quad D_2 = 2\mu_2 = 2.24 \text{ a.u.}$$

227

Using the values (81), the generalized correlations between the electrons are

$$\text{gencorr}_{cal}(\mu_1, \mu_1) = \frac{6c_1^2}{1 - 4c_1 + 18c_1^2 + 3c_3^2} = 0.217095,$$

$$\text{gencorr}_{cal}(\mu_1, \mu_2) =$$

$$= \frac{6c_1 c_2}{\sqrt{1 - 4c_1 + 18c_1^2 + 3c_3^2}\sqrt{1 - 4c_3 + 15c_3^2 + 6c_1^2}} = 0.198947,$$

showing that the dependence between the two electrons belonging to the inner subshell is stronger.

Replacing the values (81) and (82) into the expression

$$\psi = \psi(1, 2, 3) = \varphi(r_1, \mu_1)\varphi(r_2, \mu_1)\varphi(r_3, \mu_2) \times$$

$$\times \left[1 + c_1\left(3 - \frac{r_1}{\mu_1}\right) + c_2\left(3 - \frac{r_2}{\mu_1}\right) + c_3\left(3 - \frac{r_3}{\mu_2}\right)\right], \qquad (83)$$

we get $< \psi^2 >_1 = 1.29873$ and, using (74)-(76), we get the standard correlations

$$\text{stcorr}_{cal}(\mu_1, \mu_1) = 0.219793, \quad \text{stcorr}_{cal}(\mu_1, \mu_2) = -0.114556.$$

Introducing (81) and (82) into (83), and using $< \psi^2 >_1 = 1.29873$, the mean radial distance of an arbitrary electron from the nucleus during its random motion around the nucleus is

$$\bar{r} = \frac{1}{3}\frac{< \psi\left(r_1 + r_2 + r_3\right)\psi >_1}{< \psi^2 >_1} = 1.30384 \text{ a.u.},$$

and the mean diameter of the atom induced by the individual random fluctuations of its electrons is $D_3 = 2\bar{r} = 2.60768$ a.u. The mean index of instability of the atom is $\bar{r} - \bar{\mu} = 0.74384$.

In order to calculate the global correlation between two arbitrary electrons of the lithium atom we introduce again (81) and (82) into (83), and using (70)-(73) we get

$$\text{globcorr} = -0.256382.$$

Introducing again (81) and (82) into (83), let

$$d(\mu_1, \mu_1, \mu_2) = \frac{\psi^2}{< \psi^2 >_1}$$

be the joint probability density of the system of the three electrons and $d(\mu_1)$, $d(\mu_2)$, $d(\mu_1, \mu_1)$ three of its marginal densities, describing the behavior of an arbitrary electron of the first subshell, the behavior of the electron of the second subshell, and the behavior of the first subshell, respectively. Thus, $d(\mu_1, \mu_1)$, for instance, is obtained by integrating $d(\mu_1, \mu_1, \mu_2)$ with respect to the space variables (r_3, ν_3, ω_3), whereas $d(\mu_2)$ is obtained by integrating $d(\mu_1, \mu_1, \mu_2)$ with respect to the measure

$$r_1^2 \, dr_1 \, d\nu_1 \, d\omega_1 \, r_2^2 \, dr_2 \, d\nu_2 \, d\omega_2.$$

We obtain the following degrees of stability for subsystems of electrons inside the lithium atom:

$$< d^2(\mu_1, \mu_1, \mu_2) >_1^{1/2} = 0.00919999, \quad < d^2(\mu_1) >_1^{1/2} = 0.265726,$$

$$< d^2(\mu_2) >_1^{1/2} = 0.131323, \quad < d^2(\mu_1, \mu_1) >_1^{1/2} = 0.0704919,$$

$$< d^2(\mu_1, \mu_2) >_1^{1/2} = 0.0348341,$$

and the following levels of interdependence among subsystems of electrons

$$I(\{\mu_1, \mu_1, \mu_2\} : \{\mu_1\}, \{\mu_1\}, \{\mu_2\}) = 7.276389 \times 10^{-5},$$

$$I(\{\mu_1, \mu_1, \mu_2\} : \{\mu_1, \mu_1\}, \{\mu_2\}) = 5.71775 \times 10^{-5},$$

$$I(\{\mu_1, \mu_1\} : \{\mu_1\}, \{\mu_1\}) = 1.18687 \times 10^{-4},$$

$$I(\{\mu_1, \mu_2\} : \{\mu_1\}, \{\mu_2\}) = 6.18263 \times 10^{-5},$$

where the first number is the amount of interdependence among all three electrons of the lithium atom, the second is the amount of interdependence between the two subshells of the atom, the third number gives the amount of interdependence between the two electrons making up the first subshell of the atom, and the last number gives the amount of interdependence between an electron from the inner shell and the electron of the second subshell.

2^0. For the *nebulous model* of the atom, we take again

$$\mu_1 = \frac{r}{3}, \quad \mu_2 = \frac{4r}{3},$$

into (83) and we have

$$\rho_{neb}(1, 2, 3) = \psi(1, 2, 3) - \psi(3, 2, 1),$$

229

$$< \Psi_{neb}^* \Psi_{neb} >_2 = < \psi(1,2,3) \rho_{neb}(1,2,3) >_1 .$$

Minimizing the mean energy

$$E(r, c_1, c_2, c_3) = \frac{< \Psi_{neb}^* \hat{H} \Psi_{neb} >_2}{< \Psi_{neb}^* \Psi_{neb} >_2} = \frac{< \psi \hat{H} \rho_{neb} >_1}{< \psi \rho_{neb} >_1},$$

we obtain

$$r = 0.554633 \text{ a.u.}, \quad c_1 = c_2 = -0.00465165, \quad c_3 = -0.386276. \qquad (84)$$

The mean orbits of the two subshells of the lithium atom in the nebulous model are

$$\mu_1 = \frac{r}{3} = 0.184878 \text{ a.u.}, \quad \mu_2 = \frac{4r}{3} = 0.739511 \text{ a.u}, \qquad (85)$$

and the weighted mean orbit of the atom is equal to

$$\bar{\mu} = \frac{2\mu_1 + \mu_2}{3} = 0.369756 \text{ a.u.}$$

Therefore, the mean diameters of the atom induced by the statistical equilibrium and by the external subshell are

$$D_1 = 2\bar{\mu} = 0.739512 \text{ a.u.}, \quad D_2 = 2\mu_2 = 1.479022 \text{ a.u.}$$

The minimum mean energy is -7.41911 a.u. $= -201.886$ eV, and its absolute value represents 99.2136% of the absolute value of the experimental value -7.47791 a.u. $= -203.486$ eV obtained for the energy of the ground state of the lithium atom.

Introducing the values (84) into the formula (65), we obtain the generalized correlations:

$$\text{gencorr}_{neb}(\mu_1, \mu_1) = 8.85211 \times 10^{-5}, \quad \text{gencorr}_{neb}(\mu_1, \mu_2) = 4.07033 \times 10^{-3}.$$

Introducing the values (84) and (85) into the expression of the probability wave function (83), we get

$$< \psi \rho_{neb} >_1 = 1.42024,$$

and evaluating the formulas (74)-(76), the standard correlations are

$$\text{stcorr}_{neb}(\mu_1, \mu_1) = 0.170344, \quad \text{stcorr}_{neb}(\mu_1, \mu_2) = 0.0283357.$$

Introducing (84) and (85) into (83), and using $< \psi\rho_{neb} >_1 = 1.42024$, the mean radial distance of an arbitrary electron from the nucleus during its random motion around the nucleus is

$$\bar{r} = \frac{1}{3} \frac{< \psi(r_1 + r_2 + r_3)\rho_{neb} >_1}{< \psi\rho_{neb} >_1} = 1.66991 \text{ a.u.},$$

and the mean diameter of the atom induced by the random fluctuations of its electrons is $D_3 = 2\bar{r} = 3.33982$ a.u. The mean index of instability of the atom is $\bar{r} - \bar{\mu} = 1.300154$.

In order to calculate the global correlation between two arbitrary electrons of the lithium atom we introduce again (84) and (85) into (83), and using (70)-(73) we get

$$\text{globcorr} = -0.35968,$$

which shows that the individual electrons of the lithium atom are more correlated, or more dependent, than the subshells of the atom, taken as being subsystems of electrons.

Introducing again (84) and (85) into (83), let

$$d(\mu_1, \mu_1, \mu_2) = \frac{\psi\rho_{neb}}{< \psi\rho_{neb} >_1}$$

be the joint density of the system of the three electrons and $d(\mu_1), d(\mu_2), d(\mu_1, \mu_1)$ three of its marginal densities, describing the behavior of an arbitrary electron of the first subshell, the behavior of the electron of the second subshell, and the behavior of the first subshell, respectively. We obtain the following degrees of stability for subsystems of electrons inside the lithium atom in the nebulous model:

$$< d^2(\mu_1, \mu_1, \mu_2) >_1^{1/2} = 0.0098994,$$

$$< d^2(\mu_1) >_1^{1/2} = 0.28689, \quad < d^2(\mu_2) >_1^{1/2} = 0.122801,$$

$$< d^2(\mu_1, \mu_1) >_1^{1/2} = 0.0812495, \quad < d^2(\mu_1, \mu_2) >_1^{1/2} = 0.0349557,$$

and the following levels of interdependence among subsystems of electrons

$$I(\{\mu_1, \mu_1, \mu_2\} : \{\mu_1\}, \{\mu_1\}, \{\mu_2\}) = 2.07831 \times 10^{-4},$$

$$I(\{\mu_1, \mu_1, \mu_2\} : \{\mu_1, \mu_1\}, \{\mu_2\}) = 7.81317 \times 10^{-5},$$

$$I(\{\mu_1, \mu_1, \mu_2\} : \{\mu_1, \mu_2\}, \{\mu_1\}) = 1.29032 \times 10^{-4},$$

$$I(\{\mu_1, \mu_1\} : \{\mu_1\}, \{\mu_1\}) = 1.05617 \times 10^{-3},$$

$$I(\{\mu_1, \mu_2\} : \{\mu_1\}, \{\mu_2\}) = 2.74666 \times 10^{-4}.$$

Remark: As the experimental value of the diameter of the lithium atom is $D_{exp} = 3.13$, this shows that, both in the calm model of the atom and in the nebulous model, D_3 is the closest approximation of the diameter and the structure of the lithium atom depends more on the random fluctuations of its electrons than on a regular evolution along orbits of statistical equilibrium.

d) Beryllium Atom ($Z = 4$).

This atom has two subshells. The four electrons have the quantum numbers (n, ℓ, m, s) given by

$$(1, 0, 0, -\frac{1}{2}), \quad (1, 0, 0, +\frac{1}{2}), \quad (2, 0, 0, -\frac{1}{2}), \quad (2, 0, 0, +\frac{1}{2}),$$

respectively. The Hamiltonian is given by (63) with $Z = 4$. Using the weighted Bohr approximation (69), the mean orbits of the two subshells of the atom are

$$\mu_1 = \frac{r}{4}, \qquad \mu_2 = \frac{2^2 r}{4} = r. \tag{86}$$

As there are $4! = 24$ permutations of four electrons, the antisymmetric wave function is

$$\Psi = \frac{1}{\sqrt{4!}}[\tilde{\psi}(1, 2, 3, 4) - \tilde{\psi}(1, 2, 4, 3) - \tilde{\psi}(1, 3, 2, 4) + \tilde{\psi}(1, 3, 4, 2)+$$

$$+\tilde{\psi}(1, 4, 2, 3) - \tilde{\psi}(1, 4, 3, 2) - \tilde{\psi}(2, 1, 3, 4) + \tilde{\psi}(2, 1, 4, 3) + \tilde{\psi}(2, 3, 1, 4)-$$

$$-\tilde{\psi}(2, 3, 4, 1) - \tilde{\psi}(2, 4, 1, 3) + \tilde{\psi}(2, 4, 3, 1) + \tilde{\psi}(3, 1, 2, 4) - \tilde{\psi}(3, 1, 4, 2)-$$

$$-\tilde{\psi}(3, 2, 1, 4) + \tilde{\psi}(3, 2, 4, 1) + \tilde{\psi}(3, 4, 1, 2) - \tilde{\psi}(3, 4, 2, 1) - \tilde{\psi}(4, 1, 2, 3)+$$

$$+\tilde{\psi}(4, 1, 3, 2) + \tilde{\psi}(4, 2, 1, 3) - \tilde{\psi}(4, 2, 3, 1) - \tilde{\psi}(4, 3, 1, 2) + \tilde{\psi}(4, 3, 2, 1)],$$

where

$$\tilde{\psi}(1, 2, 3, 4) =$$

$$= \psi(1,2,3,4)\beta_{1,0,0}(\tau_1)\alpha_-(\sigma_1)\beta_{1,0,0}(\tau_2)\alpha_+(\sigma_2)\beta_{2,0,0}(\tau_3)\alpha_-(\sigma_3), \beta_{2,0,0}(\tau_4)\alpha_+(\sigma_4),$$

in the calm model of the atom, and

$$\tilde{\psi}(1,2,3,4) = \psi(1,2,3,4)\alpha_-(\sigma_1)\alpha_+(\sigma_2)\alpha_-(\sigma_3)\alpha_+(\sigma_4),$$

in the nebulous model of the atom, where

$$\psi = \psi(1,2,3,4) = \varphi(r_1,\mu_1)\varphi(r_2,\mu_1)\varphi(r_3,\mu_2)\varphi(r_4,\mu_2)\times$$

$$\times \left[1 + c_1 \left(3 - \frac{r_1}{\mu_1} \right) + c_2 \left(3 - \frac{r_2}{\mu_1} \right) + c_3 \left(3 - \frac{r_3}{\mu_2} \right) + c_4 \left(3 - \frac{r_4}{\mu_2} \right) \right], \quad (87)$$

For the beryllium atom the nervous model, the turbulent model, and the nebulous model are the same. With the notations from Section 4, we have

$$\rho_{neb}(1,2,3,4) = \rho_{tur}(1,2,3,4) = \rho_{ner}(1,2,3,4) =$$

$$= \psi(1,2,3,4) - \psi(1,4,3,2) - \psi(3,2,1,4) + \psi(3,4,1,2), \quad (88)$$

$$\rho_{cal}(1,2,3,4) = \psi(1,2,3,4). \quad (89)$$

1^0. For the *calm model* of the atom,

$$< \Psi_{cal}^* \Psi_{cal} >_3 = \frac{1}{4!} \sum_p < \psi^2(p(1),p(2),p(3),p(4)) >_1 =$$

$$= \frac{1}{4!} 24 < \psi^2(1,2,3,4) >_1 = < \psi^2(1,2,3,4) >_1 = 1 + \sum_{i=1}^4 3c_i^2,$$

and for the Hamiltonian we have, similarly,

$$< \Psi_{cal}^* \hat{H} \Psi_{cal} >_3 =$$

$$= \frac{1}{4!} \sum_p < \psi(p(1),p(2),p(3),p(4)) \, \hat{H}\psi(p(1),p(2),p(3),p(4)) >_1 =$$

$$= < \psi(1,2,3,4) \, \hat{H}\psi(1,2,3,4) >_1,$$

where the sums above are taken with respect to all 4! permutations of the set $\{1,2,3,4\}$, denoted by $p = (p(1),p(2),p(3),p(4))$, and the mean energy is

$$E(r,c_1,c_2,c_3,c_4) = \frac{< \Psi_{cal}^* \hat{H} \Psi_{cal} >_3}{< \Psi_{cal}^* \Psi_{cal} >_3} = \frac{< \psi \hat{H} \psi >_1}{< \psi^2 >_1}.$$

Minimizing $E(r, c_1, c_2, c_3, c_4)$ we get

$$r = 0.45934 \text{ a.u.}, \quad a = c_1 = c_2 = -0.0809111, \quad b = c_3 = c_4 = 0.226936,$$

and the corresponding ground state mean energy is

$$E^* = -16.9392 \text{ a.u.} = -460.943 \text{ eV},$$

which is much smaller than the experimental value

$$-14.6683 \text{ a.u.} = -399.149 \text{ eV},$$

contradicting the variational theorem and showing that the Hamiltonian (63), while taking into account the interactions between pairs of electrons, ignores the interaction among three and all four electrons taken together. This insufficiency may be corrected, allowing the electrons of the beryllium atom to repel each other and move, on average, farther from the nucleus, by taking r to be a correction factor in the expression of the mean energy $E(r, c_1, c_2, c_3, c_4)$. In fact, the real problem is not only to get a value as close as possible to the experimental ground state energy, but rather to use its known value in order to better approximate the model of the atom and get a good approximation for the probability wave function of the corresponding atom, allowing to make predictions about the behavior of the atom. Denoting by $a = c_1 = c_2$, and $b = c_3 = c_4$, the mean energy becomes a function of three variational parameters $E(r, a, b)$. Taking different values for r, we minimize $E(r, a, b)$ with respect to a and b. Taking the weight $r = 1$, which is equivalent to taking just Bohr's values $1/4$ and 1 as the mean orbits of the two subshells, minimizing $E(1, a, b)$ we get $a = 0.245856$, $b = 0.12620$, and the minimum mean energy -14.133 a.u.$= -384.582$ eV, whose absolute value represents 96.3504% of the absolute value of the experimental ground state energy of the atom. Therefore, the weight r belongs to the interval $0.45934 < r < 1$. When the weight r increases, the minimum value of $E(r, a, b)$ with respect to a and b increases as well. Trying different values of r, we find that for $r = 0.917$, the mean ground energy $E(0.917, a, b)$ has the minimum value

$$E^* = -14.6657 \text{ a.u.} = -399.077 \text{ eV},$$

corresponding to the coefficients

$$c_1 = c_2 = a = 0.220935, \quad c_3 = c_4 = b = 0.145214, \tag{90}$$

and whose absolute value represents 99.982% of the absolute value of the experimental value of the ground state mean energy

$$-14.6683 \text{ a.u.} = -399.149 \text{ eV}.$$

On average, the first subshell of the beryllium atom has a mean orbit equal to $\mu_1 = r/4 = 0.22925$ a.u., while the second subshell has the mean orbit equal to $\mu_2 = 4r/4 = 0.917$ a.u. The weighted mean orbit of the atom is equal to

$$\bar{\mu} = \frac{2\mu_1 + 2\mu_2}{4} = 0.573125 \text{ a.u.}$$

Therefore, the mean diameters of the atom induced by the statistical equilibrium and by the external subshell are

$$D_1 = 2\bar{\mu} = 1.14625 \text{ a.u.}, \quad D_2 = 2\mu_2 = 1.834 \text{ a.u.}$$

Using the values (90), the generalized correlations (65) become

$$\text{gencorr}_{cal}(\mu_1, \mu_1) = 0.261167, \quad \text{gencorr}_{cal}(\mu_1, \mu_2) = 0.173986,$$

$$\text{gencorr}_{cal}(\mu_2, \mu_2) = 0.115907,$$

showing that the interdependence between the two electrons belonging to the inner subshell is stronger.

Introducing the values (90) and $\mu_1 = 0.22925$, $\mu_2 = 0.917$ into the expression (87) of $\psi = \psi(1, 2, 3, 4)$, we obtain $< \psi^2 >_1 = 1.4194$ and, using (74)-(76), we get the standard correlations

$$\text{stcorr}_{cal}(\mu_1, \mu_1) = 0.366083, \quad \text{stcorr}_{cal}(\mu_1, \mu_2) = 0.127801,$$

$$\text{stcorr}_{cal}(\mu_2, \mu_2) = -0.0120384.$$

Introducing the values (90) and $\mu_1 = 0.22925$, $\mu_2 = 0.917$ into (87), and using $< \psi^2 >_1 = 1.4194$, the mean radial distance of an arbitrary electron from the nucleus during its random motion around the nucleus is

$$\bar{r} = \frac{1}{4} \frac{< \psi \, (r_1 + r_2 + r_3 + r_4) \, \psi >_1}{< \psi^2 >_1} = 1.3954 \text{ a.u.},$$

and the mean diameter of the atom induced by the individual random fluctuations of its electrons is $D_3 = 2\bar{r} = 2.7908$ a.u. The mean index of instability of the atom is $\bar{r} - \bar{\mu} = 0.822275$.

In order to calculate the global correlation between two arbitrary electrons of the beryllium atom we introduce again (90) and $\mu_1 = 0.22925$, $\mu_2 = 0.917$ into (87), and using (70)-(73) we get

$$\text{globcorr} = -0.150958,$$

which shows that the individual electrons of the beryllium atom are less correlated, or more independent, than the subshells of the atom, taken as being subsystems of electrons.

Introducing the values (90) and $\mu_1 = 0.22925$, $\mu_2 = 0.917$ into (87), let

$$d(\mu_1, \mu_1, \mu_2, \mu_2) = \frac{\psi^2}{<\psi^2>_1}$$

be the joint probability density of the system of four electrons and

$$d(\mu_1), \quad d(\mu_2), \quad d(\mu_1, \mu_1), \quad d(\mu_2, \mu_2),$$

be four of its marginal densities, describing the behavior of an arbitrary electron of the first subshell, the behavior of an arbitrary electron of the second subshell, and the behavior of each of the two subshells, respectively. Thus, $d(\mu_2, \mu_2)$, for instance, is obtained by integrating $d(\mu_1, \mu_1, \mu_2, \mu_2)$ with respect to the space variables $(r_1, \nu_1, \omega_1, r_2, \nu_2, \omega_2)$. We obtain the following degrees of stability for subsystems of electrons inside the beryllium atom:

$$< d^2(\mu_1, \mu_1, \mu_2, \mu_2) >_1^{1/2} = 0.00172626,$$

$$< d^2(\mu_1) >_1^{1/2} = 0.29582, \quad < d^2(\mu_2) >_1^{1/2} = 0.140871,$$

$$< d^2(\mu_1, \mu_1) >_1^{1/2} = 0.0877194, \quad < d^2(\mu_2, \mu_2) >_1^{1/2} = 0.0198379,$$

$$< d^2(\mu_1, \mu_2) >_1^{1/2} = 0.0416916,$$

and the following levels of interdependence among subsystems of electrons

$$I(\{\mu_1, \mu_1, \mu_2, \mu_2\} : \{\mu_1\}, \{\mu_1\}, \{\mu_2\}, \{\mu_2\}) = 1.0322 \times 10^{-5},$$

$$I(\{\mu_1, \mu_1, \mu_2, \mu_2\} : \{\mu_1, \mu_1\}, \{\mu_2, \mu_2\}) = 1.39035 \times 10^{-5},$$

$$I(\{\mu_1, \mu_1, \mu_2, \mu_2\} : \{\mu_1, \mu_2\}, \{\mu_1, \mu_2\}) = 1.19295 \times 10^{-5},$$

$$I(\{\mu_1, \mu_1\} : \{\mu_1\}, \{\mu_1\}) = 2.09783 \times 10^{-4},$$

$$I(\{\mu_2, \mu_2\} : \{\mu_2\}, \{\mu_2\}) = 6.62991 \times 10^{-6},$$

$$I(\{\mu_1, \mu_2\} : \{\mu_1\}, \{\mu_2\}) = 1.92439 \times 10^{-5},$$

where the last number is the amount of interdependence between an arbitrary electron from the first subshell and an arbitrary electron from the second subshell of the atom.

2^0. For the *nebulous model* of the atom, introducing (86) into (87) and taking into account (88), by minimizing the mean energy

$$E(r, c_1, c_2, c_3, c_4) = \frac{< \Psi^*_{neb} \hat{H} \Psi_{neb} >_2}{< \Psi^*_{neb} \Psi_{neb} >_2} = \frac{< \psi \hat{H} \rho_{neb} >_1}{< \psi \rho_{neb} >_1},$$

we obtain the values

$$r = 0.558516 \text{ a.u.}, \quad c_1 = c_2 = 0.0174535, \quad c_3 = c_4 = -0.369507. \quad (91)$$

The mean orbits of the two subshells of the beryllium atom in the nebulous model are

$$\mu_1 = \frac{r}{4} = 0.139629 \text{ a.u.}, \quad \mu_2 = r = 0.558516 \text{ a.u.} \quad (92)$$

The weighted mean orbit of the atom is equal to

$$\bar{\mu} = \frac{2\mu_1 + 2\mu_2}{4} = 0.349073 \text{ a.u.}$$

Therefore, the mean diameters of the atom induced by the statistical equilibrium and by the external subshell are

$$D_1 = 2\bar{\mu} = 0.698146 \text{ a.u.}, \quad D_2 = 2\mu_2 = 1.117038 \text{ a.u.}$$

The minimum energy is

$$-14.53 \text{ a.u.} = -395.385 \text{ eV},$$

whose absolute value represents 99.0569% of the absolute value of the experimental value

$$-14.6683 \text{ a.u.} = -399.149 \text{ eV},$$

and this surprisingly good result has been obtained using only three variational parameters, r, a, b, where $a = c_1 = c_2$ and $b = c_3 = c_4$.

Introducing the values (91) into the formula (65) we get

$$\text{gencorr}_{neb}(\mu_1, \mu_1) = 0.00104152, \quad \text{gencorr}_{neb}(\mu_1, \mu_2) = -0.0131456,$$

$$\text{gencorr}_{neb}(\mu_2, \mu_2) = 0.165917.$$

Introducing the values (91) and (92) into the expression of the probability wave function (87), we get

$$< \Psi^*_{neb} \Psi_{neb} >_2 = < \psi \rho_{neb} >_1 = 1.57776,$$

and, evaluating the formulas (74)-(76), the standard correlations are

$$\text{stcorr}_{neb}(\mu_1, \mu_1) = 0.368474, \quad \text{stcorr}_{neb}(\mu_1, \mu_2) = 0.256560,$$

$$\text{stcorr}_{neb}(\mu_2, \mu_2) = 0.159988.$$

Introducing (91) and (92) into (87), and using $< \psi \rho_{neb} >_1 = 1.57776$, the mean radial distance of an arbitrary electron from the nucleus during its random motion around the nucleus is

$$\bar{r} = \frac{1}{4} \frac{< \psi \left(r_1 + r_2 + r_3 + r_4 \right) \rho_{neb} >_1}{< \psi \rho_{neb} >_1} = 1.53467 \text{ a.u.},$$

and the mean diameter of the atom induced by the individual random fluctuations of its electrons is $D_3 = 2\bar{r} = 3.06934$ a.u. The mean index of instability of the atom is $\bar{r} - \bar{\mu} = 1.18559$.

In order to calculate the global correlation between two arbitrary electrons of the beryllium atom we introduce again (91) and (92) into (87), and using (70)-(73) we get

$$\text{globcorr} = -0.229352.$$

Introducing the values (91) and (92) into (87), let

$$d(\mu_1, \mu_1, \mu_2, \mu_2) = \frac{\psi^2}{< \psi^2 >_1}$$

be the joint probability density of the system of the three electrons and

$$d(\mu_1), \quad d(\mu_2), \quad d(\mu_1, \mu_1), \quad d(\mu_2, \mu_2),$$

be four of its marginal densities, describing the behavior of an arbitrary electron of the first subshell, the behavior of an arbitrary electron of the second subshell, and the behavior of each of the two subshells, respectively. Thus, $d(\mu_2, \mu_2)$, for instance, is obtained by integrating $d(\mu_1, \mu_1, \mu_2, \mu_2)$ with respect to the space variables $(r_1, \nu_1, \omega_1, r_2, \nu_2, \omega_2)$. We obtain the following degrees of stability for subsystems of electrons inside the beryllium atom in the nebulous model of the atom:

$$< d^2(\mu_1, \mu_1, \mu_2, \mu_2) >_1^{1/2} = 0.00234454,$$

$$< d^2(\mu_1) >_1^{1/2} = 0.340054, \quad < d^2(\mu_2) >_1^{1/2} = 0.14446,$$

$$< d^2(\mu_1, \mu_1) >_1^{1/2} = 0.115473, \quad < d^2(\mu_2, \mu_2) >_1^{1/2} = 0.0209757,$$

$$< d^2(\mu_1, \mu_2) >_1^{1/2} = 0.0483306,$$

and the following levels of interdependence among subsystems of electrons

$$I(\{\mu_1, \mu_1, \mu_2, \mu_2\} : \{\mu_1\}, \{\mu_1\}, \{\mu_2\}, \{\mu_2\}) = 6.86474 \times 10^{-5},$$

$$I(\{\mu_1, \mu_1, \mu_2, \mu_2\} : \{\mu_1, \mu_1\}, \{\mu_2, \mu_2\}) = 7.75961 \times 10^{-5},$$

$$I(\{\mu_1, \mu_1, \mu_2, \mu_2\} : \{\mu_1, \mu_2\}, \{\mu_1, \mu_2\}) = 8.69555 \times 10^{-6},$$

$$I(\{\mu_1, \mu_1\} : \{\mu_1\}, \{\mu_1\}) = 1.63682 \times 10^{-4},$$

$$I(\{\mu_2, \mu_2\} : \{\mu_2\}, \{\mu_2\}) = 1.07077 \times 10^{-4},$$

$$I(\{\mu_1, \mu_2\} : \{\mu_1\}, \{\mu_2\}) = 7.93629 \times 10^{-4},$$

where the last number is the amount of interdependence between an arbitrary electron from the first subshell and an arbitrary electron from the second subshell of the atom.

Remark 1: If we do not impose the weighted Bohr's rule for the mean orbits of the subshells of the beryllium atom and we minimize the mean ground energy as we did before, we get

$$\mu_1 = \mu_2 = 0.1633 \text{ a.u.}, \quad c_1 = c_2 = c_3 = c_4 = -0.012762,$$

and the mean energy of the ground state equal to -18.786 a.u.$= -511.197$ eV, a value smaller than the mean value obtained for the calm model of the atom and much smaller than the experimental value -399.149 eV. This result and the analysis performed in this section show that taking subshells of the atom into account is appropriate and that the nebulous model of the atom is more accurate than the calm model.

Remark 2: As the experimental value of the diameter of the beryllium atom is $D_{exp} = 2.25$, this shows that in the calm model of the atom D_2 is the closest approximation of the diameter of the atom whereas in the nebulous model D_3 is the closest approximation of the diameter. This shows that the structure of the beryllium atom depends more on the behavior of the electrons from the external subshell, in the calm model, and on the random fluctuations of its electrons, in the nebulous atom, than on a regular evolution along orbits of statistical equilibrium.

e) Boron Atom $(Z = 5)$.

This atom has three subshells. The five electrons have the quantum numbers (n, ℓ, m, s) given by

$$(1, 0, 0, -\tfrac{1}{2}), \quad (1, 0, 0, +\tfrac{1}{2}), \quad (2, 0, 0, -\tfrac{1}{2}), \quad (2, 0, 0, +\tfrac{1}{2}), \quad (2, 1, -1, -\tfrac{1}{2}),$$

respectively. The Hamiltonian is given by (63) with $Z = 5$. Using the weighted Bohr approximation (69), the mean orbits of the three subshells of the atom are

$$\mu_1 = \frac{r}{5}, \quad \mu_2 = \frac{2^2 r}{5} = \frac{4r}{5}, \quad \mu_3 = \frac{3^2 r}{5} = \frac{9r}{5}. \tag{93}$$

There are $5! = 120$ permutations of five electrons, the nervous model of the atom coincides with the turbulent model, but the turbulent model of the atom differs from the nebulous model. Thus, with the notations from Section 4, we have

$$\rho_{cal}(1, 2, 3, 4, 5) = \psi(1, 2, 3, 4, 5), \tag{94}$$

$$\rho_{ner}(1, 2, 3, 4, 5) = \rho_{tur}(1, 2, 3, 4, 5) = \psi(1, 2, 3, 4, 5) -$$

$$- \psi(3, 2, 1, 4, 5) - \psi(1, 4, 3, 2, 5) + \psi(3, 4, 1, 2, 5), \tag{95}$$

$$\rho_{neb}(1, 2, 3, 4, 5) =$$

$$= \psi(1, 2, 3, 4, 5) - \psi(3, 2, 1, 4, 5) - \psi(5, 2, 3, 4, 1) - \psi(1, 2, 5, 4, 3) +$$

$$+ \psi(3, 2, 5, 4, 1) + \psi(5, 2, 1, 4, 3) - \psi(1, 4, 3, 2, 5) + \psi(3, 4, 1, 2, 5) +$$

$$+ \psi(5, 4, 3, 2, 1) + \psi(1, 4, 5, 2, 3) - \psi(3, 4, 5, 2, 1) - \psi(5, 4, 1, 2, 3). \tag{96}$$

where the probability wave function is

$$\psi = \psi(1,2,3,4,5) = \varphi(r_1,\mu_1)\varphi(r_2,\mu_1)\varphi(r_3,\mu_2)\varphi(r_4,\mu_2)\varphi(r_5,\mu_3) \times$$

$$\times \left[1 + \sum_{i=1}^{2} c_i \left(3 - \frac{r_i}{\mu_1} \right) + \sum_{i=3}^{4} c_i \left(3 - \frac{r_i}{\mu_2} \right) + c_5 \left(3 - \frac{r_5}{\mu_3} \right) \right]. \qquad (97)$$

1^0. For the *calm model* of the atom,

$$< \Psi_{cal}^* \Psi_{cal} >_3 = \frac{1}{5!} \sum_p < \psi^2(p(1),p(2),p(3),p(4),p(5)) >_1 =$$

$$= \frac{1}{5!} 120 < \psi^2(1,2,3,4,5) >_1 = < \psi^2(1,2,3,4,5) >_1 = 1 + \sum_{i=1}^{5} 3c_i^2,$$

and for the Hamiltonian we have, similarly,

$$< \Psi_{cal}^* \hat{H} \Psi_{cal} >_3 =$$

$$= \frac{1}{5!} \sum_p < \psi(p(1),p(2),p(3),p(4),p(5)) \, \hat{H} \psi(p(1),p(2),p(3),p(4),p(5)) >_1 =$$

$$= < \psi(1,2,3,4,5) \, \hat{H} \psi(1,2,3,4,5) >_1,$$

where the sums above are taken with respect to all 5! permutations of the set $\{1,2,3,4,5\}$, denoted by $p = (p(1),p(2),p(3),p(4),p(5))$, and the mean value of energy is

$$E(r,c_1,c_2,c_3,c_4,c_5) = \frac{< \Psi_{cal}^* \hat{H} \Psi_{cal} >_3}{< \Psi_{cal}^* \Psi_{cal} >_3} = \frac{< \psi \hat{H} \psi >_1}{< \psi^2 >_1}.$$

Minimizing $E(r,c_1,c_2,c_3,c_4,c_5)$ we get

$$r = 0.415865 \text{ a.u.},$$

$$a = c_1 = c_2 = -0.106008, \quad b = c_3 = c_4 = 0.194017, \quad c = c_5 = 0.0855266,$$

and the corresponding ground state mean energy is

$$E^* = -29.8375 \text{ a.u.} = -811.926 \text{ eV},$$

241

which is much smaller than the experimental value

$$-24.658 \text{ a.u.} = -670.98452 \text{ eV},$$

contradicting the variational theorem and showing that the Hamiltonian (63), while taking into account the interactions between pairs of electrons, ignores the interaction among three, four, and all five electrons taken together. This insufficiency may be corrected, allowing the electrons of the boron atom to repel each other and move, on average, farther from the nucleus, by taking r to be a correction factor in the expression of the mean energy $E(r, c_1, c_2, c_3, c_4, c_5)$. Here we want to use the known experimental value of the mean ground energy in order to better approximate the model of the atom. Denoting by $a = c_1 = c_2, b = c_3 = c_4$, and $c = c_5$, the mean energy becomes a function of four variational parameters $E(r, a, b, c)$. Taking different values for r, and minimizing $E(r, a, b, c)$ with respect to a, b, and c, we find that for $r = 0.923$, the mean ground energy $E(0.923, a, b, c)$ has the minimum value

$$E^* = -24.6577 \text{ a.u.} = -670.975 \text{ eV},$$

corresponding to the coefficients

$$c_1 = c_2 = a = 0.224809, \quad c_3 = c_4 = b = 0.16172, \quad c_5 = c = 0.0484195, \tag{98}$$

and its absolute value represents 99.9987% of the absolute value of the experimental value of the ground state mean energy. On average, the two electrons belonging to the first subshell of the boron atom move randomly around the nucleus at a mean orbit equal to $\mu_1 = r/5 = 0.1846$ a.u., the two electrons from the second subshell move around the nucleus at a mean orbit equal to $\mu_2 = 4r/5 = 0.7384$ a.u., while the electron from the external subshell moves randomly at a mean radial distance $\mu_3 = 9r/5 = 1.6614$ a.u. from the nucleus. The weighted mean orbit of the boron atom is equal to

$$\bar{\mu} = \frac{2\mu_1 + 2\mu_2 + \mu_3}{5} = 0.70148 \text{ a.u.}$$

Therefore, the mean diameters of the atom induced by the statistical equilibrium and by the external subshell are

$$D_1 = 2\bar{\mu} = 1.40296 \text{ a.u.}, \quad D_2 = 2\mu_3 = 3.3228 \text{ a.u.}$$

Using the values (98), the generalized correlations (65) become

$$\text{gencorr}_{cal}(\mu_1, \mu_1) = 0.258199, \quad \text{gencorr}_{cal}(\mu_1, \mu_2) = 0.189009,$$
$$\text{gencorr}_{cal}(\mu_1, \mu_3) = 0.052824, \quad \text{gencorr}_{cal}(\mu_2, \mu_2) = 0.138359,$$

$$\text{gencorr}_{cal}(\mu_2, \mu_3) = 0.0386682,$$

showing that the interdependence between the two electrons belonging to the inner subshell is stronger.

Replacing the values (98) and

$$\mu_1 = 0.1846 \text{ a.u.}, \quad \mu_2 = 0.7384 \text{ a.u.}, \quad \mu_3 = 1.6614 \text{ a.u.},$$

into the expression (97) of $\psi = \psi(1, 2, 3, 4, 5)$, we get $< \psi^2 >_1 = 1.46719$ and, using (74)-(76), we get the standard correlations

$$\text{stcorr}_{cal}(\mu_1, \mu_1) = 0.298314, \quad \text{stcorr}_{cal}(\mu_1, \mu_2) = 0.190518,$$
$$\text{stcorr}_{cal}(\mu_1, \mu_3) = -0.059906, \quad \text{stcorr}_{cal}(\mu_2, \mu_2) = 0.105692,$$

$$\text{stcorr}_{cal}(\mu_2, \mu_3) = -0.0862038.$$

Introducing the values (98) and $\mu_1 = 0.1846$, $\mu_2 = 0.7384$, $\mu_3 = 1.6614$ into (97), and using $< \psi^2 >_1 = 1.46719$, the mean radial distance from the nucleus of an arbitrary electron during its random motion around the nucleus is

$$\bar{r} = \frac{1}{5} \frac{< \psi \left(r_1 + r_2 + r_3 + r_4 + r_5 \right) \psi >_1}{< \psi^2 >_1} = 1.82546 \text{ a.u.},$$

and the mean diameter of the atom induced by the individual random fluctuations of its electrons is $D_3 = 2\bar{r} = 3.65092$ a.u. The mean index of instability of the atom is $\bar{r} - \bar{\mu} = 1.12398$.

In order to calculate the global correlation between two arbitrary electrons of the boron atom we introduce again (98) and

$$\mu_1 = 0.1846 \text{ a.u.}, \quad \mu_2 = 0.7384 \text{ a.u.}, \quad \mu_3 = 1.6614 \text{ a.u.}$$

into (97), and using (70)-(73) we get

$$\text{globcorr} = -0.138849.$$

Introducing the values (98) and

$$\mu_1 = 0.1846 \text{ a.u.}, \quad \mu_2 = 0.7384 \text{ a.u.}, \quad \mu_3 = 1.6614 \text{ a.u.}$$

into (97), let

$$d(\mu_1, \mu_1, \mu_2, \mu_2, \mu_3) = \frac{\psi^2}{<\psi^2>_1}$$

be the joint probability density of the system of five electrons and

$$d(\mu_i), \quad d(\mu_i, \mu_j), \quad d(\mu_i, \mu_j, \mu_k), \quad d(\mu_i, \mu_j, \mu_k, \mu_\ell),$$

be its marginal densities, describing the behavior of different subsystems of electrons from the same shell or from different subshells. Thus, $d(\mu_2, \mu_2)$, for instance, is obtained by integrating $d(\mu_1, \mu_1, \mu_2, \mu_2, \mu_3)$ with respect to the measure

$$r_1^2 dr_1 d\nu_1 d\omega_1 r_2^2 dr_2 d\nu_2 d\omega_2 r_5^2 dr_5 d\nu_5 d\omega_5.$$

We obtain the following degrees of stability for subsystems of electrons inside the boron atom:

$$<d^2(\mu_1, \mu_1, \mu_2, \mu_2, \mu_3)>_1^{1/2} = 0.000262477,$$

$$<d^2(\mu_1)>_1^{1/2} = 0.32852, \quad <d^2(\mu_2)>_1^{1/2} = 0.158029,$$

$$<d^2(\mu_3)>_1^{1/2} = 0.097918, \quad <d^2(\mu_1, \mu_1)>_1^{1/2} = 0.108322,$$

$$<d^2(\mu_2, \mu_2)>_1^{1/2} = 0.0249864, \quad <d^2(\mu_1, \mu_1, \mu_3)>_1^{1/2} = 0.0105915,$$

$$<d^2(\mu_2, \mu_2, \mu_3)>_1^{1/2} = 0.00244408, \quad <d^2(\mu_1, \mu_1, \mu_2, \mu_2)>_1^{1/2} = 0.00269097,$$

and the following levels of interdependence among subsystems of electrons

$$I(\{\mu_1, \mu_1, \mu_2, \mu_2, \mu_3\} : \{\mu_1\}, \{\mu_1\}, \{\mu_2\}, \{\mu_2\}, \{\mu_3\}) = 1.43752 \times 10^{-6},$$

$$I(\{\mu_1, \mu_1, \mu_2, \mu_2, \mu_3\} : \{\mu_1, \mu_1\}, \{\mu_2, \mu_2\}, \{\mu_3\}) = 2.54762 \times 10^{-6},$$

$$I(\{\mu_1, \mu_1, \mu_2, \mu_2\} : \{\mu_1, \mu_1\}, \{\mu_2, \mu_2\}) = 1.56075 \times 10^{-5},$$

$$I(\{\mu_1, \mu_1, \mu_3\} : \{\mu_1, \mu_1\}, \{\mu_3\}) = 1.52078 \times 10^{-5},$$

$$I(\{\mu_2, \mu_2, \mu_3\} : \{\mu_2, \mu_2\}, \{\mu_3\}) = 2.54132 \times 10^{-6},$$

$$I(\{\mu_1, \mu_1\} : \{\mu_1\}, \{\mu_1\}) = 3.96756 \times 10^{-4},$$

$$I(\{\mu_2, \mu_2\} : \{\mu_2\}, \{\mu_2\}) = 1.31896 \times 10^{-5}.$$

2^0. For the *turbulent model* of the atom, introducing (93) into (97) and taking into account (95), by minimizing the mean energy

$$E(r, c_1, c_2, c_3, c_4, c_5) = \frac{< \Psi_{tur}^* \hat{H} \Psi_{tur} >_3}{< \Psi_{tur}^* \Psi_{tur} >_3} = \frac{< \psi \hat{H} \rho_{tur} >_1}{< \psi \rho_{tur} >_1},$$

we obtain the values
$$r = 0.539869 \text{ a.u.},$$

$$c_1 = c_2 = -0.00159598, \quad c_3 = c_4 = -0.265834, \quad c_5 = 0.542492,$$

and the corresponding ground state mean energy is

$$E^* = -25.1097 \text{ a.u.} = -683.275 \text{ eV},$$

which is much smaller than the experimental value

$$-24.658 \text{ a.u.} = -670.98452 \text{ eV},$$

contradicting the variational theorem and showing that the Hamiltonian (63), while taking into account the interactions between pairs of electrons, ignores the interaction among three, four, and all five electrons taken together. This insufficiency may be corrected, allowing the electrons of the boron atom to repel each other and move, on average, farther from the nucleus, by taking r to be a correction factor in the expression of the mean energy $E(r, c_1, c_2, c_3, c_4, c_5)$. Here we want to use the known experimental value of the mean ground energy in order to better approximate the model of the atom. Denoting by $a = c_1 = c_2, b = c_3 = c_4$, and $c = c_5$, the mean energy becomes a function of four variational parameters $E(r, a, b, c)$. Taking different values for r, and minimizing $E(r, a, b, c)$ with respect to a, b, and c, we find that for $r = 0.7$, the mean ground energy $E(0.7, a, b, c)$ has the minimum value
$$E^* = -24.637 \text{ a.u.} = -670.434 \text{ eV},$$

corresponding to the following values of the four variational parameters r, a, b, c, namely,

$$r = 0.7 \text{ a.u.}, \quad a = c_1 = c_2 = 0.180996,$$

$$b = c_3 = c_4 = -0.0762672, \quad c = c_5 = 0.285448, \tag{99}$$

and its absolute value represents 99.9179% of the absolute value of the experimental value of the ground state mean energy. The mean orbits of the three subshells of the atom are

$$\mu_1 = \frac{r}{5} = 0.14 \text{ a.u.}, \quad \mu_2 = \frac{4r}{5} = 0.56 \text{ a.u.}, \quad \mu_3 = \frac{9r}{5} = 1.26 \text{ a.u.}, \tag{100}$$

and the weighted mean orbit of the boron atom in the turbulent model is equal to

$$\bar{\mu} = \frac{2\mu_1 + 2\mu_2 + \mu_3}{5} = 0.532 \text{ a.u.}$$

Therefore, the mean diameters of the atom induced by the statistical equilibrium and by the external subshell are

$$D_1 = 2\bar{\mu} = 1.064 \text{ a.u.}, \quad D_2 = 2\mu_3 = 2.52 \text{ a.u.}$$

Introducing the values (99) into the formula (65) we get

$$\text{gencorr}_{tur}(\mu_1, \mu_1) = 0.171661, \quad \text{gencorr}_{tur}(\mu_1, \mu_2) = -0.056895,$$
$$\text{gencorr}_{tur}(\mu_1, \mu_3) = 0.252926, \quad \text{gencorr}_{tur}(\mu_2, \mu_2) = 0.018857,$$

$$\text{gencorr}_{tur}(\mu_2, \mu_3) = -0.0838289.$$

Introducing the values (99) and (100) into the expression of the probability wave function (97), we get

$$< \Psi_{tur}^* \Psi_{tur} >_3 = < \psi \, \rho_{tur} >_1 = 1.1736,$$

and evaluating the formulas (74)-(76), the standard correlations are

$$\text{stcorr}_{tur}(\mu_1, \mu_1) = 0.425271, \quad \text{stcorr}_{tur}(\mu_1, \mu_2) = 0.316365,$$
$$\text{stcorr}_{tur}(\mu_1, \mu_3) = 0.024934, \quad \text{stcorr}_{tur}(\mu_2, \mu_2) = 0.225731,$$

$$\text{stcorr}_{tur}(\mu_2, \mu_3) = -0.0119539.$$

Introducing the values (99) and (100) into the expression of the probability wave function (97), the mean radial distance from the nucleus of an arbitrary electron during its random motion around the nucleus is

$$\bar{r} = \frac{1}{5} \frac{< \psi \, (r_1 + r_2 + r_3 + r_4 + r_5) \, \rho_{tur} >_1}{< \psi \rho_{tur} >_1} = 1.48475 \text{ a.u.},$$

and the mean diameter of the atom induced by the individual random fluctuations of its electrons is $D_3 = 2\bar{r} = 2.9695$ a.u. The mean index of instability of the atom is $\bar{r} - \bar{\mu} = 0.95275$.

In order to calculate the global correlation between two arbitrary electrons of the boron atom we introduce again (99) and (100) into (97), and using (70)-(73) we get

$$\text{globcorr} = -0.12455.$$

Introducing the values (99) and (100) into the expression of the probability wave function (97), let

$$d(\mu_1, \mu_1, \mu_2, \mu_2, \mu_3) = \frac{\psi \rho_{tur}}{< \psi \rho_{tur} >_1}$$

be the joint probability density of the system of five electrons and

$$d(\mu_i), \quad d(\mu_i, \mu_j), \quad d(\mu_i, \mu_j, \mu_k), \quad d(\mu_i, \mu_j, \mu_k, \mu_\ell),$$

be its marginal densities, describing the behavior of different subsystems of electrons from the same shell or from different subshells. Thus, $d(\mu_2, \mu_2)$, for instance, is obtained by integrating $d(\mu_1, \mu_1, \mu_2, \mu_2, \mu_3)$ with respect to the measure

$$r_1^2 dr_1 d\nu_1 d\omega_1 r_2^2 dr_2 d\nu_2 d\omega_2 r_5^2 dr_5 d\nu_5 d\omega_5.$$

We obtain the following degrees of stability for subsystems of electrons inside the boron atom:

$$< d^2(\mu_1, \mu_1, \mu_2, \mu_2, \mu_3) >_1^{1/2} = 0.000494781,$$

$$< d^2(\mu_1) >_1^{1/2} = 0.380616, \quad < d^2(\mu_2) >_1^{1/2} = 0.164166,$$

$$< d^2(\mu_3) >_1^{1/2} = 0.129356, \quad < d^2(\mu_1, \mu_1) >_1^{1/2} = 0.144691,$$

$$< d^2(\mu_2, \mu_2) >_1^{1/2} = 0.0267799, \quad < d^2(\mu_1, \mu_1, \mu_3) >_1^{1/2} = 0.0187016,$$

$$< d^2(\mu_2, \mu_2, \mu_3) >_1^{1/2} = 0.00351419, \quad < d^2(\mu_1, \mu_1, \mu_2, \mu_2) >_1^{1/2} = 0.00379447,$$

and the following levels of interdependence among subsystems of electrons

$$I(\{\mu_1, \mu_1, \mu_2, \mu_2, \mu_3\} : \{\mu_1\}, \{\mu_1\}, \{\mu_2\}, \{\mu_2\}, \{\mu_3\}) = 1.02608 \times 10^{-5},$$

$$I(\{\mu_1, \mu_1, \mu_2, \mu_2, \mu_3\} : \{\mu_1, \mu_1\}, \{\mu_2, \mu_2\}, \{\mu_3\}) = 6.44671 \times 10^{-6},$$

$$I(\{\mu_1, \mu_1, \mu_2, \mu_2\} : \{\mu_1, \mu_1\}, \{\mu_2, \mu_2\}) = 8.03381 \times 10^{-5},$$

$$I(\{\mu_1, \mu_1, \mu_3\} : \{\mu_1, \mu_1\}, \{\mu_3\}) = 1.50674 \times 10^{-5},$$

$$I(\{\mu_2, \mu_2, \mu_3\} : \{\mu_2, \mu_2\}, \{\mu_3\}) = 5.00473 \times 10^{-5},$$

$$I(\{\mu_1, \mu_1\} : \{\mu_1\}, \{\mu_1\}) = 1.77516 \times 10^{-4},$$

$$I(\{\mu_2, \mu_2\} : \{\mu_2\}, \{\mu_2\}) = 1.70716 \times 10^{-4}.$$

3^0. For the *nebulous model* of the atom, introducing (93) into (97), taking into account (96), and minimizing the mean energy

$$E(r, c_1, c_2, c_3, c_4, c_5) = \frac{< \Psi^*_{neb}\, \hat{H}\Psi_{neb} >_2}{< \Psi^*_{neb}\, \Psi_{neb} >_2} = \frac{< \psi\, \hat{H}\rho_{neb} >_1}{< \psi\, \rho_{neb} >_1},$$

we obtain the values

$$r = 0.554896 \text{ a.u.},$$

$$c_1 = c_2 = 0.0147055, \quad c_3 = c_4 = -0.258471, \quad c_5 = -0.132429. \quad (101)$$

The mean orbits of the three subshells of the boron atom in the nebulous model are (in a.u.)

$$\mu_1 = \frac{r}{5} = 0.110979, \quad \mu_2 = \frac{4r}{5} = 0.443917, \quad \mu_3 = \frac{9r}{5} = 0.998813, \quad (102)$$

and the weighted mean orbit of the boron atom in the nebulous model is

$$\bar{\mu} = \frac{2\mu_1 + 2\mu_2 + \mu_3}{5} = 0.421721 \text{ a.u.}$$

Therefore, the mean diameters of the atom induced by the statistical equilibrium and by the external subshell are

$$D_1 = 2\bar{\mu} = 0.843442 \text{ a.u.}, \quad D_2 = 2\mu_3 = 1.99763 \text{ a.u.}$$

The minimum energy is

$$-24.2281 \text{ a.u.} = -659.285 \text{ eV},$$

whose absolute value represents 98.2564% of the experimental value -24.658 a.u. $= -670.98452$ eV, and this result has been obtained using only four variational parameters, r, a, b, c, where $a = c_1 = c_2$, $b = c_3 = c_4$, and $c = c_5$.

Introducing the values (101) into the formula (65) we get

$$\text{gencorr}_{neb}(\mu_1, \mu_1) = 0.00092777, \qquad \text{gencorr}_{neb}(\mu_1, \mu_2) = -0.0106313,$$
$$\text{gencorr}_{neb}(\mu_1, \mu_3) = -0.0066691, \qquad \text{gencorr}_{neb}(\mu_2, \mu_2) = 0.12182500,$$

$$\text{gencorr}_{neb}(\mu_2, \mu_3) = 0.0764219.$$

Introducing the values (101) and (102) into the expression of the probability wave function (97), we get

$$< \Psi^*_{neb} \Psi_{neb} >_2 = < \psi \rho_{neb} >_1 = 0.317023,$$

and evaluating the formulas (74)-(76), the standard correlations are

$$\text{stcorr}_{neb}(\mu_1, \mu_1) = 0.310425, \qquad \text{stcorr}_{neb}(\mu_1, \mu_2) = 0.253306,$$
$$\text{stcorr}_{neb}(\mu_1, \mu_3) = 0.132242, \qquad \text{stcorr}_{neb}(\mu_2, \mu_2) = 0.184092,$$

$$\text{stcorr}_{neb}(\mu_2, \mu_3) = 0.0912202.$$

Introducing the values (101) and (102) into the expression of the probability wave function (97), the mean radial distance from the nucleus of an arbitrary electron during its random motion around the nucleus is

$$\bar{r} = \frac{1}{5} \frac{< \psi \left(r_1 + r_2 + r_3 + r_4 + r_5 \right) \rho_{neb} >_1}{< \psi \rho_{neb} >_1} = 1.98798 \text{ a.u.,}$$

and the mean diameter of the atom induced by the individual random fluctuations of its electrons is $D_3 = 2\bar{r} = 3.97596$ a.u. The mean index of instability of the atom is $\bar{r} - \bar{\mu} = 1.566259$.

In order to calculate the global correlation between two arbitrary electrons of the boron atom we introduce again (101) and (102) into (97), and using (70)-(73) we get

$$\text{globcorr} = -0.195797.$$

Introducing the values (101) and (102) into the expression of the probability wave function (97), let

$$d(\mu_1, \mu_1, \mu_2, \mu_2, \mu_3) = \frac{\psi \rho_{neb}}{< \psi \rho_{neb} >_1}$$

be the joint probability density of the system of five electrons and

$$d(\mu_i), \quad d(\mu_i, \mu_j), \quad d(\mu_i, \mu_j, \mu_k), \quad d(\mu_i, \mu_j, \mu_k, \mu_\ell),$$

be its marginal densities, describing the behavior of different subsystems of electrons from the same shell or from different subshells. Thus, $d(\mu_2, \mu_2)$, for instance, is obtained by integrating $d(\mu_1, \mu_1, \mu_2, \mu_2, \mu_3)$ with respect to the measure

$$r_1^2 dr_1 d\nu_1 d\omega_1 r_2^2 dr_2 d\nu_2 d\omega_2 r_5^2 dr_5 d\nu_5 d\omega_5.$$

We obtain the following degrees of stability for subsystems of electrons inside the boron atom:

$$< d^2(\mu_1, \mu_1, \mu_2, \mu_2, \mu_3) >_1^{1/2} = 0.000804841,$$

$$< d^2(\mu_1) >_1^{1/2} = 0.395819, \quad < d^2(\mu_2) >_1^{1/2} = 0.265887,$$

$$< d^2(\mu_3) >_1^{1/2} = 0.172900, \quad < d^2(\mu_1, \mu_1) >_1^{1/2} = 0.151938,$$

$$< d^2(\mu_2, \mu_2) >_1^{1/2} = 0.043970, \quad < d^2(\mu_1, \mu_1, \mu_3) >_1^{1/2} = 0.0260006,$$

$$< d^2(\mu_2, \mu_2, \mu_3) >_1^{1/2} = 0.00561169, \quad < d^2(\mu_1, \mu_1, \mu_2, \mu_2) >_1^{1/2} = 0.00632538,$$

and the following levels of interdependence among subsystems of electrons

$$I(\{\mu_1, \mu_1, \mu_2, \mu_2, \mu_3\} : \{\mu_1\}, \{\mu_1\}, \{\mu_2\}, \{\mu_2\}, \{\mu_3\}) = 1.11021 \times 10^{-3},$$

$$I(\{\mu_1, \mu_1, \mu_2, \mu_2, \mu_3\} : \{\mu_1, \mu_1\}, \{\mu_2, \mu_2\}, \{\mu_3\}) = 3.50249 \times 10^{-4},$$

$$I(\{\mu_1, \mu_1, \mu_2, \mu_2\} : \{\mu_1, \mu_1\}, \{\mu_2, \mu_2\}) = 3.55332 \times 10^{-4},$$

$$I(\{\mu_1, \mu_1, \mu_3\} : \{\mu_1, \mu_1\}, \{\mu_3\}) = 2.69474 \times 10^{-4},$$

$$I(\{\mu_2, \mu_2, \mu_3\} : \{\mu_2, \mu_2\}, \{\mu_3\}) = 1.99073 \times 10^{-3},$$

$$I(\{\mu_1, \mu_1\} : \{\mu_1\}, \{\mu_1\}) = 4.73483 \times 10^{-3},$$

$$I(\{\mu_2, \mu_2\} : \{\mu_2\}, \{\mu_2\}) = 2.67257 \times 10^{-2}.$$

Remark 1: In the calm model of the boron atom, if we ignore the subshells and the electrons are allowed to be completely free, the mean ground energy of the atom is minimized for

$$\mu_1 = \mu_2 = \mu_3 = 0.134845 \text{ a.u.}, \quad c_1 = c_2 = c_3 = c_4 = c_5 = 0.00576931,$$

and its minimum value is -35.1562 a.u. $= -956.656$ eV, a value much smaller than the experimental value -670.975 eV. On the other side, if we assume that the atom has three subshells and their mean orbits were given just by the Bohr's values $k^2/Z = k^2/5$ for $k = 1, 2, 3$, then the minimum mean ground energy would be -646.088 eV, whose absolute value represents 96.2895% of the absolute value of the experimental value of the mean ground energy. This result and the analysis performed in this section show that taking subshells of the atom into account with weighted Bohr's mean radial distances from the nucleus is appropriate and that the turbulent and nebulous models of the atom are more accurate than the calm model in describing the stucture of the boron atom.

Remark 2: In the turbulent model of the atom, if we take Bohr's values as mean orbits of the shells, which means to take $r = 1$ in (69), then by minimizing the mean energy $E(1, c_1, c_2, c_3, c_4, c_5)$, we obtain

$$c_1 = c_2 = 0.477744, \quad c_3 = c_4 = -0.0554296, \quad c_5 = 0.0906082,$$

and the minimum value of the energy is -22.7618 a.u. $= -619.385$ eV, whose absolute value represents 92.3099% of the experimental value -24.658 a.u.$= -670.98452$ eV.

Remark 3: In the turbulent model of the atom, if we suppose that the electrons of the boron atom are independent, i.e., $c_1 = c_2 = c_3 = c_4 = c_5 = 0$, and in statistical equilibrium, then the mean energy $E(r, 0, 0, 0, 0, 0)$ is minimum for $r = 0.568694$ a.u., and the minimum energy is -24.5143 a.u. $= -667.073$ eV, whose absolute value represents 99.4171% of the experimental value -24.658 a.u. $= -670.98452$ eV, and this surprisingly good result has been obtained using only one variational parameter, r. In such a case, the expression of the mean ground energy of the boron atom in the turbulent model, using the only variational parameter r, is

$$E(r) = \frac{7.9282294}{r^2} - \frac{27.882236}{r}.$$

f) Carbon Atom ($Z = 6$).

This atom has three subshells. The six electrons have the quantum numbers (n, ℓ, m, s) given by

$$\left(1, 0, 0, -\frac{1}{2}\right), \quad \left(1, 0, 0, +\frac{1}{2}\right), \quad \left(2, 0, 0, -\frac{1}{2}\right),$$

$$(2, 0, 0, +\frac{1}{2}), \quad (2, 1, -1, -\frac{1}{2}), \quad (2, 1, -1, +\frac{1}{2}),$$

respectively. The Hamiltonian is given by (63) with $Z = 6$. Using the weighted Bohr approximation (69), the mean orbits of the three subshells of the atom are

$$\mu_1 = \frac{r}{6}, \quad \mu_2 = \frac{2^2 r}{6} = \frac{2r}{3}, \quad \mu_3 = \frac{3^2 r}{6} = \frac{3r}{2}. \tag{103}$$

There are $6! = 720$ permutations of six electrons, the nervous model of the atom coincides with the turbulent model but the turbulent model of the atom differs from the nebulous model. Thus, with the notations from Section 4, we have:

$$\rho_{cal}(1, 2, 3, 4, 5, 6) = \psi(1, 2, 3, 4, 5, 6), \tag{104}$$

$$\rho_{ner}(1, 2, 3, 4, 5, 6) = \rho_{tur}(1, 2, 3, 4, 5, 6) =$$
$$= \psi(1, 2, 3, 4, 5, 6) - \psi(3, 2, 1, 4, 5, 6) - \psi(1, 4, 3, 2, 5, 6) + \psi(3, 4, 1, 2, 5, 6), \tag{105}$$

$$\rho_{neb}(1, 2, 3, 4, 5, 6) =$$
$$= \psi(1, 2, 3, 4, 5, 6) - \psi(3, 2, 1, 4, 5, 6) - \psi(5, 2, 3, 4, 1, 6) -$$
$$- \psi(1, 2, 5, 4, 3, 6) + \psi(3, 2, 5, 4, 1, 6) + \psi(5, 2, 1, 4, 3, 6) -$$
$$- \psi(1, 4, 3, 2, 5, 6) + \psi(3, 4, 1, 2, 5, 6) + \psi(5, 4, 3, 2, 1, 6) +$$
$$+ \psi(1, 4, 5, 2, 3, 6) - \psi(3, 4, 5, 2, 1, 6) - \psi(5, 4, 1, 2, 3, 6) -$$
$$- \psi(1, 6, 3, 4, 5, 2) + \psi(3, 6, 1, 4, 5, 2) + \psi(5, 6, 3, 4, 1, 2) +$$
$$+ \psi(1, 6, 5, 4, 3, 2) - \psi(3, 6, 5, 4, 1, 2) - \psi(5, 6, 1, 4, 3, 2) -$$
$$- \psi(1, 2, 3, 6, 5, 4) + \psi(3, 2, 1, 6, 5, 4) + \psi(5, 2, 3, 6, 1, 4) +$$
$$+ \psi(1, 2, 5, 6, 3, 4) - \psi(3, 2, 5, 6, 1, 4) - \psi(5, 2, 1, 6, 3, 4) +$$
$$+ \psi(1, 4, 3, 6, 5, 2) - \psi(3, 4, 1, 6, 5, 2) - \psi(5, 4, 3, 6, 1, 2) -$$
$$- \psi(1, 4, 5, 6, 3, 2) + \psi(3, 4, 5, 6, 1, 2) + \psi(5, 4, 1, 6, 3, 2) +$$
$$+ \psi(1, 6, 3, 2, 5, 4) - \psi(3, 6, 1, 2, 5, 4) - \psi(5, 6, 3, 2, 1, 4) -$$
$$- \psi(1, 6, 5, 2, 3, 4) + \psi(3, 6, 5, 2, 1, 4) + \psi(5, 6, 1, 2, 3, 4), \tag{106}$$

where

$$\psi = \psi(1, 2, 3, 4, 5, 6) =$$

$$= \varphi(r_1, \mu_1)\varphi(r_2, \mu_1)\varphi(r_3, \mu_2)\varphi(r_4, \mu_2)\varphi(r_5, \mu_3)\varphi(r_6, \mu_3) \times$$

$$\times \left[1 + c_1 \left(3 - \frac{r_1}{\mu_1} \right) + c_2 \left(3 - \frac{r_2}{\mu_1} \right) + c_3 \left(3 - \frac{r_3}{\mu_2} \right) + \right.$$

$$\left. + c_4 \left(3 - \frac{r_4}{\mu_2} \right) + c_5 \left(3 - \frac{r_5}{\mu_3} \right) + c_6 \left(3 - \frac{r_6}{\mu_3} \right) \right]. \tag{107}$$

1^0. For the *calm model* of the atom,

$$< \Psi_{cal}^* \Psi_{cal} >_3 = \frac{1}{6!} \sum_p < \psi^2(p(1), p(2), p(3), p(4), p(5), p(6)) >_1 =$$

$$= \frac{1}{6!} 720 < \psi^2(1, 2, 3, 4, 5, 6) >_1 = < \psi^2(1, 2, 3, 4, 5, 6) >_1 = 1 + \sum_{i=1}^{6} 3c_i^2,$$

and for the Hamiltonian we have, similarly,

$$< \Psi_{cal}^* \hat{H} \Psi_{cal} >_3 =$$

$$= \frac{1}{6!} \sum_p < \psi(p(1), p(2), \dots, p(6)) \, \hat{H} \psi(p(1), p(2), \dots, p(6)) >_1 =$$

$$= < \psi(1, 2, 3, 4, 5, 6) \, \hat{H} \psi(1, 2, 3, 4, 5, 6) >_1,$$

where the sums above are taken with respect to all $6!$ permutations of the set $\{1, 2, 3, 4, 5, 6\}$, denoted by $p = (p(1), p(2), p(3), p(4), p(5), p(6))$, and the mean energy is

$$E(r, c_1, c_2, c_3, c_4, c_5, c_6) = \frac{< \Psi_{cal}^* \hat{H} \Psi_{cal} >_3}{< \Psi_{cal}^* \Psi_{cal} >_3} = \frac{< \psi \hat{H} \psi >_1}{< \psi^2 >_1}.$$

Minimizing $E(r, c_1, c_2, c_3, c_4, c_5, c_6)$ we get

$$r = 0.382723 \text{ a.u.},$$

$$a = c_1 = c_2 = -0.125765, \quad b = c_3 = c_4 = 0.16227, \quad c = c_5 = c_6 = 0.074773,$$

and the corresponding ground state mean energy is

$$E^* = -47.4994 \text{ a.u.} = -1292.53 \text{ eV},$$

which is much smaller than the experimental value

$$-37.8555 \text{ a.u.} = -1030.11 \text{ eV},$$

contradicting the variational theorem and showing that the Hamiltonian (63), while taking into account the interactions between pairs of electrons, ignores the interaction among three, four, five, and all six electrons taken together. This insufficiency may be corrected, allowing the electrons of the carbon atom to repel each other and move, on average, farther from the nucleus, by taking r to be a correction factor in the expression of the mean energy

$$E(r, c_1, c_2, c_3, c_4, c_5, c_6).$$

Here we want to use the known experimental value of the mean ground energy in order to better approximate the model of the atom. Denoting by $a = c_1 = c_2, b = c_3 = c_4$, and $c = c_5 = c_6$, the mean energy becomes a function of four variational parameters $E(r, a, b, c)$. We have seen that the minimum value of $E(r, a, b, c)$ is much smaller than the experimental value and it corresponds to $r = 0.382723$. On the other hand, taking the weighting factor $r = 1$, in which case (103) are just the Bohr's orbits, we obtain the minimum value of $E(1, a, b, c)$ to be -36.2464 a.u. $= -986.323$ eV, whose absolute value represents only 95.7492% of the absolute value of the experimental value of the mean ground energy. Therefore, the weighting factor r has to be located somewhere between 0.38 and 1. The search for the right value is not difficult and for $r = 0.918$, the minimum value of the mean ground energy $E(0.918, a, b, c)$ is

$$E^* = -37.8375 \text{ a.u.} = -1029.62 \text{ eV},$$

corresponding to the coefficients

$$c_1 = c_2 = a = 0.223918, \quad c_3 = c_4 = b = 0.172499,$$

$$c_5 = c_6 = c = 0.0549091, \tag{108}$$

and its absolute value represents 99.9523% of the absolute value of the experimental value of the mean ground state energy. On average, the two electrons belonging to the first subshell of the carbon atom move around the nucleus with a mean orbit $\mu_1 = r/6 = 0.153$ a.u., the two electrons from the second subshell move around the nucleus with a mean orbit $\mu_2 = 4r/6 = 0.612$ a.u., while the two electrons from the external subshell have the mean orbit $\mu_3 = 9r/6 = 1.377$ a.u. around the nucleus. The weighted mean orbit of the electrons of the carbon atom is

$$\bar{\mu} = \frac{1}{6}(2\mu_1 + 2\mu_2 + 2\mu_3) = 0.714 \text{ a.u.}$$

Therefore, the mean diameters of the atom induced by the statistical equilibrium and by the external subshell are

$$D_1 = 2\bar{\mu} = 1.428 \text{ a.u.,} \quad D_2 = 2\mu_3 = 2.754 \text{ a.u.}$$

Using the values (108), the generalized correlations (65) become

$$\text{gencorr}_{cal}(\mu_1, \mu_1) = 0.249976, \quad \text{gencorr}_{cal}(\mu_1, \mu_2) = 0.195765,$$
$$\text{gencorr}_{cal}(\mu_1, \mu_3) = 0.058664, \quad \text{gencorr}_{cal}(\mu_2, \mu_2) = 0.153310,$$
$$\text{gencorr}_{cal}(\mu_2, \mu_3) = 0.045942, \quad \text{gencorr}_{cal}(\mu_3, \mu_3) = 0.013767,$$

showing that the interdependence between the two electrons belonging to the inner subshell is stronger, followed by the interdependence between the two electrons from the second subshell.

Replacing the values (108) and

$$\mu_1 = 0.153 \text{ a.u.,} \quad \mu_2 = 0.612 \text{ a.u.,} \quad \mu_3 = 1.377 \text{ a.u.,}$$

into the expression (107) of $\psi = \psi(1, 2, 3, 4, 5, 6)$, we get $< \psi^2 >_1 = 1.49746$ and, using (74)-(76), we get the standard correlations

$$\text{stcorr}_{cal}(\mu_1, \mu_1) = 0.370184, \quad \text{stcorr}_{cal}(\mu_1, \mu_2) = 0.286566,$$
$$\text{stcorr}_{cal}(\mu_1, \mu_3) = 0.087047, \quad \text{stcorr}_{cal}(\mu_2, \mu_2) = 0.215813,$$
$$\text{stcorr}_{cal}(\mu_2, \mu_3) = 0.048925, \quad \text{stcorr}_{cal}(\mu_3, \mu_3) = -0.036116.$$

Replacing the values (108) and

$$\mu_1 = 0.153 \text{ a.u.,} \quad \mu_2 = 0.612 \text{ a.u.,} \quad \mu_3 = 1.377 \text{ a.u.,}$$

into the expression (107) of $\psi = \psi(1, 2, 3, 4, 5, 6)$, and using $< \psi^2 >_1 = 1.49746$, the mean radial distance from the nucleus of an arbitrary electron during its random motion around the nucleus is

$$\bar{r} = \frac{1}{6} \frac{< \psi \, (r_1 + r_2 + r_3 + r_4 + r_5 + r_6) \, \psi >_1}{< \psi^2 >_1} = 1.89437 \text{ a.u.,}$$

and the mean diameter of the atom induced by the individual random fluctuations of its electrons is $D_3 = 2\bar{r} = 3.78874$ a.u. The mean index of instability of the atom is $\bar{r} - \bar{\mu} = 1.18037$.

In order to calculate the global correlation between two arbitrary electrons of the carbon atom we introduce again (108) and

$$\mu_1 = 0.153 \text{ a.u.}, \quad \mu_2 = 0.612 \text{ a.u.}, \quad \mu_3 = 1.377 \text{ a.u.},$$

into (107), and using (70)-(73) we get

$$\text{globcorr} = -0.105142.$$

Replacing the values (108) and

$$\mu_1 = 0.153 \text{ a.u.}, \quad \mu_2 = 0.612 \text{ a.u.}, \quad \mu_3 = 1.377 \text{ a.u.},$$

into (107), let

$$d(\mu_1, \mu_1, \mu_2, \mu_2, \mu_3, \mu_3) = \frac{\psi^2}{< \psi^2 >_1},$$

be the joint probability density of the system of six electrons and

$$d(\mu_i), \quad d(\mu_i, \mu_j), \quad d(\mu_i, \mu_j, \mu_k),$$

$$d(\mu_i, \mu_j, \mu_k, \mu_\ell), \quad d(\mu_i, \mu_j, \mu_k, \mu_\ell, \mu_t),$$

be its marginal densities, describing the behavior of different subsystems of electrons from the same shell or from different subshells. Thus, $d(\mu_2, \mu_2)$, for instance, is obtained by integrating $d(\mu_1, \mu_1, \mu_2, \mu_2, \mu_3, \mu_3)$ with respect to the measure

$$r_1^2 dr_1 d\nu_1 d\omega_1 r_2^2 dr_2 d\nu_2 d\omega_2 r_5^2 dr_5 d\nu_5 d\omega_5 r_6^2 dr_6 d\nu_6 d\omega_6.$$

We obtain the following degrees of stability for subsystems of electrons inside the carbon atom:

$$< d^2(\mu_1, \mu_1, \mu_2, \mu_2, \mu_3, \mu_3) >_1^{1/2} = 0.0000453824,$$

$$< d^2(\mu_1) >_1^{1/2} = 0.359396, \quad < d^2(\mu_2) >_1^{1/2} = 0.174286,$$

$$< d^2(\mu_3) >_1^{1/2} = 0.107936, \quad < d^2(\mu_1, \mu_1) >_1^{1/2} = 0.129706,$$

$$< d^2(\mu_2, \mu_2) >_1^{1/2} = 0.0304128, \quad < d^2(\mu_3, \mu_3) >_1^{1/2} = 0.0116479,$$

$$< d^2(\mu_1, \mu_1, \mu_2, \mu_2) >_1^{1/2} = 0.00392836, \quad < d^2(\mu_1, \mu_1, \mu_3, \mu_3) >_1^{1/2} = 0.00150665,$$

$$< d^2(\mu_2, \mu_2, \mu_3, \mu_3) >_1^{1/2} = 0.000353491,$$

and the following levels of interdependence among subsystems of electrons

$$I(\{\mu_1, \mu_1, \mu_2, \mu_2, \mu_3, \mu_3\} : \{\mu_1\}, \{\mu_1\}, \{\mu_2\}, \{\mu_2\}, \{\mu_3\}, \{\mu_3\}) = 3.26655 \times 10^{-7},$$

$$I(\{\mu_1, \mu_1, \mu_2, \mu_2, \mu_3, \mu_3\} : \{\mu_1, \mu_1\}, \{\mu_2, \mu_2\}, \{\mu_3, \mu_3\}) = 5.65362 \times 10^{-7},$$

$$I(\{\mu_1, \mu_1, \mu_2, \mu_2\} : \{\mu_1, \mu_1\}, \{\mu_2, \mu_2\}) = 1.63672 \times 10^{-5},$$

$$I(\{\mu_1, \mu_1, \mu_3, \mu_3\} : \{\mu_1, \mu_1\}, \{\mu_3, \mu_3\}) = 4.15322 \times 10^{-6},$$

$$I(\{\mu_2, \mu_2, \mu_3, \mu_3\} : \{\mu_2, \mu_2\}, \{\mu_3, \mu_3\}) = 7.54161 \times 10^{-7},$$

$$I(\{\mu_1, \mu_1\} : \{\mu_1\}, \{\mu_1\}) = 5.40956 \times 10^{-4},$$

$$I(\{\mu_2, \mu_2\} : \{\mu_2\}, \{\mu_2\}) = 3.73831 \times 10^{-5},$$

$$I(\{\mu_3, \mu_3\} : \{\mu_3\}, \{\mu_3\}) = 2.33643 \times 10^{-6}.$$

2^0. For the *turbulent model* of the atom, introducing (103) into (107) and taking into account (105), by minimizing the mean energy

$$E(r, c_1, c_2, c_3, c_4, c_5, c_6) = \frac{< \Psi^*_{tur} \hat{H} \Psi_{tur} >_3}{< \Psi^*_{tur} \Psi_{tur} >_3} = \frac{< \psi \hat{H} \rho_{tur} >_1}{< \psi \rho_{tur} >_1},$$

we get $r = 0.507274$ a.u., and

$$a = c_1 = c_2 = -0.0348155 \quad b = c_3 = c_4 = -0.21055, \quad c = c_5 = c_6 = 0.363363,$$

for which the corresponding ground state mean energy is

$$E^* = -39.499 \text{ a.u.} = -1074.83 \text{ eV},$$

which is smaller than the experimental value

$$-37.8555 \text{ a.u.} = -1030.11 \text{ eV},$$

contradicting the variational theorem and showing that the Hamiltonian (63), while taking into account the interactions between pairs of electrons, ignores the interaction among three, four, five, and all six electrons taken together.

This insufficiency may be corrected, allowing the electrons of the carbon atom to repel each other and move, on average, farther from the nucleus, by taking r to be a correction factor in the expression of the mean energy

$$E(r, c_1, c_2, c_3, c_4, c_5, c_6).$$

Here we want to use the known experimental value of the mean ground energy in order to better approximate the model of the atom. Denoting by $a = c_1 = c_2, b = c_3 = c_4$, and $c = c_5 = c_6$, the mean energy becomes a function of four variational parameters $E(r, a, b, c)$. We have seen that the minimum value of $E(r, a, b, c)$ is much smaller than the experimental value and it corresponds to $r = 0.382723$. On the other hand, taking the weighting factor $r = 1$, in which case (103) are just the Bohr's orbits, we obtain for $a = b = c = 0$ the value of $E(1, 0, 0, 0)$ to be -30.4201 a.u. $= -827.78$ eV, whose absolute value represents only 80.3584% of the absolute value of the experimental value of the mean ground energy of the carbon atom. Therefore, the weighting factor r has to be located somewhere between 0.50 and 1. The search for the right value is not difficult and for $r = 0.74$, the minimum value of the mean ground energy $E(0.74, a, b, c)$ is

$$E^* = -37.8295 \text{ a.u.} = -1029.4 \text{ eV},$$

corresponding to the coefficients

$$r = 0.74, \quad c_1 = c_2 = a = 0.219355, \quad c_3 = c_4 = b = -0.0234488,$$

$$c_5 = c_6 = c = 0.196749, \tag{109}$$

and its absolute value represents 99.9312% of the absolute value of the experimental value of the mean ground energy -1030.11 eV. The mean orbits of the three subshells of the carbon atom in the turbulent model are (in a.u.)

$$\mu_1 = \frac{r}{6} = 0.123333, \quad \mu_2 = \frac{4r}{6} = 0.493333, \quad \mu_3 = \frac{9r}{6} = 1.11. \tag{110}$$

The weighted mean radial distance from the nucleus of the electrons of the carbon atom is

$$\bar{\mu} = \frac{1}{6}(2\mu_1 + 2\mu_2 + 2\mu_3) = 0.575555 \text{ a.u.}$$

Therefore, the mean diameters of the atom induced by the statistical equilibrium and by the external subshell are

$$D_1 = 2\bar{\mu} = 1.15111 \text{ a.u.}, \quad D_2 = 2\mu_3 = 2.22 \text{ a.u.}$$

Introducing the values (109) into the formula (65) we get

$$\text{gencorr}_{tur}(\mu_1, \mu_1) = 0.235820, \quad \text{gencorr}_{tur}(\mu_1, \mu_2) = -0.021883,$$
$$\text{gencorr}_{tur}(\mu_1, \mu_3) = 0.213484, \quad \text{gencorr}_{tur}(\mu_2, \mu_2) = 0.002031,$$
$$\text{gencorr}_{tur}(\mu_2, \mu_3) = -0.019810, \quad \text{gencorr}_{tur}(\mu_3, \mu_3) = 0.193263.$$

Introducing the values (109) and (110) into the expression of the probability wave function (107), we get

$$< \Psi_{tur}^* \Psi_{tur} >_3 = < \psi \, \rho_{tur} >_1 = 1.18448,$$

and evaluating the formulas (74)-(76), the standard correlations are

$$\text{stcorr}_{tur}(\mu_1, \mu_1) = 0.454117, \quad \text{stcorr}_{tur}(\mu_1, \mu_2) = 0.366872,$$
$$\text{stcorr}_{tur}(\mu_1, \mu_3) = 0.138384, \quad \text{stcorr}_{tur}(\mu_2, \mu_2) = 0.291443,$$
$$\text{stcorr}_{tur}(\mu_2, \mu_3) = 0.096068, \quad \text{stcorr}_{tur}(\mu_3, \mu_3) = -0.007853.$$

Introducing the values (109) and (110) into (107), and using $< \psi \rho_{tur} >_1 = 1.18448$, the mean radial distance from the nucleus of an arbitrary electron during its random motion around the nucleus is

$$\bar{r} = \frac{1}{6} \frac{< \psi \, (r_1 + r_2 + r_3 + r_4 + r_5 + r_6) \, \rho_{tur} >_1}{< \psi \rho_{tur} >_1} = 1.53524 \text{ a.u.},$$

and the mean diameter of the atom induced by the individual random fluctuations of its electrons is $D_3 = 2\bar{r} = 3.07048$ a.u. The mean index of instability of the atom is $\bar{r} - \bar{\mu} = 0.959685$.

In order to calculate the global correlation between two arbitrary electrons of the carbon atom we introduce again (109) and (110) into (107), and using (70)-(73) we get

$$\text{globcorr} = -0.100786.$$

Introducing the values (109) and (110) into (107), let

$$d(\mu_1, \mu_1, \mu_2, \mu_2, \mu_3, \mu_3) = \frac{\psi^2}{< \psi^2 >_1},$$

be the joint probability density of the system of six electrons and

$$d(\mu_i), \quad d(\mu_i, \mu_j), \quad d(\mu_i, \mu_j, \mu_k),$$

$$d(\mu_i, \mu_j, \mu_k, \mu_\ell), \quad d(\mu_i, \mu_j, \mu_k, \mu_\ell, \mu_t),$$

be its marginal densities, describing the behavior of different subsystems of electrons from the same shell or from different subshells. We obtain the following degrees of stability for subsystems of electrons inside the carbon atom:

$$< d^2(\mu_1, \mu_1, \mu_2, \mu_2, \mu_3, \mu_3) >_1^{1/2} = 0.0000931701,$$

$$< d^2(\mu_1) >_1^{1/2} = 0.414591, \quad < d^2(\mu_2) >_1^{1/2} = 0.181531,$$

$$< d^2(\mu_3) >_1^{1/2} = 0.130323, \quad < d^2(\mu_1, \mu_1) >_1^{1/2} = 0.171803,$$

$$< d^2(\mu_2, \mu_2) >_1^{1/2} = 0.0328125, \quad < d^2(\mu_3, \mu_3) >_1^{1/2} = 0.0169439,$$

$$< d^2(\mu_1, \mu_1, \mu_2, \mu_2) >_1^{1/2} = 0.00549015, \quad < d^2(\mu_1, \mu_1, \mu_3, \mu_3) >_1^{1/2} = 0.00290126,$$

$$< d^2(\mu_2, \mu_2, \mu_3, \mu_3) >_1^{1/2} = 0.000564232,$$

and the following levels of interdependence among subsystems of electrons

$$I(\{\mu_1, \mu_1, \mu_2, \mu_2, \mu_3, \mu_3\} : \{\mu_1\}, \{\mu_1\}, \{\mu_2\}, \{\mu_2\}, \{\mu_3\}, \{\mu_3\}) = 3.03238 \times 10^{-6},$$

$$I(\{\mu_1, \mu_1, \mu_2, \mu_2, \mu_3, \mu_3\} : \{\mu_1, \mu_1\}, \{\mu_2, \mu_2\}, \{\mu_3, \mu_3\}) = 2.34738 \times 10^{-6},$$

$$I(\{\mu_1, \mu_1, \mu_2, \mu_2\} : \{\mu_1, \mu_1\}, \{\mu_2, \mu_2\}) = 1.47136 \times 10^{-4},$$

$$I(\{\mu_1, \mu_1, \mu_3, \mu_3\} : \{\mu_1, \mu_1\}, \{\mu_3, \mu_3\}) = 9.75386 \times 10^{-6},$$

$$I(\{\mu_2, \mu_2, \mu_3, \mu_3\} : \{\mu_2, \mu_2\}, \{\mu_3, \mu_3\}) = 8.26059 \times 10^{-6},$$

$$I(\{\mu_1, \mu_1\} : \{\mu_1\}, \{\mu_1\}) = 8.28401 \times 10^{-5},$$

$$I(\{\mu_2, \mu_2\} : \{\mu_2\}, \{\mu_2\}) = 1.40989 \times 10^{-4},$$

$$I(\{\mu_3, \mu_3\} : \{\mu_3\}, \{\mu_3\}) = 4.03107 \times 10^{-5}.$$

3^0. For the *nebulous model* of the atom, introducing (103) into (107) and taking into account (106), by minimizing the mean energy

$$E(r, c_1, c_2, c_3, c_4, c_5, c_6) = \frac{< \Psi^*_{neb} \hat{H} \Psi_{neb} >_2}{< \Psi^*_{neb} \Psi_{neb} >_2} = \frac{< \psi \hat{H} \rho_{neb} >_1}{< \psi \rho_{neb} >_1},$$

we obtain the values

$$r = 0.5462 \text{ a.u.},$$

$$c_1 = c_2 = 0.0057583985, \quad c_3 = c_4 = -0.166646437,$$

$$c_5 = c_6 = 0.00161467167. \tag{111}$$

The mean orbits of the three subshells of the carbon atom in the nebulous model are (in a.u.)

$$\mu_1 = \frac{r}{6} = 0.0910333, \quad \mu_2 = \frac{4r}{6} = 0.364133, \quad \mu_3 = \frac{9r}{6} = 0.8193. \tag{112}$$

The weighted mean radial distance of the electrons from the nucleus is

$$\bar{\mu} = \frac{1}{6}(2\mu_1 + 2\mu_2 + 2\mu_3) = 0.424822 \text{ a.u.}$$

Therefore, the mean diameters of the atom induced by the statistical equilibrium and by the external subshell are

$$D_1 = 2\bar{\mu} = 0.849644 \text{ a.u.}, \quad D_2 = 2\mu_3 = 1.6386 \text{ a.u.}$$

The minimum energy is -36.7488068 a.u. $= -999.994$ eV, whose absolute value represents 97.0764% of the absolute value of the experimental value obtained for the mean ground energy of the carbon atom -1030.11 eV, and this result has been obtained using only four variational parameters, r, a, b, c, where $a = c_1 = c_2$, $b = c_3 = c_4$, and $c = c_5 = c_6$. Introducing the values (111) into the formula (65) we get

$$\text{gencorr}_{neb}(\mu_1, \mu_1) = 0.00017388, \quad \text{gencorr}_{neb}(\mu_1, \mu_2) = -0.00365679,$$
$$\text{gencorr}_{neb}(\mu_1, \mu_3) = 0.00004841, \quad \text{gencorr}_{neb}(\mu_2, \mu_2) = 0.07690400,$$
$$\text{gencorr}_{neb}(\mu_2, \mu_3) = -0.00101819, \quad \text{gencorr}_{neb}(\mu_3, \mu_3) = 0.00001348.$$

Introducing the values (111) and (112) into the expression of the probability wave function (107), we get

$$< \Psi_{neb}^* \Psi_{neb} >_2 = < \psi \, \rho_{neb} >_1 = 0.0280023,$$

and evaluating the formulas (74)-(76), the standard correlations are

$$\text{stcorr}_{neb}(\mu_1, \mu_1) = 0.410798, \quad \text{stcorr}_{neb}(\mu_1, \mu_2) = 0.364799,$$
$$\text{stcorr}_{neb}(\mu_1, \mu_3) = 0.266225, \quad \text{stcorr}_{neb}(\mu_2, \mu_2) = 0.321713,$$
$$\text{stcorr}_{neb}(\mu_2, \mu_3) = 0.229745, \quad \text{stcorr}_{neb}(\mu_3, \mu_3) = 0.152656.$$

Introducing the values (111) and (112) into (107), and using $< \psi \rho_{neb} >_1 = 0.0280023$ the mean radial distance from the nucleus of an arbitrary electron during its random motion around the nucleus is

$$\bar{r} = \frac{1}{6} \frac{< \psi \left(r_1 + r_2 + r_3 + r_4 + r_5 + r_6 \right) \rho_{neb} >_1}{< \psi \rho_{neb} >_1} = 1.97577 \text{ a.u.,}$$

and the mean diameter of the atom induced by the individual random fluctuations of its electrons is $D_3 = 2\bar{r} = 3.95154$ a.u. The mean index of instability of the atom is $\bar{r} - \bar{\mu} = 1.550948$.

In order to calculate the global correlation between two arbitrary electrons of the carbon atom we introduce again (111) and (112) into (107), and using (70)-(73) we get

$$globcorr = -0.152165.$$

Introducing the values (111) and (112) into (107), let

$$d(\mu_1, \mu_1, \mu_2, \mu_2, \mu_3, \mu_3) = \frac{\psi^2}{< \psi^2 >_1},$$

be the joint probability density of the system of six electrons and

$$d(\mu_i), \quad d(\mu_i, \mu_j), \quad d(\mu_i, \mu_j, \mu_k),$$

$$d(\mu_i, \mu_j, \mu_k, \mu_\ell), \quad d(\mu_i, \mu_j, \mu_k, \mu_\ell, \mu_t),$$

be its marginal densities, describing the behavior of different subsystems of electrons from the same shell or from different subshells. We obtain the following degrees of stability for subsystems of electrons inside the carbon atom in the nebulous model:

$$< d^2(\mu_1, \mu_1, \mu_2, \mu_2, \mu_3, \mu_3) >_1^{1/2} = 0.00167018,$$

$$< d^2(\mu_1) >_1^{1/2} = 0.433622, \quad < d^2(\mu_2) >_1^{1/2} = 0.336182,$$

$$< d^2(\mu_3) >_1^{1/2} = 0.239188, \quad < d^2(\mu_1, \mu_1) >_1^{1/2} = 0.187509,$$

$$< d^2(\mu_2, \mu_2) >_1^{1/2} = 0.110001, \quad < d^2(\mu_3, \mu_3) >_1^{1/2} = 0.0525022,$$

$$< d^2(\mu_1, \mu_1, \mu_2, \mu_2) >_1^{1/2} = 0.019455, \quad < d^2(\mu_1, \mu_1, \mu_3, \mu_3) >_1^{1/2} = 0.00980571,$$

$$< d^2(\mu_2, \mu_2, \mu_3, \mu_3) >_1^{1/2} = 0.0026334,$$

and the following levels of interdependence among subsystems of electrons

$$I(\{\mu_1, \mu_1, \mu_2, \mu_2, \mu_3, \mu_3\} : \{\mu_1\}, \{\mu_1\}, \{\mu_2\}, \{\mu_2\}, \{\mu_3\}, \{\mu_3\}) = 4.54408 \times 10^{-4},$$

$$I(\{\mu_1, \mu_1, \mu_2, \mu_2, \mu_3, \mu_3\} : \{\mu_1, \mu_1\}, \{\mu_2, \mu_2\}, \{\mu_3, \mu_3\}) = 5.87265 \times 10^{-4},$$

$$I(\{\mu_1, \mu_1, \mu_2, \mu_2\} : \{\mu_1, \mu_1\}, \{\mu_2, \mu_2\}) = 1.17122 \times 10^{-3},$$

$$I(\{\mu_1, \mu_1, \mu_3, \mu_3\} : \{\mu_1, \mu_1\}, \{\mu_3, \mu_3\}) = 3.89254 \times 10^{-5},$$

$$I(\{\mu_2, \mu_2, \mu_3, \mu_3\} : \{\mu_2, \mu_2\}, \{\mu_3, \mu_3\}) = 3.1419 \times 10^{-3},$$

$$I(\{\mu_1, \mu_1\} : \{\mu_1\}, \{\mu_1\}) = 5.19528 \times 10^{-4},$$

$$I(\{\mu_2, \mu_2\} : \{\mu_2\}, \{\mu_2\}) = 3.01785 \times 10^{-3},$$

$$I(\{\mu_3, \mu_3\} : \{\mu_3\}, \{\mu_3\}) = 4.7088 \times 10^{-3}.$$

Remark 1: In the calm model of the carbon atom, if we ignore the sub-shells and the electrons are allowed to be completely free, the mean ground energy of the atom is minimized for

$$\mu_1 = \mu_2 = \mu_3 = 0.114102 \text{ a.u.}, \quad c_1 = c_2 = c_3 = c_4 = c_5 = c_6 = 0.00632201,$$

and its minimum value is -59.0742 a.u. $= -1607.5$ eV, a value much smaller than the experimental value -1030.11 eV. On the other side, if we assume that the atom has three subshells and their mean radial distances from the nucleus were given just by the Bohr's values $k^2/Z = k^2/6$ for $k = 1, 2, 3$, then the minimum mean ground energy would be -986.323 eV, whose absolute value represents 95.7492% of the absolute value of the experimental value of the mean ground energy. This result and the analysis performed in this section show that taking subshells of the atom into account with weighted Bohr's mean radial distances from the nucleus is appropriate and that the turbulent and nebulous models of the atom are more accurate than the calm model in describing the structure of the carbon atom.

Remark 2: Let us notice that in the turbulent model of the carbon atom, if we take $a = b = c = 0$ and use the weighting factor r as being the only variational parameter, then the expression of the mean ground energy $E(r, 0, 0, 0)$ becomes

$$11.47220596\frac{1}{r^2} - 41.89232001\frac{1}{r},$$

which, for $r = 0.61$ becomes

$$-37.845 \text{ a.u.} = -1029.82 \text{ eV},$$

whose absolute value represents 99.972% of the absolute value of the experimental value -1030.22 eV of the mean ground energy of the carbon atom.

Remark 3: In the nebulous model of the carbon atom, if we assume that the electrons are in statistical equilibrium, i.e. $a = b = c = 0$, then the minimum mean energy of the atom is

$$-36.2966 \text{ a.u.} = -987.689 \text{ eV},$$

whose absolute value represents 95.8819% of the absolute value of the experimental value of the mean energy of the atom.

Remark 4: Let us notice that in the nebulous model of the carbon atom, replacing the linear probability wave function given by (107) by a quadratic function containing only four variational parameters, we obtain the minimum mean ground energy

$$-36.8247 \text{ a.u.} = -1002.06 \text{ eV}.$$

Remark 5: As the experimental value of the diameter of the carbon atom is $D_{exp} = 1.54$, this shows that in the calm and turbulent models of the atom D_1 is the closest approximation of the diameter of the atom whereas, in the nebulous model, D_2 is the closest approximation of the diameter. This shows that the structure of the carbon atom is more stable, being closer to statistical equilibrium in the calm and turbulent models, and depends more on the behavior of the electrons from the external subshell in the nebulous model. The atom is less influenced by the random fluctuations of its electrons.

g) Nitrogen Atom $(Z = 7)$.

This atom has three subshells. The seven electrons have the quantum numbers (n, ℓ, m, s) given by

$$(1, 0, 0, -\tfrac{1}{2}), \quad (1, 0, 0, +\tfrac{1}{2}), \quad (2, 0, 0, -\tfrac{1}{2}), \quad (2, 0, 0, +\tfrac{1}{2}),$$

$$(2, 1, -1, -\tfrac{1}{2}), \quad (2, 1, -1, +\tfrac{1}{2}), \quad (2, 1, 0, -\tfrac{1}{2}).$$

respectively. The Hamiltonian is given by (63) with $Z = 7$. Using the weighted Bohr approximation (69), the mean orbits of the three subshells of the nitrogen atom are

$$\mu_1 = \frac{r}{7}, \quad \mu_2 = \frac{2^2 r}{7} = \frac{4r}{7}, \quad \mu_3 = \frac{3^2 r}{7} = \frac{9r}{7}. \tag{113}$$

There are $7! = 5040$ permutations of seven electrons, and all four models (calm, nervous, turbulent, and nebulous) of the atom are different. Thus, with the notations from Section 4, we have

$$\rho_{cal}(1,2,3,4,5,6,7) = \psi(1,2,3,4,5,6,7), \tag{114}$$

$$\rho_{tur}(1,2,3,4,5,6,7) =$$
$$= \psi(1,2,3,4,5,6,7) - \psi(1,2,7,4,5,6,3) - \psi(3,2,1,4,5,6,7) +$$
$$+ \psi(3,2,7,4,5,6,1) + \psi(7,2,1,4,5,6,3) - \psi(7,2,3,4,5,6,1) -$$
$$- \psi(1,4,3,2,5,6,7) + \psi(1,4,7,2,5,6,3) + \psi(3,4,1,2,5,6,7) -$$
$$- \psi(3,4,7,2,5,6,1) - \psi(7,4,1,2,5,6,3) + \psi(7,4,3,2,5,6,1), \tag{115}$$

where the probability wave function

$$\psi = \psi(1,2,3,4,5,6,7) =$$

$$= \varphi(r_1,\mu_1)\varphi(r_2,\mu_1)\varphi(r_3,\mu_2)\varphi(r_4,\mu_2)\varphi(r_5,\mu_3)\varphi(r_6,\mu_3)\varphi(r_7,\mu_3) \times$$

$$\times \left[1 + c_1\left(3 - \frac{r_1}{\mu_1}\right) + c_2\left(3 - \frac{r_2}{\mu_1}\right) + c_3\left(3 - \frac{r_3}{\mu_2}\right) + \right.$$

$$\left. + c_4\left(3 - \frac{r_4}{\mu_2}\right) + c_5\left(3 - \frac{r_5}{\mu_3}\right) + c_6\left(3 - \frac{r_6}{\mu_3}\right) + c_7\left(3 - \frac{r_7}{\mu_3}\right) \right]. \tag{116}$$

1^0. For the *calm model* of the atom,

$$< \Psi^*_{cal}\Psi_{cal} >_3 = \frac{1}{7!}\sum_p < \psi^2(p(1),p(2),p(3),p(4),p(5),p(6),p(7)) >_1 =$$

$$= \frac{1}{7!}5040 < \psi^2(1,2,3,4,5,6,7) >_1 = < \psi^2(1,2,3,4,5,6,7) >_1 = 1 + \sum_{i=1}^{7} 3c_i^2,$$

and for the Hamiltonian we have, similarly,

$$< \Psi_{cal}^* \hat{H} \Psi_{cal} >_3 =$$

$$= \frac{1}{7!} \sum_p < \psi(p(1), p(2), \dots, p(7)) \, \hat{H} \psi(p(1), p(2), \dots, p(7)) >_1 =$$

$$= < \psi(1, 2, 3, 4, 5, 6, 7) \, \hat{H} \psi(1, 2, 3, 4, 5, 6, 7) >_1,$$

where the sums above are taken with respect to all 7! permutations of the set $\{1, 2, 3, 4, 5, 6, 7\}$, denoted by $p = (p(1), p(2), p(3), p(4), p(5), p(6), p(7))$, and the mean energy is

$$E(r, c_1, c_2, c_3, c_4, c_5, c_6, c_7) = \frac{< \Psi_{cal}^* \hat{H} \Psi_{cal} >_3}{< \Psi_{cal}^* \Psi_{cal} >_3} = \frac{< \psi \hat{H} \psi >_1}{< \psi^2 >_1}.$$

Minimizing $E(r, c_1, c_2, c_3, c_4, c_5, c_6, c_7)$ we get

$$r = 0.357273 \text{ a.u.},$$

$$a = c_1 = c_2 = -0.140627, \quad b = c_3 = c_4 = 0.138106,$$

$$c = c_5 = c_6 = c_7 = 0.065886,$$

and the corresponding ground state mean energy is

$$E^* = -70.6434 \text{ a.u.} = -1922.32 \text{ eV},$$

which is much smaller than the experimental value

$$-54.6115 \text{ a.u.} = -1486.07 \text{ eV},$$

contradicting the variational theorem and showing that the Hamiltonian (63), while taking into account the interactions between pairs of electrons, ignores the interaction among three, four, five, six, and all seven electrons taken together. This insufficiency may be corrected, allowing the electrons of the nitrogen atom to repel each other and move, on average, farther from the nucleus, by taking r to be a correction factor in the expression of the mean energy

$$E(r, c_1, c_2, c_3, c_4, c_5, c_6, c_7).$$

Here we want to use the known experimental value of the mean ground energy in order to better approximate the model of the atom. Denoting by $a = c_1 = c_2$, $b = c_3 = c_4$, and $c = c_5 = c_6 = c_7$, the mean energy becomes a function of four variational parameters $E(r, a, b, c)$. We have seen that the minimum value of $E(r, a, b, c)$ is much smaller than the experimental value and it corresponds to $r = 0.357273$. On the other hand, taking the weighting factor $r = 1$, in which case (113) are just the Bohr's orbits, we obtain the minimum value of $E(1, a, b, c)$ to be -51.8713 a.u. $= -1411.5$ eV, whose absolute value represents only 94.9821% of the absolute value of the experimental value of the mean ground energy. Therefore, the weighting factor r has to be located somewhere between 0.35 and 1. The search for the right value is not difficult and for $r = 0.91$, the minimum value of the mean ground energy $E(0.91, a, b, c)$ is

$$E^* = -54.508 \text{ a.u.} = -1483.25 \text{ eV},$$

corresponding to the coefficients

$$c_1 = c_2 = a = 0.221506, \quad c_3 = c_4 = b = 0.179827,$$

$$c_5 = c_6 = c_7 = c = 0.0592934, \tag{117}$$

and its absolute value represents 99.8102% of the absolute value of the experimental value of the mean ground state energy. On average, the two electrons belonging to the first subshell of the nitrogen atom move randomly around the nucleus with a mean orbit distance equal to $\mu_1 = r/7 = 0.13$ a.u., the two electrons from the second subshell move randomly around the nucleus at a mean orbit distance equal to $\mu_2 = 4r/7 = 0.52$ a.u., while the three electrons from the external subshell have the mean orbit $\mu_3 = 9r/7 = 1.17$ a.u.. The weighted mean orbit of the nitrogen atom is

$$\bar{\mu} = \frac{1}{7}(2\mu_1 + 2\mu_2 + 3\mu_3) = 0.687143 \text{ a.u.}$$

Therefore, the mean diameters of the atom induced by the statistical equilibrium and by the external subshell are

$$D_1 = 2\bar{\mu} = 1.37429 \text{ a.u.}, \quad D_2 = 2\mu_3 = 2.34 \text{ a.u.}$$

Using the values (117), the generalized correlations (65) become

$$\text{gencorr}_{cal}(\mu_1, \mu_1) = 0.240748, \qquad \text{gencorr}_{cal}(\mu_1, \mu_2) = 0.198224,$$
$$\text{gencorr}_{cal}(\mu_1, \mu_3) = 0.061908, \qquad \text{gencorr}_{cal}(\mu_2, \mu_2) = 0.163212,$$
$$\text{gencorr}_{cal}(\mu_2, \mu_3) = 0.050973, \qquad \text{gencorr}_{cal}(\mu_3, \mu_3) = 0.015919,$$

showing that the interdependence between the two electrons belonging to the inner subshell is stronger, followed by the interdependence between an electron from the first subshell and an electron from the second subshell.

Replacing the values (117) and $\mu_1 = 0.13$ a.u., $\mu_2 = 0.52$ a.u., and $\mu_3 = 1.17$ a.u. into the expression (116) of $\psi = \psi(1, 2, 3, 4, 5, 6, 7)$, we get $< \psi^2 >_1 = 1.52006$ and, using (74)-(76), we get the standard correlations

$$\text{stcorr}_{cal}(\mu_1, \mu_1) = 0.426100, \qquad \text{stcorr}_{cal}(\mu_1, \mu_2) = 0.354342,$$
$$\text{stcorr}_{cal}(\mu_1, \mu_3) = 0.180795, \qquad \text{stcorr}_{cal}(\mu_2, \mu_2) = 0.291580,$$
$$\text{stcorr}_{cal}(\mu_2, \mu_3) = 0.140797, \qquad \text{stcorr}_{cal}(\mu_3, \mu_3) = 0.047175.$$

Replacing the values (117) and $\mu_1 = 0.13$ a.u., $\mu_2 = 0.52$ a.u., and $\mu_3 = 1.17$ a.u. into the expression (116) and using $< \psi^2 >_1 = 1.52006$ we get the mean radial distance from the nucleus of an arbitrary electron during its random motion around the nucleus:

$$\bar{r} = \frac{1}{7} \frac{< \psi \left(r_1 + r_2 + r_3 + r_4 + r_5 + r_6 + r_7 \right) \psi >_1}{< \psi^2 >_1} = 1.83925 \text{ a.u.},$$

and the mean diameter of the atom induced by the individual random fluctuations of its electrons is $D_3 = 2\bar{r} = 3.6785$ a.u. The mean index of instability of the atom is $\bar{r} - \bar{\mu} = 1.152107$.

In order to calculate the global correlation between two arbitrary electrons of the nitrogen atom we introduce again (117) and $\mu_1 = 0.13$ a.u., $\mu_2 = 0.52$ a.u., and $\mu_3 = 1.17$ a.u. into (116), and using (70)-(73) we get

$$\text{globcorr} = -0.0822074.$$

Replacing the values (117) and $\mu_1 = 0.13$ a.u., $\mu_2 = 0.52$ a.u., and $\mu_3 = 1.17$ a.u. into the expression (116), let

$$d(\mu_1, \mu_1, \mu_2, \mu_2, \mu_3, \mu_3, \mu_3) = \frac{\psi^2}{< \psi^2 >_1},$$

be the joint probability density of the system of seven electrons and

$$d(\mu_i), \quad d(\mu_i, \mu_j), \quad d(\mu_i, \mu_j, \mu_k),$$

$$d(\mu_i, \mu_j, \mu_k, \mu_\ell), \quad d(\mu_i, \mu_j, \mu_k, \mu_\ell, \mu_t),$$

$$d(\mu_i, \mu_j, \mu_k, \mu_\ell, \mu_t, \mu_s),$$

be its marginal densities, describing the behavior of different subsystems of electrons from the same shell or from different subshells. Thus, $d(\mu_3, \mu_3, \mu_3)$, for instance, is obtained by integrating $d(\mu_1, \mu_1, \mu_2, \mu_2, \mu_3, \mu_3, \mu_3)$ with respect to the measure

$$r_1^2 dr_1 d\nu_1 d\omega_1 r_2^2 dr_2 d\nu_2 d\omega_2 r_3^2 dr_3 d\nu_3 d\omega_3 r_4^2 dr_4 d\nu_4 d\omega_4.$$

We obtain the following degrees of stability for subsystems of electrons inside the nitrogen atom:

$$< d^2(\mu_1, \mu_1, \mu_2, \mu_2, \mu_3, \mu_3, \mu_3) >_1^{1/2} = 0.00000866978,$$

$$< d^2(\mu_1) >_1^{1/2} = 0.388371, \quad < d^2(\mu_2) >_1^{1/2} = 0.189532,$$

$$< d^2(\mu_3) >_1^{1/2} = 0.117357, \quad < d^2(\mu_1, \mu_1) >_1^{1/2} = 0.151496,$$

$$< d^2(\mu_2, \mu_2) >_1^{1/2} = 0.0359871, \quad < d^2(\mu_3, \mu_3, \mu_3) >_1^{1/2} = 0.00161532,$$

$$< d^2(\mu_1, \mu_1, \mu_2, \mu_2) >_1^{1/2} = 0.00543618,$$

$$< d^2(\mu_1, \mu_1, \mu_3, \mu_3, \mu_3) >_1^{1/2} = 0.000243722,$$

$$< d^2(\mu_2, \mu_2, \mu_3, \mu_3, \mu_3) >_1^{1/2} = 0.0000579383,$$

and the following levels of interdependence among subsystems of electrons

$$I(\{\mu_1, \mu_1, \mu_2, \mu_2, \mu_3, \mu_3, \mu_3\} : \{\mu_1\}, \{\mu_1\}, \{\mu_2\}, \{\mu_2\}, \{\mu_3\}, \{\mu_3\}, \{\mu_3\}) =$$

$$= 8.77163 \times 10^{-8},$$

$$I(\{\mu_1, \mu_1, \mu_2, \mu_2, \mu_3, \mu_3, \mu_3\} : \{\mu_1, \mu_1\}, \{\mu_2, \mu_2\}, \{\mu_3, \mu_3, \mu_3\}) = 1.36728 \times 10^{-7},$$

$$I(\{\mu_1, \mu_1, \mu_2, \mu_2\} : \{\mu_1, \mu_1\}, \{\mu_2, \mu_2\}) = 1.5696 \times 10^{-5},$$

$$I(\{\mu_1, \mu_1, \mu_3, \mu_3, \mu_3\} : \{\mu_1, \mu_1\}, \{\mu_3, \mu_3, \mu_3\}) = 9.91008 \times 10^{-7},$$

$$I(\{\mu_2, \mu_2, \mu_3, \mu_3, \mu_3\} : \{\mu_2, \mu_2\}, \{\mu_3, \mu_3, \mu_3\}) = 1.92195 \times 10^{-7},$$

$$I(\{\mu_1, \mu_1\} : \{\mu_1\}, \{\mu_1\}) = 6.63184 \times 10^{-4},$$

$$I(\{\mu_2, \mu_2\} : \{\mu_2\}, \{\mu_2\}) = 6.47911 \times 10^{-5},$$

$$I(\{\mu_3, \mu_3, \mu_3\} : \{\mu_3\}, \{\mu_3\}, \{\mu_3\}) = 9.82802 \times 10^{-7}.$$

2^0. For the *turbulent model* of the atom, introducing (113) into (116) and taking into account (115), by minimizing the mean energy

$$E(r, c_1, c_2, c_3, c_4, c_5, c_6, c_7) = \frac{< \Psi_{tur}^* \, \hat{H} \Psi_{tur} >_3}{< \Psi_{tur}^* \, \Psi_{tur} >_3} = \frac{< \psi \, \hat{H} \rho_{tur} >_1}{< \psi \, \rho_{tur} >_1},$$

we obtain the values

$$r = 0.504479 \text{ a.u.},$$

$$a = c_1 = c_2 = -0.0172181, \quad b = c_3 = c_4 = -0.144738,$$

$$c = c_5 = c_6 = c_7 = 0.141009,$$

and the corresponding ground state mean energy is

$$E^* = -56.0554 \text{ a.u.} = -1525.36 \text{ eV},$$

which is smaller than the experimental value

$$-54.6115 \text{ a.u.} = -1486.07 \text{ eV},$$

contradicting the variational theorem and showing that the Hamiltonian (63), while taking into account the interactions between pairs of electrons, ignores the interaction among three, four, five, six, and all seven electrons taken together. This insufficiency may be corrected, allowing the electrons of the nitrogen atom to repel each other and move, on average, farther from the nucleus, by taking r to be a correction factor in the expression of the mean energy

$$E(r, c_1, c_2, c_3, c_4, c_5, c_6, c_7).$$

Here we want to use the known experimental value of the mean ground energy in order to better approximate the model of the atom. Denoting by $a = c_1 = c_2$, $b = c_3 = c_4$, and $c = c_5 = c_6 = c_7$, the mean energy becomes a function of four variational parameters $E(r, a, b, c)$. We have seen that the minimum value of $E(r, a, b, c)$ is much smaller than the experimental value and it corresponds to $r = 0.504479$. On the other hand, taking the weighting factor $r = 1$, in which case (113) are just the Bohr's orbits, we obtain for

$$a = 0.526248, \quad b = -0.0417114, \quad c = 0.0025695,$$

the minimum value of $E(1, a, b, c)$ equal to

$$-54.4139 \text{ a.u.} = -1480.69 \text{ eV},$$

whose absolute value represents only 90.0947% of the absolute value of the experimental value of the mean ground energy of the nitrogen atom. Therefore, the weighting factor r has to be located somewhere between 0.50 and 1. The search for the right value is not difficult and, for $r = 0.673$, the minimum value of the mean ground energy $E(0.673, a, b, c)$ is

$$E^* = -54.6028 \text{ a.u.} = -1485.83 \text{ eV},$$

corresponding to the coefficients

$$c_1 = c_2 = a = 0.0960587, \quad c_3 = c_4 = b = -0.0725801,$$

$$c_5 = c_6 = c_7 = c = 0.154955, \tag{118}$$

and its absolute value represents 99.9838% of the absolute value of the experimental value of the mean ground energy -54.6115 a.u. $= -1486.07$ eV. The mean orbits of the three subshells of the nitrogen atom in the turbulent model are (in a.u.)

$$\mu_1 = \frac{r}{7} = 0.0961429, \quad \mu_2 = \frac{4r}{7} = 0.384571, \quad \mu_3 = \frac{9r}{7} = 0.865286. \tag{119}$$

The weighted mean orbit of the nitrogen atom is

$$\bar{\mu} = \frac{1}{7}(2\mu_1 + 2\mu_2 + 3\mu_3) = 0.508184 \text{ a.u.}$$

Therefore, the mean diameters of the atom induced by the statistical equilibrium and by the external subshell are

$$D_1 = 2\bar{\mu} = 1.01637 \text{ a.u.}, \quad D_2 = 2\mu_3 = 1.73057 \text{ a.u.}$$

271

Introducing the values (118) into the formula (65) we get

$$\text{gencorr}_{tur}(\mu_1, \mu_1) = 0.0537739, \qquad \text{gencorr}_{tur}(\mu_1, \mu_2) = -0.0320309,$$
$$\text{gencorr}_{tur}(\mu_1, \mu_3) = 0.0893042, \qquad \text{gencorr}_{tur}(\mu_2, \mu_2) = 0.0190795,$$
$$\text{gencorr}_{tur}(\mu_2, \mu_3) = -0.0531949, \qquad \text{gencorr}_{tur}(\mu_3, \mu_3) = 0.1483110.$$

Introducing the values (118) and (119) into the expression of the probability wave function (116), we get

$$< \Psi_{tur}^* \Psi_{tur} >_3 = < \psi \rho_{tur} >_1 = 0.150411,$$

and evaluating the formulas (74)-(76), the standard correlations are

$$\text{stcorr}_{tur}(\mu_1, \mu_1) = 0.471962, \qquad \text{stcorr}_{tur}(\mu_1, \mu_2) = 0.406346,$$
$$\text{stcorr}_{tur}(\mu_1, \mu_3) = 0.245165, \qquad \text{stcorr}_{tur}(\mu_2, \mu_2) = 0.347280,$$
$$\text{stcorr}_{tur}(\mu_2, \mu_3) = 0.203033, \qquad \text{stcorr}_{tur}(\mu_3, \mu_3) = 0.102183.$$

Introducing the values (118) and (119) into (116) and using $< \psi \rho_{tur} >_1 = 0.150411$, we get the mean radial distance from the nucleus of an arbitrary electron during its random motion around the nucleus:

$$\bar{r} = \frac{1}{7} \frac{< \psi (r_1 + r_2 + r_3 + r_4 + r_5 + r_6 + r_7) \rho_{tur} >_1}{< \psi \rho_{tur} >_1} = 1.49589 \text{ a.u.},$$

and the mean diameter of the atom induced by the individual random fluctuations of its electrons is $D_3 = 2\bar{r} = 2.99178$ a.u. The mean index of instability of the atom is $\bar{r} - \bar{\mu} = 0.987706$.

In order to calculate the global correlation between two arbitrary electrons of the nitrogen atom we introduce again (118) and (119) into (116), and using (70)-(73) we get

$$\text{globcorr} = -0.0926667.$$

Introducing the values (118) and (119) into (116), let

$$d(\mu_1, \mu_1, \mu_2, \mu_2, \mu_3, \mu_3, \mu_3) = \frac{\psi^2}{< \psi^2 >_1},$$

be the joint probability density of the system of seven electrons and

$$d(\mu_i), \quad d(\mu_i, \mu_j), \quad d(\mu_i, \mu_j, \mu_k),$$

272

$$d(\mu_i, \mu_j, \mu_k, \mu_\ell), \quad d(\mu_i, \mu_j, \mu_k, \mu_\ell, \mu_t),$$

$$d(\mu_i, \mu_j, \mu_k, \mu_\ell, \mu_t, \mu_s),$$

be its marginal densities, describing the behavior of different subsystems of electrons from the same shell or from different subshells. We obtain the following degrees of stability for subsystems of electrons inside the nitrogen atom in the turbulent model of the atom:

$$< d^2(\mu_1, \mu_1, \mu_2, \mu_2, \mu_3, \mu_3, \mu_3) >_1^{1/2} = 0.0000607381,$$

$$< d^2(\mu_1) >_1^{1/2} = 0.472506, \quad < d^2(\mu_2) >_1^{1/2} = 0.397013,$$

$$< d^2(\mu_3) >_1^{1/2} = 0.151219, \quad < d^2(\mu_1, \mu_1) >_1^{1/2} = 0.212351,$$

$$< d^2(\mu_2, \mu_2) >_1^{1/2} = 0.0779563, \quad < d^2(\mu_3, \mu_3, \mu_3) >_1^{1/2} = 0.00610043,$$

$$< d^2(\mu_1, \mu_1, \mu_2, \mu_2) >_1^{1/2} = 0.0155892,$$

$$< d^2(\mu_1, \mu_1, \mu_3, \mu_3, \mu_3) >_1^{1/2} = 0.00126481,$$

$$< d^2(\mu_2, \mu_2, \mu_3, \mu_3, \mu_3) >_1^{1/2} = 0.000308991,$$

and the following levels of interdependence among subsystems of electrons

$$I(\{\mu_1, \mu_1, \mu_2, \mu_2, \mu_3, \mu_3, \mu_3\} : \{\mu_1\}, \{\mu_1\}, \{\mu_2\}, \{\mu_2\}, \{\mu_3\}, \{\mu_3\}, \{\mu_3\}) =$$

$$= 6.09482 \times 10^{-5},$$

$$I(\{\mu_1, \mu_1, \mu_2, \mu_2, \mu_3, \mu_3, \mu_3\} : \{\mu_1, \mu_1\}, \{\mu_2, \mu_2\}, \{\mu_3, \mu_3, \mu_3\}) = 4.02492 \times 10^{-5},$$

$$I(\{\mu_1, \mu_1, \mu_2, \mu_2\} : \{\mu_1, \mu_1\}, \{\mu_2, \mu_2\}) = 9.64968 \times 10^{-4},$$

$$I(\{\mu_1, \mu_1, \mu_3, \mu_3, \mu_3\} : \{\mu_1, \mu_1\}, \{\mu_3, \mu_3, \mu_3\}) = 3.06232 \times 10^{-5},$$

$$I(\{\mu_2, \mu_2, \mu_3, \mu_3, \mu_3\} : \{\mu_2, \mu_2\}, \{\mu_3, \mu_3, \mu_3\}) = 1.66576 \times 10^{-4},$$

$$I(\{\mu_1, \mu_1\} : \{\mu_1\}, \{\mu_1\}) = 1.09108 \times 10^{-2},$$

$$I(\{\mu_2, \mu_2\} : \{\mu_2\}, \{\mu_2\}) = 7.96628 \times 10^{-2},$$

$$I(\{\mu_3, \mu_3, \mu_3\} : \{\mu_3\}, \{\mu_3\}, \{\mu_3\}) = 2.64249 \times 10^{-3}.$$

Remark 1: In the calm model of the nitrogen atom, if we ignore the subshells and the electrons are allowed to be completely free, the mean ground energy of the atom is minimized for

$$\mu_1 = \mu_2 = \mu_3 = 0.117021 \text{ a.u.}, \quad c_1 = c_2 = c_3 = c_4 = c_5 = c_6 = c_7 = 0.0847102,$$

and its minimum value is -91.5466 a.u. $= -2491.13$ eV, a value much smaller than the experimental value -1486.07 eV. On the other side, if we assume that the atom has three subshells and their mean radial distances from the nucleus were given just by the Bohr's values $k^2/Z = k^2/7$ for $k = 1, 2, 3$, then the minimum mean ground energy would be -1411.5 eV, whose absolute value represents 94.9821% of the absolute value of the experimental value of the mean ground energy. This result and the analysis performed in this section show that taking subshells of the atom into account with weighted Bohr's mean radial distances from the nucleus is appropriate and that the turbulent model of the atom is more accurate than the calm model in describing the structure of the nitrogen atom.

Remark 2: Let us notice that in the turbulent model of the nitrogen atom, if we take $a = b = c = 0$ and use the weighting factor r as being the only variational parameter, then the expression of the mean ground energy $E(r, 0, 0, 0)$ becomes

$$5.49714062463582686866573812 \times$$

$$\times \left(2.979271532759831973463213 5\frac{1}{r^2} - 10.8610444464640366987291 49\frac{1}{r} \right),$$

which becomes minimum for $r = 0.548616$ and equal to

$$-54.4139 \text{ a.u.} = -1480.69 \text{ eV},$$

whose absolute value represents 99.6379% of the absolute value of the experimental value -1486.07 eV of the mean ground energy of the nitrogen atom. This result shows that the nitrogen atom is close to statistical equilibrium.

Remark 3: As the experimental value of the diameter of the nitrogen atom is $D_{exp} = 1.06$, this shows that, both in the calm model of the atom and in the turbulent model, D_1 is the closest approximation of the diameter and the structure of the nitrogen atom is quite stable, being closer to statistical equilibrium and being less influenced by the random fluctuations of its individual electrons or by the random behavior of its external subshell.

h) Argon Atom $(Z = 18)$.

We jump now to a complex atom which has 18 electrons structured in five subshells. There are two electrons in the first, second, and fourth subshells, and six electrons in the third and fifth subshells. The 18 electrons have the quantum numbers (n, ℓ, m, s) given by

$$(1,0,0,-\tfrac{1}{2}), \quad (1,0,0,+\tfrac{1}{2}), \quad (2,0,0,-\tfrac{1}{2}), \quad (2,0,0,+\tfrac{1}{2}), \quad (2,1,-1,-\tfrac{1}{2}),$$

$$(2,1,-1,+\tfrac{1}{2}), \quad (2,1,0,-\tfrac{1}{2}), \quad (2,1,0,+\tfrac{1}{2}), \quad (2,1,1,-\tfrac{1}{2}), \quad (2,1,1,+\tfrac{1}{2}),$$

$$(3,0,0,-\tfrac{1}{2}), \quad (3,0,0,+\tfrac{1}{2}), \quad (3,1,-1,-\tfrac{1}{2}), \quad (3,1,-1,+\tfrac{1}{2}), \quad (3,1,0,-\tfrac{1}{2}),$$

$$(3,1,0,+\tfrac{1}{2}), \quad (3,1,1,-\tfrac{1}{2}), \quad (3,1,1,+\tfrac{1}{2}),$$

respectively. The Hamiltonian is given by (63) with $Z = 18$. Using the weighted Bohr approximation (69), the mean radial distances from the nucleus of the five subshells of the atom are

$$\mu_1 = \frac{r}{18}, \quad \mu_2 = \frac{4r}{18}, \quad \mu_3 = \frac{9r}{18}, \quad \mu_4 = \frac{16r}{18}, \quad \mu_5 = \frac{25r}{18}. \tag{120}$$

There are $18! = 6,402,373,705,728,000$ permutations of 18 electrons, and all four models (calm, nervous, turbulent, and nebulous) of the atom are different. As the computations from this paper have been done using an IBM ThinkPad 380 personal computer with 80MB RAM, and 400MB available on the hard disk, the complexity of the argon atom allows us to deal only with the calm model of the atom which is only a rough approximation of the probabilistic structure of the atom. In such a case, we have

$$\rho_{cal}(1, 2, \ldots, 18) = \psi(1, 2, \ldots, 18),$$

where the probability wave function is

$$\psi = \psi(1, 2, \ldots, 18) =$$

$$= \prod_{i=1}^{2} \varphi(r_i, \mu_1) \prod_{i=3}^{4} \varphi(r_i, \mu_2) \prod_{i=5}^{10} \varphi(r_i, \mu_3) \prod_{i=11}^{12} \varphi(r_i, \mu_4) \prod_{i=13}^{18} \varphi(r_i, \mu_5) \times$$

$$\times \left[1 + \sum_{i=1}^{2} c_i \left(3 - \frac{r_i}{\mu_1} \right) + \sum_{i=3}^{4} c_i \left(3 - \frac{r_i}{\mu_2} \right) + \right.$$

275

$$+ \sum_{i=5}^{10} c_i \left(3 - \frac{r_i}{\mu_3} \right) + \sum_{i=11}^{12} c_i \left(3 - \frac{r_i}{\mu_4} \right) + \sum_{i=13}^{18} c_i \left(3 - \frac{r_i}{\mu_5} \right) \Bigg]. \qquad (121)$$

We have

$$< \Psi_{cal}^* \Psi_{cal} >_3 = \frac{1}{18!} \sum_p < \psi^2(p(1), p(2), \ldots, p(18)) >_1 =$$

$$=< \psi^2(1, 2, \ldots, 18) >_1 = 1 + \sum_{i=1}^{18} 3c_i^2,$$

and for the Hamiltonian we have, similarly,

$$< \Psi_{cal}^* \hat{H} \Psi_{cal} >_3 =$$

$$= \frac{1}{18!} \sum_p < \psi(p(1), p(2), \ldots, p(18)) \, \hat{H} \psi(p(1), p(2), \ldots, p(18)) >_1 =$$

$$=< \psi(1, 2, \ldots, 18) \, \hat{H} \psi(1, 2, \ldots, 18) >_1,$$

where the sums above are taken with respect to all 18! permutations of the set $\{1, 2, \ldots, 18\}$, denoted by $p = (p(1), p(2), \ldots, p(18))$, and the mean energy is

$$E(r, c_1, c_2, \ldots, c_{18}) =$$

$$= \frac{< \Psi_{cal}^* \hat{H} \Psi_{cal} >_3}{< \Psi_{cal}^* \Psi_{cal} >_3} = \frac{< \psi \, \hat{H} \psi >_1}{< \psi^2 >_1}.$$

The mean energy $E(r, c_1, c_2, \ldots, c_{18})$ becomes minimum for

$$r = 0.256714 \text{ a.u.},$$

$$c_1 = c_2 = -0.190955, \quad c_3 = c_4 = 0.065532,$$

$$c_5 = c_6 = c_7 = c_8 = c_9 = c_{10} = 0.0436731, \quad c_{11} = c_{12} = 0.0216786,$$

$$c_{13} = c_{14} = c_{15} = c_{16} = c_{17} = c_{18} = 0.0108941,$$

and its minimum value is -794.318 a.u. $= -21614.7$ eV, much smaller than the experimental value

$$- 529.108 \text{ a.u.} = -14397.8666 \text{ eV}, \qquad (122)$$

which shows that the Hamiltonian, on one side, does not take into account instant, global interdependences among more than two electrons and, on the other side, that the calm model of the atom is only a rough approximation of the probabilistic structure of the atom. This insufficiency may be corrected, however, by allowing the electrons of the argon atom to repel each other and move, on average, farther from the nucleus. But how far can the electrons move from the nucleus? We can use the known experimental value (122) of the ground energy of the argon atom in order to find the right weighting factor r and the values of the coefficients $c_1, c_2, \ldots, c_{18}$ in order to approximate better the probability wave function of the atom and to calculate different correlations between the electrons of this atom.

As far as the spatial coordinates are concerned, the probability wave function is symmetric with respect to the electrons belonging to the same subshell. Let us denote by

$$a = c_1 = c_2, \quad b = c_3 = c_4, \quad c = c_5 = c_6 = c_7 = c_8 = c_9 = c_{10},$$

$$d = c_{11} = c_{12}, \quad e = c_{13} = c_{14} = c_{15} = c_{16} = c_{17} = c_{18},$$

in which case the mean energy becomes a function depending on six variational parameters $E(r, a, b, c, d, e)$. Taking the weighting factor $r = 1$, for which (120) are just Bohr's orbits, the minimum value of the mean energy $E(1, a, b, c, d, e)$ is -452.585 a.u. $= -12315.6$ eV, whose absolute value represents 85.5374% of the absolute value of the experimental value (122). Therefore the weighting factor r has to be somewhere between $0.256714 < r < 1$. The search for the right value of r is not difficult at all, and for $r = 0.779$ a.u. the mean energy $E(0.779, a, b, c, d, e)$ is minimum for

$$c_1 = c_2 = a = 0.168735, \quad c_3 = c_4 = b = 0.223709,$$

$$c_5 = c_6 = c_7 = c_8 = c_9 = c_{10} = c = 0.096383, \quad c_{11} = c_{12} = d = 0.0432255,$$

$$c_{13} = c_{14} = c_{15} = c_{16} = c_{17} = c_{18} = e = 0.0208763, \tag{123}$$

and its minimum value is

$$-529.022 \text{ a.u.} = -14395.5 \text{ eV},$$

whose absolute value represents 99.9838% of the absolute value of the experimental value (122). For the weighting factor $r = 0.779$, the mean orbits of the five subshells of the argon atom are (in a.u.)

$$\mu_1 = 0.0432778, \quad \mu_2 = 0.173111, \quad \mu_3 = 0.3895,$$

$$\mu_4 = 0.69244, \quad \mu_5 = 1.08194. \tag{124}$$

The weighted mean orbit of the argon atom is

$$\bar{\mu} = \frac{1}{18}(2\mu_1 + 2\mu_2 + 6\mu_3 + 2\mu_4 + 6\mu_5) = 0.591461 \text{ a.u.}$$

Therefore, the mean diameters of the atom induced by the statistical equilibrium and by the external subshell are

$$D_1 = 2\bar{\mu} = 1.182922 \text{ a.u.}, \quad D_2 = 2\mu_5 = 2.16388 \text{ a.u.}$$

Using the values (123), the generalized correlations (65) become

$$\text{gencorr}_{cal}(\mu_1, \mu_1) = 0.129016, \quad \text{gencorr}_{cal}(\mu_1, \mu_2) = 0.168585,$$
$$\text{gencorr}_{cal}(\mu_1, \mu_3) = 0.072100, \quad \text{gencorr}_{cal}(\mu_1, \mu_4) = 0.030981,$$
$$\text{gencorr}_{cal}(\mu_1, \mu_5) = 0.014617, \quad \text{gencorr}_{cal}(\mu_2, \mu_2) = 0.220290,$$
$$\text{gencorr}_{cal}(\mu_2, \mu_3) = 0.094213, \quad \text{gencorr}_{cal}(\mu_2, \mu_4) = 0.040483,$$
$$\text{gencorr}_{cal}(\mu_2, \mu_5) = 0.019099, \quad \text{gencorr}_{cal}(\mu_3, \mu_3) = 0.040293,$$
$$\text{gencorr}_{cal}(\mu_3, \mu_4) = 0.017314, \quad \text{gencorr}_{cal}(\mu_3, \mu_5) = 0.008168,$$
$$\text{gencorr}_{cal}(\mu_4, \mu_4) = 0.007440, \quad \text{gencorr}_{cal}(\mu_4, \mu_5) = 0.003510,$$

$$\text{gencorr}_{cal}(\mu_5, \mu_5) = 0.001656,$$

showing that the interdependence inside and between the most interior subshells of the atom is stronger whereas the external subshells have more intra and internal independence.

Replacing the values (123) and (124) into (121), we get

$$< \psi^2 >_1 = 1.65737$$

and, using (74)-(76), we get the standard correlations

$$\text{stcorr}_{cal}(\mu_1, \mu_1) = 0.481939, \quad \text{stcorr}_{cal}(\mu_1, \mu_2) = 0.460319,$$
$$\text{stcorr}_{cal}(\mu_1, \mu_3) = 0.418272, \quad \text{stcorr}_{cal}(\mu_1, \mu_4) = 0.344646,$$
$$\text{stcorr}_{cal}(\mu_1, \mu_5) = 0.221243, \quad \text{stcorr}_{cal}(\mu_2, \mu_2) = 0.439566,$$
$$\text{stcorr}_{cal}(\mu_2, \mu_3) = 0.399218, \quad \text{stcorr}_{cal}(\mu_2, \mu_4) = 0.328593,$$
$$\text{stcorr}_{cal}(\mu_2, \mu_5) = 0.210282, \quad \text{stcorr}_{cal}(\mu_3, \mu_3) = 0.362192,$$
$$\text{stcorr}_{cal}(\mu_3, \mu_4) = 0.297434, \quad \text{stcorr}_{cal}(\mu_3, \mu_5) = 0.189068,$$
$$\text{stcorr}_{cal}(\mu_4, \mu_4) = 0.243030, \quad \text{stcorr}_{cal}(\mu_4, \mu_5) = 0.152195,$$

$$\mathrm{stcorr}_{cal}(\mu_5, \mu_5) = 0.091009.$$

Replacing the values (123) and (124) into (121), and using $< \psi^2 >_1 = 1.65737$, we get the mean radial distance from the nucleus of an arbitrary electron during its random motion around the nucleus:

$$\bar{r} = \frac{1}{18} \frac{< \psi \left(\sum_{i=1}^{18} r_i \right) \psi >_1}{< \psi^2 >_1} = 1.68071 \text{ a.u.,}$$

and the mean diameter of the atom induced by the individual random fluctuations of its electrons is $D_3 = 2\bar{r} = 3.36142$ a.u. The mean index of instability of the atom is $\bar{r} - \bar{\mu} = 1.089249$.

In order to calculate the global correlation between two arbitrary electrons of the argon atom we introduce again (123) and (124) into (121), and using (70)-(73) we get

$$\mathrm{globcorr} = -0.0288284.$$

Replacing the values (123) and (124) into (121), let

$$d(\mu_1, \mu_1, \mu_2, \mu_2, \mu_3, \mu_3, \mu_3, \mu_3, \mu_3, \mu_3, \mu_4, \mu_4, \mu_5, \mu_5, \mu_5, \mu_5, \mu_5, \mu_5) =$$

$$= \frac{\psi^2}{< \psi^2 >_1},$$

be the joint probability density of the system of 18 electrons and

$$d(\mu_i), \quad d(\mu_i, \mu_j), \quad d(\mu_i, \mu_j, \mu_k), \quad d(\mu_i, \mu_j, \mu_k, \mu_\ell), \quad \text{etc.,}$$

be its marginal densities, describing the behavior of different subsystems of electrons from the same shell or from different subshells. We obtain the following degrees of stability for subsystems of electrons inside the argon atom in the calm model of the atom:

$$< d^2(\mu_1, \mu_1, \mu_2, \mu_2, \mu_3, \mu_3, \mu_3, \mu_3, \mu_3, \mu_3, \mu_4, \mu_4, \mu_5, \mu_5, \mu_5, \mu_5, \mu_5, \mu_5) >_1^{1/2} =$$

$$= 0.000000000000223256,$$

$$< d^2(\mu_1) >_1^{1/2} = 0.646099, \quad < d^2(\mu_2) >_1^{1/2} = 0.332404,$$

$$< d^2(\mu_3) >_1^{1/2} = 0.207041, \quad < d^2(\mu_4) >_1^{1/2} = 0.150617,$$

$$< d^2(\mu_5) >_1^{1/2} = 0.118913,$$

279

$$< d^2(\mu_1, \mu_1) >_1^{1/2} = 0.418449, \quad < d^2(\mu_2, \mu_2) >_1^{1/2} = 0.111202,$$

$$< d^2(\mu_3, \mu_3, \mu_3, \mu_3, \mu_3, \mu_3) >_1^{1/2} = 0.0000786384,$$

$$< d^2(\mu_4, \mu_4) >_1^{1/2} = 0.0226842,$$

$$< d^2(\mu_5, \mu_5, \mu_5, \mu_5, \mu_5, \mu_5) >_1^{1/2} = 0.00000282638,$$

$$< d^2(\mu_1, \mu_1, \mu_2, \mu_2) >_1^{1/2} = 0.0466802,$$

$$< d^2(\mu_1, \mu_1, \mu_3, \mu_3, \mu_3, \mu_3, \mu_3, \mu_3) >_1^{1/2} = 0.0000326955,$$

$$< d^2(\mu_1, \mu_1, \mu_4, \mu_4) >_1^{1/2} = 0.00948914,$$

$$< d^2(\mu_1, \mu_1, \mu_5, \mu_5, \mu_5, \mu_5, \mu_5, \mu_5) >_1^{1/2} = 0.00000118137,$$

$$< d^2(\mu_2, \mu_2, \mu_3, \mu_3, \mu_3, \mu_3, \mu_3, \mu_3) >_1^{1/2} = 0.00000867568,$$

$$< d^2(\mu_2, \mu_2, \mu_4, \mu_4) >_1^{1/2} = 0.00252155,$$

$$< d^2(\mu_2, \mu_2, \mu_5, \mu_5, \mu_5, \mu_5, \mu_5, \mu_5) >_1^{1/2} = 0.000000313845,$$

$$< d^2(\mu_3, \mu_3, \mu_3, \mu_3, \mu_3, \mu_3, \mu_4, \mu_4) >_1^{1/2} = 0.00000177904,$$

$$< d^2(\mu_3, \mu_3, \mu_3, \mu_3, \mu_3, \mu_3, \mu_5, \mu_5, \mu_5, \mu_5, \mu_5, \mu_5) >_1^{1/2} =$$
$$= 0.000000000221233,$$

$$< d^2(\mu_4, \mu_4, \mu_5, \mu_5, \mu_5, \mu_5, \mu_5, \mu_5) >_1^{1/2} = 0.0000000640813,$$

and the following levels of interdependence among subsystems of electrons

$$I(\{\mu_1, \mu_1, \mu_2, \mu_2, \mu_3, \mu_3, \mu_3, \mu_3, \mu_3, \mu_3, \mu_4, \mu_4, \mu_5, \mu_5, \mu_5, \mu_5, \mu_5, \mu_5\} :$$

$$\{\mu_1\}, \{\mu_1\}, \{\mu_2\}, \{\mu_2\}, \{\mu_3\}, \{\mu_3\}, \{\mu_3\}, \{\mu_3\}, \{\mu_3\}, \{\mu_3\}, \{\mu_4\}, \{\mu_4\},$$
$$\{\mu_5\}, \{\mu_5\}, \{\mu_5\}, \{\mu_5\}, \{\mu_5\}, \{\mu_5\}) = 9.76358 \times 10^{-15},$$

$$I(\{\mu_1, \mu_1, \mu_2, \mu_2, \mu_3, \mu_3, \mu_3, \mu_3, \mu_3, \mu_3, \mu_4, \mu_4, \mu_5, \mu_5, \mu_5, \mu_5, \mu_5, \mu_5\} :$$

$$\{\mu_1, \mu_1\}, \{\mu_2, \mu_2\}, \{\mu_3, \mu_3, \mu_3, \mu_3, \mu_3, \mu_3\}, \{\mu_4, \mu_4\}, \{\mu_5, \mu_5, \mu_5, \mu_5, \mu_5, \mu_5\}) =$$
$$= 1.13524 \times 10^{-14},$$

$$I(\{\mu_1, \mu_1, \mu_2, \mu_2\} : \{\mu_1, \mu_1\}, \{\mu_2, \mu_2\}) = 1.48040 \times 10^{-4},$$

$$I(\{\mu_1, \mu_1, \mu_3, \mu_3, \mu_3, \mu_3, \mu_3, \mu_3\} : \{\mu_1, \mu_1\}, \{\mu_3, \mu_3, \mu_3, \mu_3, \mu_3, \mu_3\}) =$$

$$= 2.10618 \times 10^{-7},$$

$$I(\{\mu_1, \mu_1, \mu_4, \mu_4\} : \{\mu_1, \mu_1\}, \{\mu_4, \mu_4\}) = 3.04072 \times 10^{-6},$$

$$I(\{\mu_1, \mu_1, \mu_5, \mu_5, \mu_5, \mu_5, \mu_5, \mu_5\} : \{\mu_1, \mu_1\}, \{\mu_5, \mu_5, \mu_5, \mu_5, \mu_5, \mu_5\}) =$$
$$= 1.32183 \times 10^{-9},$$

$$I(\{\mu_2, \mu_2, \mu_3, \mu_3, \mu_3, \mu_3, \mu_3, \mu_3\} : \{\mu_2, \mu_2\}, \{\mu_3, \mu_3, \mu_3, \mu_3, \mu_3, \mu_3\}) =$$
$$= 6.90358 \times 10^{-8},$$

$$I(\{\mu_2, \mu_2, \mu_4, \mu_4\} : \{\mu_2, \mu_2\}, \{\mu_4, \mu_4\}) = 9.73548 \times 10^{-7},$$

$$I(\{\mu_2, \mu_2, \mu_5, \mu_5, \mu_5, \mu_5, \mu_5, \mu_5\} : \{\mu_2, \mu_2\}, \{\mu_5, \mu_5, \mu_5, \mu_5, \mu_5\}) =$$
$$= 4.52981 \times 10^{-10},$$

$$I(\{\mu_3, \mu_3, \mu_3, \mu_3, \mu_3, \mu_3, \mu_4, \mu_4\} :$$
$$\{\mu_3, \mu_3, \mu_3, \mu_3, \mu_3, \mu_3\}, \{\mu_4, \mu_4\}) = 4.81268 \times 10^{-9},$$

$$I(\{\mu_3, \mu_3, \mu_3, \mu_3, \mu_3, \mu_3, \mu_5, \mu_5, \mu_5, \mu_5, \mu_5, \mu_5\} :$$
$$\{\mu_3, \mu_3, \mu_3, \mu_3, \mu_3, \mu_3\}, \{\mu_5, \mu_5, \mu_5, \mu_5, \mu_5, \mu_5\}) = 1.02931 \times 10^{-12},$$

$$I(\{\mu_4, \mu_4, \mu_5, \mu_5, \mu_5, \mu_5, \mu_5, \mu_5\} : \{\mu_4, \mu_4\}, \{\mu_5, \mu_5, \mu_5, \mu_5, \mu_5, \mu_5\}) =$$
$$= 3.29921 \times 10^{-11},$$

$$I(\{\mu_1, \mu_1\} : \{\mu_1\}, \{\mu_1\}) = 1.00409 \times 10^{-3},$$

$$I(\{\mu_2, \mu_2\} : \{\mu_2\}, \{\mu_2\}) = 7.09545 \times 10^{-4},$$

$$I(\{\mu_3, \mu_3, \mu_3, \mu_3, \mu_3, \mu_3\} : \{\mu_3\}, \{\mu_3\}, \{\mu_3\}, \{\mu_3\}, \{\mu_3\}, \{\mu_3\}) =$$
$$= 1.2801 \times 10^{-7},$$

$$I(\{\mu_4, \mu_4\} : \{\mu_4\}, \{\mu_4\}) = 1.14869 \times 10^{-6},$$

$$I(\{\mu_5, \mu_5, \mu_5, \mu_5, \mu_5, \mu_5\} : \{\mu_5\}, \{\mu_5\}, \{\mu_5\}, \{\mu_5\}, \{\mu_5\}) =$$
$$= 9.36049 \times 10^{-10}.$$

Remark: As the experimental value of the diameter of the argon atom is $D_{exp} = 3.82$, this shows that in the calm model of the atom D_3 is the closest approximation of the diameter and the structure of the argon atom depends more on the random fluctuations of its electrons than on a regular evolution along orbits of statistical equilibrium or on the exclusive behavior of its external subshell.

6. COMPUTER PROGRAMS.

a) Generalities.

From computational point of view, this monograph contains the numerical results obtained by using the excellent symbolic mathematics software package MATHEMATICA, created by Wolfram (1991), and the fast subroutine 'integExp' from Feagin's (1994) package "Quantum" which may be attached to MATHEMATICA. In order to save space, the open computer programs are given for the nebulous model of the beryllium atom only but they may be easily adapted to other models and atoms. The corresponding programs were run on a personal computer IBM ThinkPad 380 with Intel Pentium processor, 80MB RAM and 400MB available on the hard disk. The examples and numerical results from this paper refer to hydrogen, helium, lithium, beryllium, boron, carbon, nitrogen, and argon atoms. Similar applications to more complex atoms require more powerful computers.

Each command has a number given by the computer. We can refer to the output of such a command, let say command numbered [54], by typing %54, and we can continue performing operations on this output. The symbol % alone represents the output of the previous command. We can call back a previous command by using the back arrow key ↑ and run it again or change it correspondingly. This saves us a lot of time when repetitive identical or similar loops are involved. If a command ends with a semicolon ; then the computer's answer is not displayed on the screen, saving time. This is very useful when we deal with commands that refer to intermediary steps, whose partial results are very cumbersome, in a long computation. Finally, MATHEMATICA is case sensitive. Its commands start with a capital letter, except the special command "integExp", intensively used in the programs from this monograph, which is a very fast subroutine for calculating exponential integrals of the form

$$\int_0^{+\infty} r^n e^{-a\,r} dr.$$

b) Minimizing the ground state mean energy of the atom.

In what follows: 'phi' is φ, 'psi' is ψ, 'ppsi' is ρ, 'm' is μ, 'invri' is the integral $< \psi(1/r_i)\rho >_1$, 'invrij' is $< \psi(1/r_{ij})\rho >_1$, 'laplacei' is the integral of the laplacian $< \psi \nabla_i^2 \rho >_1$, 'norm' is $< \psi \rho >_1$, and 'energy' is the mean value of the energy $< \psi \hat{H} \rho >_1 / < \psi \rho >_1$. After each command press the 'Enter' key.

math
[1] Needs["Quantum'integExp'"]
[2] phi[r_,m_]:=((8*m^3*Pi)^(−1/2))*Exp[−r/(2*m)];
[3] psi[r1_,r2_,r3_,r4_]: =phi[r1,r/4]*phi[r2,r/4]*phi[r3,r] *phi[r4,r]*
    (1+a*(3 − 4*r1/r)+a*(3 − 4*r2/r) +b*(3−r3/r)+b*(3−r4/r));
[4] ppsi[r1_,r2_,r3_,r4_]:=psi[r1,r2,r3,r4]−psi[r1,r4,r3,r2]−psi[r3,r2,r1,r4]+
    psi[r3,r4,r1,r2];
[5] integExp[4*Pi*r1^2*ppsi[r1,r2,r3,r4]* psi[r1,r2,r3,r4],{r1,0,Infinity}];
[6] integExp[4*Pi*r2^2*%,{r2,0,Infinity}];
[7] integExp[4*Pi*r3^2*%,{r3,0,Infinity}];
[8] integExp[4*Pi*r4^2*%,{r4,0,Infinity}];
[9] norm=Together[%]
[10] integExp[4*Pi*r1*ppsi[r1,r2,r3,r4]* psi[r1,r2,r3,r4],{r1,0,Infinity}];
[11] integExp[4*Pi*r2^2*%,{r2,0,Infinity}];
[12] integExp[4*Pi*r3^2*%,{r3,0,Infinity}];
[13] integExp[4*Pi*r4^2*%,{r4,0,Infinity}];
[14] invr1=Together[%]
[15] integExp[4*Pi*r2*ppsi[r1,r2,r3,r4]* psi[r1,r2,r3,r4],{r2,0,Infinity}];
[16] integExp[4*Pi*r1^2*%,{r1,0,Infinity}];
[17] integExp[4*Pi*r3^2*%,{r3,0,Infinity}];
[18] integExp[4*Pi*r4^2*%,{r4,0,Infinity}];
[19] invr2=Together[%]
[20] integExp[4*Pi*r3*ppsi[r1,r2,r3,r4]* psi[r1,r2,r3,r4],{r3,0,Infinity}];
[21] integExp[4*Pi*r1^2*%,{r1,0,Infinity}];
[22] integExp[4*Pi*r2^2*%,{r2,0,Infinity}];
[23] integExp[4*Pi*r4^2*%,{r4,0,Infinity}];
[24] invr3=Together[%]
[25] integExp[4*Pi*r4*ppsi[r1,r2,r3,r4]* psi[r1,r2,r3,r4],{r4,0,Infinity}];
[26] integExp[4*Pi*r1^2*%,{r1,0,Infinity}];
[27] integExp[4*Pi*r2^2*%,{r2,0,Infinity}];
[28] integExp[4*Pi*r3^2*%,{r3,0,Infinity}];
[29] invr4=Together[%]
[30] laplace[r1_,r2_,r3_,r4_]:=D[psi[r1,r2,r3,r4],{r1,2 }]+
    2*D[psi[r1,r2,r3,r4],r1]/r1;

283

```
[31] Expand[4*Pi*r1^2*ppsi[r1,r2,r3,r4]* laplace[r1,r2,r3,r4]];
[32] integExp[%,{r1,0,Infinity}];
[33] integExp[4*Pi*r2^2*%,{r2,0,Infinity}];
[34] integExp[4*Pi*r3^2*%,{r3,0,Infinity}];
[35] integExp[4*Pi*r4^2*%,{r4,0,Infinity}];
[36] laplace1=Together[%]
[37] laplace[r1_,r2_,r3_,r4_]:=D[psi[r1,r2,r3,r4],{r2,2 }]+
     2*D[psi[r1,r2,r3,r4],r2]/r2;
[38] Expand[4*Pi*r2^2*ppsi[r1,r2,r3,r4]* laplace[r1,r2,r3,r4]];
[39] integExp[%,{r2,0,Infinity}];
[40] integExp[4*Pi*r1^2*%,{r1,0,Infinity}];
[41] integExp[4*Pi*r3^2*%,{r3,0,Infinity}];
[42] integExp[4*Pi*r4^2*%,{r4,0,Infinity}];
[43] laplace2=Together[%]
[44] laplace[r1_,r2_,r3_,r4_]:=D[psi[r1,r2,r3,r4],{r3,2 }]+
     2*D[psi[r1,r2,r3,r4],r3]/r3;
[45] Expand[4*Pi*r3^2*ppsi[r1,r2,r3,r4]* laplace[r1,r2,r3,r4]];
[46] integExp[%,{r3,0,Infinity}];
[47] integExp[4*Pi*r1^2*%,{r1,0,Infinity}];
[48] integExp[4*Pi*r2^2*%,{r2,0,Infinity}];
[49] integExp[4*Pi*r4^2*%,{r4,0,Infinity}];
[50] laplace3=Together[%]
[51] laplace[r1_,r2_,r3_,r4_]:=D[psi[r1,r2,r3,r4],{r4,2 }]+
     2*D[psi[r1,r2,r3,r4],r4]/r4;
[52] Expand[4*Pi*r4^2*ppsi[r1,r2,r3,r4]* laplace[r1,r2,r3,r4]];
[53] integExp[%,{r4,0,Infinity}];
[54] integExp[4*Pi*r1^2*%,{r1,0,Infinity}];
[55] integExp[4*Pi*r2^2*%,{r2,0,Infinity}];
[56] integExp[4*Pi*r3^2*%,{r3,0,Infinity}];
[57] laplace4=Together[%]
[58] integExp[4*Pi*r4^2*ppsi[r1,r2,r3,r4]* psi[r1,r2,r3,r4],{r4,0,Infinity}];
[59] integExp[4*Pi*r3^2*%,{r3,0,Infinity}];
[60] integExp[4*Pi*r2*%,{r2,0,Infinity}];
[61] integExp[4*Pi*r1^2*%,{r1,0,Infinity}];
[62] Integrate[4*Pi*r2*(r2-r1) * %%%, {r2,0,r1}];
[63] integExp[4*Pi*r1*%,{r1,0,Infinity}];
[64] % + %%%;
```

[65] invr12=Together[%]
[66] integExp[4⋆Pi⋆r4^2⋆ppsi[r1,r2,r3,r4]⋆ psi[r1,r2,r3,r4],{r4,0,Infinity}];
[67] integExp[4⋆Pi⋆r2^2⋆%,{r2,0,Infinity}];
[68] integExp[4⋆Pi⋆r3⋆%,{r3,0,Infinity}];
[69] integExp[4⋆Pi⋆r1^2⋆%,{r1,0,Infinity}];
[70] Integrate[4⋆Pi⋆r3⋆(r3−r1) ⋆ %%%, {r3,0,r1}];
[71] integExp[4⋆Pi⋆r1⋆%,{r1,0,Infinity}];
[72] % + %%%;
[73] invr13=Together[%]
[74] integExp[4⋆Pi⋆r3^2⋆ppsi[r1,r2,r3,r4]⋆ psi[r1,r2,r3,r4],{r3,0,Infinity}];
[75] integExp[4⋆Pi⋆r2^2⋆%,{r2,0,Infinity}];
[76] integExp[4⋆Pi⋆r4⋆%,{r4,0,Infinity}];
[77] integExp[4⋆Pi⋆r1^2⋆%,{r1,0,Infinity}];
[78] Integrate[4⋆Pi⋆r4⋆(r4−r1) ⋆ %%%, {r4,0,r1}];
[79] integExp[4⋆Pi⋆r1⋆%,{r1,0,Infinity}];
[80] % + %%%;
[81] invr14=Together[%]
[82] integExp[4⋆Pi⋆r4^2⋆ppsi[r1,r2,r3,r4]⋆ psi[r1,r2,r3,r4],{r4,0,Infinity}];
[83] integExp[4⋆Pi⋆r1^2⋆%,{r1,0,Infinity}];
[84] integExp[4⋆Pi⋆r3⋆%,{r3,0,Infinity}];
[85] integExp[4⋆Pi⋆r2^2⋆%,{r2,0,Infinity}];
[86] Integrate[4⋆Pi⋆r3⋆(r3−r2) ⋆ %%%, {r3,0,r2}];
[87] integExp[4⋆Pi⋆r2⋆%,{r2,0,Infinity}];
[88] % + %%%;
[89] invr23=Together[%]
[90] integExp[4⋆Pi⋆r3^2⋆ppsi[r1,r2,r3,r4]⋆ psi[r1,r2,r3,r4],{r3,0,Infinity}];
[91] integExp[4⋆Pi⋆r1^2⋆%,{r1,0,Infinity}];
[92] integExp[4⋆Pi⋆r4⋆%,{r4,0,Infinity}];
[93] integExp[4⋆Pi⋆r2^2⋆%,{r2,0,Infinity}];
[94] Integrate[4⋆Pi⋆r4⋆(r4−r2) ⋆ %%%, {r4,0,r2}];
[95] integExp[4⋆Pi⋆r2⋆%,{r2,0,Infinity}];
[96] % + %%%;
[97] invr24=Together[%]
[98] integExp[4⋆Pi⋆r2^2⋆ppsi[r1,r2,r3,r4]⋆ psi[r1,r2,r3,r4],{r2,0,Infinity}];
[99] integExp[4⋆Pi⋆r1^2⋆%,{r1,0,Infinity}];
[100] integExp[4⋆Pi⋆r4⋆%,{r4,0,Infinity}];
[101] integExp[4⋆Pi⋆r3^2⋆%,{r3,0,Infinity}];

[102] Integrate[4*Pi*r4*(r4−r3) * %%%, {r4,0,r3}];
[103] integExp[4*Pi*r3*%,{r3,0,Infinity}];
[104] % + %%%;
[105] invr34=Together[%]
[106] energy=(−(laplace1+laplace2+laplace3+laplace4)/2−
    4*(invr1+invr2+invr3+invr4) + (invr12+invr13+invr14+
    invr23+invr24+invr34))/norm;
[107] FindMinimum[energy,{r,1},{a,0},{b,0}]
[108] Save["results",phi,psi,ppsi,energy]
[109] Quit

Remark 1: The output of command [107] is the minimum value of the ground state mean energy of the atom, in atomic units, and the values of the variational parameters r, a, b.

Remark 2: For the calm model of the beryllium atom, it is enough to omit command [4] and replace 'ppsi' by 'psi' throughout the above computer program.

Remark 3: After command [108], the computer stores in the file called "results" the functions mentioned between the square brackets. If during another MATHEMATICA session we want to call this file for further computations, we type the command: <<results;

Remark 4: In the above program, command [1] calls the subroutine 'integExp'. Commands [2]–[4] define the functions φ, ψ, and ρ_{neb}, respectively. Commands [5]–[9] calculate $< \psi \rho_{neb} >_1$, which is labeled 'norm'. Commands [10]–[29] calculate the integrals

$$< \psi \frac{1}{r_i} \rho_{neb} >_1,$$

labeled 'invri'. Commands [30]–[57] calculate the integrals

$$< \psi \nabla_i^2 \rho_{neb} >_1,$$

labeled 'laplacei'. Commands [58]–[105] calculate the integrals

$$< \psi \frac{1}{r_{ij}} \rho_{neb} >_1,$$

labeled 'invrij'. Command [106] evaluates the expression of the ground state mean energy of the atom, which is minimized by command [107]. The subroutine 'FindMinimum' is a powerful tool in MATHEMATICA which locates the minimum of a function of several variables when a starting search point is given. Making a good choice for this starting search point is a difficult task in general but, fortunately, not in our context. In minimizing the mean energy of the atom it is very reasonable to start our search for a minimum by taking the mean radius equal to 1 a.u. and all variational parameters equal to zero, an assumption which is equivalent to assuming that the electrons of the atom are independent and in statistical equilibrium. Looking subsequently for a minimum of the mean energy of the atom, we find the corresponding values of the variational parameters which characterize the ground state of the respective atom.

c) Adjusting the weight of the mean orbits of the subshells.

For the calm model of the beryllium atom, running the above program with 'ppsi' replaced by 'psi', we obtain that the values of the variational parameters are:

$$r = 0.45934 \text{ a.u.}, \quad a = -0.0809111, \quad b = 0.226936,$$

and the minimum value of the ground state mean energy of the atom is

$$\text{energy } = -16.9392 \text{ a.u.} = -460.943 \text{ eV},$$

which is much smaller than the experimental value -399.149 eV. On the other side, taking the weight $r = 1$, which is equivalent to assuming that the mean orbits of the two subshells of the atom are given by Bohr's numbers $1/4$ and 1, respectively, which amounts to the use of the following MATHE-MATICA commands:

```
math
<<results;
energia=energy /. r-> 1;
FindMinimum[energia,{a,0},{b,0}];
```

we obtain that the value of the ground state mean energy of the atom is energia$= -14.133$ a.u. $= -384.582$ eV, whose absolute value represents 96.3504% of the absolute value of the experimental value. Therefore, the weight r which correspond to the experimental value is located in the interval $0.45934 < r < 1$. Repeating the above two commands with different values for r we easily locate the right value and for:

energia=energy /. r− > 0.917;
FindMinimum[energia,{a,0},{b,0}];

we obtain that the value of the ground state mean energy of the atom is energia$= -14.6657$ a.u. $= -399.077$ eV, whose absolute value represents 99.982% of the absolute value of the experimental value, and the corresponding values of the variational parameters are $a = 0.220935$ and $b = 0.145214$.

d) The computation of generalized correlations.

In order to be specific, we take into account the nebulous model of the beryllium atom and we want to calculate the generalized correlation as given by formula (65) with:

$$Z = 4, \quad a = c_1 = c_2 = 0.0174535, \quad b = c_3 = c_4 = -0.369507,$$

the last two values being obtained by minimizing the mean energy of the atom. Using MATHEMATICA, the corresponding commands are:

math
[1] a=0.0174512
[2] b=−0.369517
[3] gencorr[x_,y_]:=6⋆x⋆y/(((1 − 4⋆x+12⋆x^2+6⋆(a^2+b^2))⋆
    (1−4⋆y+ 12⋆y^2+6⋆(a^2+b^2))) ^(1/2))
[4] gencorr[x,y] /. {x− >a, y− >a}
[5] gencorr[x,y] /. {x− >a, y− >b}
[6] gencorr[x,y] /. {x− >b, y− >b}
[7] Quit

The outcomes of [4], [5], [6] are:

$$\text{gencorr}_{neb}(\mu_1, \mu_1) = 0.00104152, \quad \text{gencorr}_{neb}(\mu_1, \mu_2) = -0.0131456,$$

$$\text{gencorr}_{neb}(\mu_2, \mu_2) = 0.165917,$$

respectively.

e) The computation of the global correlation.

In order to be specific, we take into account the nebulous model of the beryllium atom and we want to calculate the global correlation between two arbitrary electrons of the atom as given by formulas (70)-(73) with:

$$Z = 4, \quad r = 0.558519 \text{ a.u.}, \quad a = 0.0174535, \quad b = -0.369507,$$

the last three values being obtained by minimizing the mean energy of the atom. Assume that the functions φ, ψ, and ρ_{neb}, called phi, psi, and ppsi in our MATHEMATICA programs, have been saved in the file "results".

```
math
[1] Needs["Quantum'integExp'"]
[2] <<results;
[3] density=psi[r1,r2,r3,r4]*ppsi[r1,r2,r3,r4] /.
    {r- >0.558519,a- >0.0174535,b- > -0.369517};
[4] integExp[4*Pi*r1^2*density, {r1,0,Infinity}];
[5] integExp[4*Pi*r2^2*%, {r2,0,Infinity}];
[6] integExp[4*Pi*r3^2*%, {r3,0,Infinity}];
[7] integExp[4*Pi*r4^2*%, {r4,0,Infinity}]
[8] integExp[4*Pi*r1^2* (r1+r2+r3+r4)*density, {r1,0,Infinity}];
[9] integExp[4*Pi*r2^2*%, {r2,0,Infinity}];
[10] integExp[4*Pi*r3^2*%, {r3,0,Infinity}];
[11] integExp[4*Pi*r4^2*%, {r4,0,Infinity}];
[12] mean= %/(4*%7)
[13] integExp[4*Pi*r1^2* ((r1-mean)^2+(r2-mean)^2+
     (r3-mean)^2+(r4-mean)^2)*density, {r1,0,Infinity}];
[14] integExp[4*Pi*r2^2*%, {r2,0,Infinity}];
[15] integExp[4*Pi*r3^2*%, {r3,0,Infinity}];
[16] integExp[4*Pi*r4^2*%, {r4,0,Infinity}];
[17] var= %/(4*%7);
[18] stdev=Sqrt[%]
[19] integExp[4*Pi*r1^2* ((r1-mean)*(r2-mean)+
     (r1-mean)*(r3-mean)+(r1-mean)*(r4-mean)+
     (r2-mean)*(r3-mean)+(r2-mean)*(r4-mean)+
     (r3-mean)*(r4-mean))*density, {r1,0,Infinity}];
[20] integExp[4*Pi*r2^2*%, {r2,0,Infinity}];
```

[21] integExp[4⋆Pi⋆r3^2⋆%, {r3,0,Infinity}];
[22] integExp[4⋆Pi⋆r4^2⋆%, {r4,0,Infinity}];
[23] cov= 2 ⋆ %/(12⋆%7)
[24] globcorr=cov/var

Remark 1: The output of [7] is $< \psi \rho_{neb} >_1 = 1.57776$.

Remark 2: The output of [12] is the mean radial distance of an arbitrary electron during its random motion around nucleus, namely $\bar{r} = 1.53467$ a.u., where

$$\bar{r} = \frac{1}{4} \frac{< \psi \left(r_1 + r_2 + r_3 + r_4 \right) \rho_{neb} >_1}{< \psi \rho_{neb} >_1}.$$

Remark 3: The output of [18] is the standard deviation of the radial distance of an arbitrary electron of the atom during its random motion around the nucleus, namely stdev=1.43382, where stdev $= \sqrt{\text{var}}$ and

$$\text{var} = \frac{1}{4} \frac{< \psi \left[\sum_{i=1}^4 (r_i - \bar{r})^2 \right] \rho_{neb} >_1}{< \psi \rho_{neb} >_1}.$$

Remark 4: The output of [23] is the covariance between two arbitrary electrons of the atom with respect to their radial distances from the nucleus during their random motion around nucleus, namely cov $= -0.47151$, where

$$\text{cov} = \frac{2}{12}(< \psi \left[(r_1 - \bar{r})(r_2 - \bar{r}) + (r_1 - \bar{r})(r_3 - \bar{r}) + (r_1 - \bar{r})(r_4 - \bar{r}) + \right.$$

$$\left. +(r_2 - \bar{r})(r_3 - \bar{r}) + (r_2 - \bar{r})(r_4 - \bar{r}) + (r_3 - \bar{r})(r_4 - \bar{r}) \right] \rho_{neb} >_1)/ < \psi \rho_{neb} >_1.$$

Remark 5: The output of [24] is the global correlation between two arbitrary electrons of the atom and is equal to -0.229352.

f) The computation of standard correlations.

In order to be specific, we take into account the nebulous model of the beryllium atom and we want to calculate the standard correlation between two arbitrary electrons of the atom as given by formulas (74)-(76) with:

$$Z = 4, \quad r = 0.558519 \text{ a.u.}, \quad a = 0.0174535, \quad b = -0.369507,$$

the last three values being obtained by minimizing the mean energy of the atom. Assume that the functions φ, ψ, and ρ_{neb}, called phi, psi, and ppsi in our MATHEMATICA programs, have been saved in the file "results".

```
math
[1] Needs["Quantum'integExp'"]
[2] <<results;
[3] density=psi[r1,r2,r3,r4]*ppsi[r1,r2,r3,r4] /.
    {r- >0.558519,a- >0.0174535,b- > -0.369507};
[4] p=(r1-m)^2+(r2-m)^2+(r3-m)^2+ (r4-m)^2;
[5] Expand[p*density];
[6] integExp[4*Pi*r1^2*%, {r1,0,Infinity}];
[7] integExp[4*Pi*r2^2*%, {r2,0,Infinity}];
[8] integExp[4*Pi*r3^2*%, {r3,0,Infinity}];
[9] integExp[4*Pi*r4^2*%, {r4,0,Infinity}];
[10] w= %/4;
[11] q=(r1-m)*((r2-n)+(r3-n)+(r4-n))+
    (r2-m)*((r1-n)+(r3-n)+(r4-n))+
    (r3-m)*((r1-n)+(r2-n)+(r4-n))+
    (r4-m)*((r1-n)+(r2-n)+(r3-n));
[12] Expand[q*density];
[13] integExp[4*Pi*r1^2*%, {r1,0,Infinity}];
[14] integExp[4*Pi*r2^2*%, {r2,0,Infinity}];
[15] integExp[4*Pi*r3^2*%, {r3,0,Infinity}];
[16] integExp[4*Pi*r4^2*%, {r4,0,Infinity}];
[17] c=%/12;
[18] r=0.558519
[19] c11=c /. {m- >r/4,n- >r/4};
[20] w1=w /. m- >r/4;
[21] stcorr11=c11/w1
[22] c12=c /. {m- >r/4,n- >r};
[23] w2=w /. m- >r;
[24] stcorr12=c12/Sqrt[w1*w2]
[25] c22=c /. {m- >r,n- >r};
[26] stcorr22=c22/w2
[27] Quit
```

The output of [21], [24], and [26] is:

$$\text{stcorr}(\mu_1, \mu_1) = 0.368474, \quad \text{stcorr}(\mu_1, \mu_2) = 0.256560,$$

$$\text{stcorr}(\mu_2, \mu_2) = 0.159988,$$

respectively.

g) The computation of degrees of stability and interdependence.

In order to be specific, we take into account the nebulous model of the beryllium atom and we want to calculate the degree of stability of different subsets of electrons and the amount of interdependence among subsets of electrons, as discussed in Subsection h of Section 4 of Part II. Assume:

$$Z = 4, \quad r = 0.558519 \text{ a.u.}, \quad a = 0.0174535, \quad b = -0.369507,$$

the last three values being obtained by minimizing the mean energy of the atom as shown in Subsection b of this Section. Assume that the functions φ, ψ, and ρ_{neb}, called phi, psi, and ppsi in our MATHEMATICA programs, have been saved in the file named "results".

```
math
[1] Needs["Quantum'integExp'"]
[2] <<results;
[3] density=psi[r1,r2,r3,r4]*ppsi[r1,r2,r3,r4] /.
    {r- >0.558519,a- >0.0174535,b- > -0.369507};
[4] integExp[4*Pi*r1^2*density, {r1,0,Infinity}];
[5] integExp[4*Pi*r2^2*%, {r2,0,Infinity}];
[6] integExp[4*Pi*r3^2*%, {r3,0,Infinity}];
[7] integExp[4*Pi*r4^2*%, {r4,0,Infinity}]
[8] d(1,2,3,4)=density/%;
[9] integExp[4*Pi*r4^2*d(1,2,3,4), {r4,0,Infinity}];
[10] d(1,2)=integExp[4*Pi*r3^2*%, {r3,0,Infinity}];
[11] integExp[4*Pi*r4^2*d(1,2,3,4), {r4,0,Infinity}];
[12] d(1,3)=integExp[4*Pi*r2^2*%, {r2,0,Infinity}];
[13] integExp[4*Pi*r2^2*d(1,2,3,4), {r2,0,Infinity}];
[14] d(3,4)=integExp[4*Pi*r1^2*%, {r1,0,Infinity}];
```

292

[15] d(1)=integExp[4⋆Pi⋆r2^2⋆d(1,2), {r2,0,Infinity}];
[16] d(3)=integExp[4⋆Pi⋆r1^2⋆d(1,3), {r1,0,Infinity}];
[17] integExp[4⋆Pi⋆r1^2⋆d(1,2,3,4)^2, {r1,0,Infinity}];
[18] integExp[4⋆Pi⋆r2^2⋆%, {r2,0,Infinity}];
[19] integExp[4⋆Pi⋆r3^2⋆%, {r3,0,Infinity}];
[20] integExp[4⋆Pi⋆r4^2⋆%, {r4,0,Infinity}]
[21] e1234=Sqrt[%]
[22] integExp[4⋆Pi⋆r1^2⋆d(1,2)^2, {r1,0,Infinity}];
[23] integExp[4⋆Pi⋆r2^2⋆%, {r2,0,Infinity}]
[24] e12=Sqrt[%]
[25] integExp[4⋆Pi⋆r1^2⋆d(1,3)^2, {r1,0,Infinity}];
[26] integExp[4⋆Pi⋆r3^2⋆%, {r3,0,Infinity}]
[27] e13=Sqrt[%]
[28] integExp[4⋆Pi⋆r3^2⋆d(3,4)^2, {r3,0,Infinity}];
[29] integExp[4⋆Pi⋆r4^2⋆%, {r4,0,Infinity}]
[30] e34=Sqrt[%]
[31] integExp[4⋆Pi⋆r1^2⋆d(1)^2, {r1,0,Infinity}];
[32] e1=Sqrt[%]
[33] integExp[4⋆Pi⋆r3^2⋆d(3)^2, {r3,0,Infinity}];
[34] e3=Sqrt[%]
[35] i(1;2;3;4)=Abs[e1⋆e1⋆e3⋆e3−e1234]
[36] i(1,2;3,4)=Abs[e12⋆e34−e1234]
[37] i(1,3;2,4)=Abs[e13⋆e13−e1234]
[38] i(1;2)=Abs[e1⋆e1−e12]
[39] i(1;3)=Abs[e1⋆e3−e13]
[40] i(3;4)=Abs[e3⋆e3−e34]
[41] Quit

In the above program, e1234, e12, e13, e34, e1, and e3 are the norms measuring the degrees of stability for subsystems of electrons belonging to the two subshells inside the beryllium atom in the nebulous model, namely,

$$< d^2(\mu_1, \mu_1, \mu_2, \mu_2) >_1^{1/2} = 0.00234454, \quad < d^2(\mu_1, \mu_1) >_1^{1/2} = 0.115473,$$

$$< d^2(\mu_1, \mu_2) >_1^{1/2} = 0.0483306, \quad < d^2(\mu_2, \mu_2) >_1^{1/2} = 0.0209757,$$

$$< d^2(\mu_1) >_1^{1/2} = 0.340054, \quad < d^2(\mu_2) >_1^{1/2} = 0.14446,$$

respectively, and i(1;2;3;4), i(1,2;3,4), i(1,3;2,4), i(1;2), i(1;3), i(3;4), measure the levels of interdependence among systems of electrons belonging to the subshells of the beryllium atom in nebulous state, namely

$$I(\{\mu_1, \mu_1, \mu_2, \mu_2\} : \{\mu_1\}, \{\mu_1\}, \{\mu_2\}, \{\mu_2\}) = 6.86474 \times 10^{-5},$$

$$I(\{\mu_1, \mu_1, \mu_2, \mu_2\} : \{\mu_1, \mu_1\}, \{\mu_2, \mu_2\}) = 7.75961 \times 10^{-5},$$

$$I(\{\mu_1, \mu_1, \mu_2, \mu_2\} : \{\mu_1, \mu_2\}, \{\mu_1, \mu_2\}) = 8.69555 \times 10^{-6},$$

$$I(\{\mu_1, \mu_1\} : \{\mu_1\}, \{\mu_1\}) = 1.63682 \times 10^{-4},$$

$$I(\{\mu_1, \mu_2\} : \{\mu_1\}, \{\mu_2\}) = 7.93629 \times 10^{-5},$$

$$I(\{\mu_2, \mu_2\} : \{\mu_2\}, \{\mu_2\}) = 1.07077 \times 10^{-4},$$

respectively.

7. CONCLUSION.

In system theory, maximum entropy probability distributions are frequently used to describe statistical equilibrium. They offer the most unbiased probabilistic models subject to given moments of some random variables. As a system of electrons moving randomly around a nucleus, an atom is not in statistical equilibrium and the maximum entropy probability distributions are not suitable to describe its behavior. In the first part of this monograph, the variational principle of minimum mean deviation from statistical equilibrium subject to generalized mixed moments was used for constructing probability wave functions of quantum systems. The parameters of such probability wave functions were determined by minimizing (in fact, by looking for stationary points of) the mean energy of the system. The new formalism was applied to the quantum systems (the free particle in a one- or three-dimensional box, the two noninteracting free particles in a one-dimensional box, the harmonic oscillator, the hydrogen atom) for which the standard approach, based on solving the Schrödinger equation, gives the exact solution. Subsequently, the new formalism was applied to the ground state of the helium atom, for which the corresponding Schrödinger equation cannot be solved exactly but there are excellent trial functions known in literature, inspired by the breakthrough made by Hylleraas (1929, 1963), and to the ground state of the lithium atom in the calm model, for which there are many difficulties in assigning similar trial functions due to its shell structure.

The main objective of Part II of the monograph is to show how the formalism induced by the minimum mean deviation from statistical equilibrium may be efectively applied to more complex atoms. First of all, the linear approximation of the probability wave function induced by this variational principle is examined. The objective is not only to recover the experimental value of the ground state mean energy of the atom, as a goodness-of-fit test for the proposed trial functions, but rather to use the known experimental values in order to better approximate the unknown parameters of these trial functions and to test different probabilistic models of the structure of the respective atom.

The second objective of Part II is to examine four types of probabilistic models of the structure of an atom. In all of them, the electrons move randomly around the nucleus but the models differ as to how many quantum numbers remain unchanged during this random motion. Denoting by n the principal quantum number, by ℓ the angular momentum quantum number, by m the magnetic quantum number, and by s the spin, in the calm model of the atom the electrons keep all four quantum numbers unchanged during their random motion around the nucleus. In the nervous model of the atom the electrons keep only three quantum numbers unchanged, namely (ℓ, m, s). In the turbulent model of the atom, the electrons keep only two quantum numbers unchanged, namely (m, s). In the nebulous model of the atom, which corresponds to what is routinely accepted today, the electrons keep only the spin unchanged during their random motion. A chaotic model of the atom would allow the electrons to change all quantum numbers. The paper focuses on the calm, turbulent, and nebulous models for the structure of the atom. The nebulous model is closer to what is believed nowadays to be the real structure of atoms, but the antisymmetrization of the corresponding probability functions raises enormous complications for complex atoms. The calm and turbulent models offer rougher approximations of the real structure of atoms but, on the other hand, the corresponding probability functions are much easier to be antisymmetrized and used for making predictions.

Even if the Schrödinger equation cannot be solved exactly for the helium atom, excellent trial functions have been proposed which give astonishingly close approximations to the mean energy of the helium atom in the ground state. It has been found to be advantageous to include terms containing

295

the electronic interdistance explicitly in the trial function and this is exactly what was done by Hylleraas in 1929. Such trial functions, however, cannot be generalized in a straightforward way to more complex atoms because, starting already with the lithium atom, it is not clear how to cope with the shell structure of such atoms, without mentioning that for atoms containing more than two electrons, neither the corresponding Hamiltonian nor any trial function including interdistances between pairs of electrons can reflect global instant interactions among more than two electrons taken together. As McQuarrie (1983,p.294) noticed in his excellent book on quantum chemistry: "These calculations (Hylleraas' and his followers') are quite difficult computationally and do not readily lend themselves to large atoms and molecules. The orbital concept has been of great use to chemists, or Hylleraas and company's approach has abandoned the orbital concept altogether." A shell picture of the atom has been in force since the time of the old quantum theory. In fact, the exclusion principle emerged along with spin to explain the picture before the coming of quantum mechanics. The shells were originally defined to be the rigid quantized electron orbits of Bohr's planetary model of the atom. According to Bohr, the orbits of an atom having Z electrons are located at the consecutive distances $k^2/Z, (k = 1, 2, \ldots)$, from the nucleus. In the present monograph, there are no rigid orbits and the electrons move randomly around the nucleus but inside successive subshells whose mean orbits are given by weighted Bohr's values. More precisely, in statistical equilibrium, all the electrons belonging to the same k-th subshell have the same mean distance from nucleus equal to

$$\mu_k = r\frac{k^2}{Z}, \qquad (k = 1, 2, \ldots),$$

where the weight r is a variational parameter. The good numerical approximations obtained in the probabilistic models of the structure of the atom analyzed in this paper somehow bring justice to Bohr and entitle us to say that he was right after all, but on average only!

Once an approximation of the probability wave function of the atom is obtained, it may be used for calculating different correlations between the electrons or subshells of the respective atom. As McQuarrie (1983, p.297) noticed: "The inclusion of electron correlations in atomic and molecular wave functions is a problem of current and active interest." In this paper three kinds of correlations are introduced. The global correlation measures the dependence between two arbitrary electrons of the atom. The generalized correlation gives the dependence between two arbitrary subshells of the atom along the directions induced by Laguerre polynomials which play an important role in the approximation of the probability wave function of the atom. Finally, the standard correlation from classical statistics measures the linear dependence between two arbitrary subshells of the atom, based on the deviations from the corresponding mean orbits of these subshells. Obviously, all these correlations depend on the numerical approximations of the variational parameters of the corresponding probability wave functions of the atom, but they are, however, useful in comparing different probabilistic models of the same atom or the same probabilistic model for different atoms if the same degree of approximation is used.

The correlations are useful but they can measure only the linear dependence between pairs of electrons or subshells. This is obviously not enough because very often the global, instant interdependence among more than two electrons or two subshells is relevant. The Hamiltonian of the atom has the same weak point because, apart from the kinetic energy part, it takes into account only interdistances between pairs of electrons and has no components refering to the instant spatial configuration of three, or more electrons taken together as subsystems. In system theory, global interdependence among the disjoint subsystems making up a more complex system may be numerically obtained by using Watanabe's (1969) entropic measure of connection which is the sum of the entropies of the subsystems minus the entropy of the whole system. Unfortunately, as the probability wave function of an atom induces continuous probability density functions in infinite multidimensional regions, the corresponding entropies cannot be effectively calculated. On the other hand, the so called information energy (Onicescu, (1966)), which is the Euclidean norm of a probability density function, may be calculated for any subsystem of electrons. In this paper, a global measure of interdependence based on the information energy, namely the absolute value of the difference between the sum of the information energies of the disjoint subsystems of a system of electrons and the information energy of the whole system, is used for measuring the global connection among the electrons making up a subshell or among different subshells of the respective atom.

The formalism allows three approximations of the diameter of atoms. The known experimental values of these diameters, when available, indirectly tell us whether the diameter of a specific atom is determined by the individual behavior of all the electrons of the atom or by the behavior of the external subshell of the atom only. A simple index of instability of the atom is also computed.

From computational point of view, the monograph contains the numerical results obtained by using the excellent symbolic mathematics software package MATHEMATICA, created by Wolfram (1991), and Feagin's (1994) package "Quantum" which is designed to be attached to the other packages of MATHEMATICA. In order to save space, the open computer programs are given for the beryllium atom only but they may be obviously adapted to the other types of atoms. The author has run the corresponding programs on a personal computer IBM ThinkPad 380 with Intel Pentium processor, 80MB RAM and 400MB available on the hard disk. The examples and numerical results refer to the hydrogen, helium, lithium, beryllium, boron, carbon, nitrogen, and argon atoms. Similar applications to more complex atoms require more powerful computers.

At the end of this monograph, let us mention something beautiful, written by J.C. Polkinghorne (1986, pp.78-79): "What does quantum mechanics really have to say about the nature of the physical world? We have to confess that we are at a stage of understanding where any answer must be to some extent tentative." He also reminds us what Niels Bohr once cautiously said: "The entire formalism is to be regarded as a tool for deriving predictions... There is no quantum world. There is only an abstract quantum physical description. It is wrong to think that the task of physics is to find out how nature is. Physics concerns what we can say about nature." Today, some people talk about the end of physics, but we still have difficulties in assigning a probability wave function to a complex atom. This monograph pleads in favour of a mathematical model of the atoms based on the assumption that the stability of a quantum system may be viewed only at the level of mean values and correlations, and that the trend towards statistical equilibrium is perturbed by random fluctuations due to the universal connection and interdependence.

References

[1] Abrahams, J. (1982). On the selection of measures of distance between probability distributions. *Information Sciences*, **26**, 109-113.

[2] Abramowitz, M. and Stegun, I.A. (eds.) (1972). *Handbook of Mathematical Functions*. Dover, New York, pp.773-783.

[3] Alhassid, Y. and Levine, R.D. (1977). Entropy and chemical change III. The maximal entropy (subject to constraints) procedure as a dynamical theory. *Journal of Chemical Physics*, **67**, 4321-4339.

[4] Allen, W. (1978). *Getting Even*. Vintage Book, New York.

[5] Ampère, A.M. (1814). Sur la détermination des proportions dans lesquelles les corps se combinent d'après le nombre et la disposition respective des molécules dont leur particules intégrantes sont composés. *Annales de Chimie et de Physique*, **90**, 40.

[6] Anderson, E.E. (1971). *Modern Physics and Quantum Mechanics*. W.B. Saunders Company, Philadelphia-London-Toronto.

[7] Baublitz Jr., M (1988). Derivation of the Schrödinger equation from a stochastic theory. *Progress in Theoretical Physics*, **80**, 232-244.

[8] Bell, E.T. (1986). *Men of Mathematics*. Simon & Schuster, New York.

[9] Ben-Shaul, A., Levine, R.D., and Bernstein, R.B. (1972). Entropy and chemical change II. Analysis of product energy distributions: temperature and entropy deficiency. *Journal of Chemical Physics*, **57**, 5427-5447.

[10] Berkson, J. (1980). Minimum chi-square, not maximum likelihood. *Annals of Statistics*, **8**, 457-487.

[11] Bernstein, R.B. and Levine, R.D. (1972a). Entropy and chemical change. I. Characterization of product (and reactant) energy distributions in reactive molecular collisions: information and entropy deficiency. *Journal of Chemical Physics*, **57**, 434-449.

[12] Bernstein, R.B. and Levine, R.D. (1972b). Entropy and chemical change. II. Analysis of product energy distributions: temperature and entropy deficiency. *Journal of Chemical Physics*, **57**, 5427-5447.

[13] Bethe, H.A. and Salpeter, E.E. (1933). Quantenmechanik der Ein- und Zwei-Elektronenprobleme. In Geiger, H. and Scheel, K. (eds.) *Handbuch der Physik*. Band 24, *Quantentheorie*. Verlag von Julius Springer, Berlin, pp.273-560.

[14] Bethe, H.A. and Salpeter, E.E. (1957). Quantum mechanics of one- and two-electron systems. In Flügge, S. (ed.) *Handbuch der Physik*. Band 35, *Atome I*. Springer-Verlag, Berlin, pp.88-439.

[15] Blacher, R. (1985). Coefficients de correlation d'ordere (I,J) et variances d'ordre I. In: Brezinski, C., Draux, A., Magnus, A.P., Maroni, P., and Ronveaux, A. (eds.) *Polynomes Orthogonaux et Applications*. Lecture Notes in Mathematics, No.1171, Berlin, Springer-Verlag, pp.475-485.

[16] Blackmore, J.T. (1972). *Ernst Mach. His Work, Life, and Influence*. University of California Press, Berkeley.

[17] Bohm, D. (1984). *Causality and Chance in Modern Physics*. (3th ed.). University of Pennsylvania Press, Philadelphia, Pennsylvania.

[18] Böhm, A. (1979). *Quantum Mechanics*. Springer-Verlag, New York-Heidelberg-Berlin.

[19] Bohr, N. (1913). On the constitution of atoms and molecules. *Philosophical Magazine* ser.6, **26**, 1-25, 476-502, 857-875.

[20] Bohr, N. (1921). Atomic structure. *Nature*, **107**, 104-107.

[21] Bohr, N. (1923). Linienspektren und Atombau. *Annalen der Physik*, **71**, 228-288.

[22] Bohr, N. (1924). *The Theory of Spectra and Atomic Constitution*. (2nd ed.). Cambridge University Press, Cambridge, England.

[23] Bohr, N. and Coster, D. (1923). Röntgenspektren und periodisches System der Elemente. *Zeitschrift für Physik*, **12**, 342-374. {The English translation in: Rud Nielsen J. (ed.) (1977) *Collected Works. Volume 4: The Periodic System (1920-1923)*. North-Holland Publishing Co., Amsterdam-New York-Oxford, pp.519-548.}

[24] Boltzmann, L. (1872). Weitere Studien über das Wärmegleichgewicht unter Gasmolekülen. *Sitzungsberichte, K. Akademie der Wissenschaften in Wien, Math.-Naturwiss. Kl.*, **66**, 275-370.

[25] Born, M. (1926). Zur Quantenmechanik der Stossvorgänge. *Zeitschrift für Physik*, **38**, 803-827.

[26] Born, M. (1963). *Ausgewählte Aghandlungen*. Vandenhoeck & Rupprecht, Göttingen.

[27] Born, M. (1969). *Atomic Physics*. (8th ed.). Blakie and Son Ltd, London-Glasgow.

[28] Broglie, L. de (1924). *Recherches sur la théorie des quanta*. Masson, Paris.

[29] Brush, S.G. (1983). *Statistical Physics and the Atomic Theory of Matter, from Boyle and Newton to Landau and Onsager*. Princeton University Press, Princeton, New Jersey.

[30] Bühler, W.K. (1981). *Gauss*. Springer-Verlag, Berlin-Heidelberg-New York.

[31] Clarke, B. and Barron, A.R. (1994). Jeffreys' prior is asymptotically least favorable under entropy risk. *Journal of Statistical Planning and Inference*, **41**, 113-147.

[32] Condon, E.U. and Shortley, G.H. (1963). *The Theory of Atomic Spectra*. Cambridge University Press, Cambridge, England.

[33] Cramér, H. (1946). *Mathematical Methods of Statistics*. Princeton University Press, Princeton, New Jersey.

[34] Csiszár, I. (1975). I-divergence geometry of probability distributions and minimization problems. *Annals of Probability*, **3**, 146-158.

[35] Csiszár, I. (1991). Why least squares and maximum entropy? An axiomatic approach to inference for linear inverse problem. *Annals of Statistics*, **19**, 2032-2066.

[36] Csiszár, I. (1996). Maxent, mathematics, and information theory. In Hanson, K.M. and Silver, R.N. (eds.), *Maximum Entropy and Bayesian Methods*. Kluwer Academic Publishers, Dordrecht-Boston-London, 35-50.

[37] D'Abro, A. (1951). *The Rise of the New Physics*. Vol.1 and Vol.2. Dover Publications, New York.

[38] Dacunha-Castelle, D. and Gamboa, F. (1990). Maximum d'entropie et problème des moments. *Annales de l' Institut H.Poincaré*, **4**, 567-596.

[39] Dean, J.A. (ed.) (1992). *Lange's Handbook of Chemistry*. (14th ed.). McGraw-Hill, New York.

[40] Dinur, U. and Levine, R.D. (1975). On the entropy of a continuous distribution. *Chemical Physics*, **9**, 17-27.

[41] Dirac, P.A.M. (1926). On the theory of quantum mechanics. *Proceedings of the Royal Society of London*, **A112**, 661-677.

[42] Ehrenfest, P. and Ehrenfest, T. (1907). Über zwei bekannte Einwande gegen das Boltzmannsche H-Theorem. *Physikalische Zeitschrift*, **8**, 311-314.

[43] Elsgolts, L.E. (1977). *Differential Equations and the Calculus of Variations*. MIR Publishers, Moscow.

[44] Fano, U. (1969). Doubly excited states of atoms. In Bederson, B., Cohen, V.W., and Pichanick, F.M.J. (eds.), *Atomic Physics*. Plenum Press, New York, 209-225.

[45] Feagin, J.M. (1994). *Quantum Methods with Mathematica*. Springer-Verlag, New York.

[46] Fechner, G.T. (1826). Über die Möglichkeit, scheinbare Abstossung auf Anziehungskräfte zurückzuführen. *Kastner's Archiv für die gesamte Naturlehre*, **9**, 257-285.

[47] Fechner, G.T. (1828). Über die Anwendung des Gravitationsgesetzes auf die Atomlehre. *Kastner's Archiv für die gesamte Naturlehre*, **15**, 257-291.

[48] Fisher, R.A. (1922). On the mathematical foundations of theoretical statistics. *Philosophical Transactions of the Royal Society of London, Ser. A*, **222**, 309-368.

[49] Fischer, R.A. (1924). The condition under which χ^2 measures the discrepancy between observations and hypothesis. *Journal of Royal Statistical Society*, **87**, 442-450.

[50] Fisher, R.A. (1925). *Statistical Methods for Research Workers*. (13th ed.). Hafner, New York.

[51] Flügge, S. (1974). *Practical Quantum Mechanics*. Springer-Verlag, Berlin-Heidelberg-New York.

[52] Fréchet, M. (1951). Sur les tableaux de corrélation dont les marges sont données. *Annales de l'Université de Lyon*, III ser., **A14**, 53-77.

[53] Frieden, B.R. (1989). Fisher information as the basis for the Schrödinger wave equation. *American Journal of Physics*, **57**(11), 1004-1008.

[54] Frieden, B.R. (1990) Fisher information, disorder, and the equilibrium distributions of physics. *Physical Review A*, **15**, 4265-4276.

[55] Frieden, B.R. (1991). Fisher information and the complex nature of the Schrödinger wave equation. *Foundations of Physics*, **21**, 757-771.

[56] Gauss, C.F. (1809). *Theoria Motus Corporum Coelestium*. Perthes, Hamburg. Translation reprinted as *Theory of the Motions of the Heavenly Bodies Moving about the Sun in Conic Sections*. Dover, New York, 1963.

[57] Gelfand, I.M. and Fomin, S.V. (1963). *Calculus of Variations*. Prentice-Hall, Englewood Cliffs, New Jersey.

[58] Ghosh, S.K., Berkowitz, M., and Parr, R.G. (1984). Transcription of ground-state density-functional theory into a local thermodynamics. *Proceedings of the National Academy of Sciences of the USA*, **81**, 8028-8031.

[59] Gibbs, J.W. (1902). *Elementary Principles in Statistical Mechanics, developed with especial reference to the rational foundation of Thermodynamics*. Scribner, New York.

[60] Goldstine, H.H. (1980). *A History of the Calculus of Variations from the 17th through the 19th Century*. Springer-Verlag, New York-Heidelberg-Berlin.

[61] Good, I.J. (1971). The probabilistic explication of information, evidence, surprise, causality, explanation, and utility. In Godambe, V.P. and Sprott, D.A. (eds.) *Foundations of Statistical Inference*. Holt, Rinehart, and Winston of Canada, Toronto, pp.108-130.

[62] Gradshteyn, I.S. and Ryzhik, I.M. (1980). *Table of Integrals, Series, and Products*. Academic Press, Orlando, Florida.

[63] Gray, D.E. (ed.) (1972). *American Institute of Physics Handbook*. (3rd ed.). McGraw-Hill, New York.

[64] Gregory, B. (1990). *Inventing Reality. Physics as Language*. Wiley, New York.

[65] Guiasu, S. (1965). Sur le théorème H de L. Boltzmann. *Comptes Rendus de l'Académie des Sciences de Paris*, **261**, 1179-1181.

[66] Guiasu, S. (1966). La mécanique statistique non conservative. *Revue Roumaine des Mathématiques Pures et Appliqués*, **11**, 541-557. Also reprinted in Schober, A. (ed.) (1984). *Irreversibility and Nonpotentiality in Statistical Mechanics*. Hadronic Press, Nonantum, Massachusetts, pp.43-59.

[67] Guiasu, S. (1971). Weighted entropy. *Reports on Mathematical Physics*, **2**, 165-179.

[68] Guiasu, S. (1977). *Information Theory with Applications*. McGraw-Hill, New York.

[69] Guiasu, S. (1986). The relative dimension of a probabilistic experiment. *Information Sciences*, **39**, 175-185.

[70] Guiasu, S. (1987). Probability space of wave functions. *Physical Review A*, **36**, 1971-1977.

[71] Guiasu, S. (1990a). A classification of the main probability distributions by minimizing the weighted logarithmic measure of deviation. *Annals of the Institute of Statistical Mathematics*, **42**, 269- 279.

[72] Guiasu, S. (1990b). Selecting relevant projections onto subsets of coordinates: a minimax dependence-based approach. *Journal of Complexity*, **6**, 24-40.

[73] Guiasu, S. (1992). Deducing the Schrödinger equation from minimum χ^2, *International Journal of Theoretical Physics*, **31**, 1153-1176.

[74] Guiasu, S. (1993). Projection pursuit based on maximum dependence. *International Journal of Uncertainty, Fuzziness and Knowledge-Based Systems*, **1**, 95-106.

[75] Guiasu, S. (1998a). Quantum Mechanics based on probability wave functions induced by the minimum mean deviation from statistical equilibrium. *International Journal of Theoretical Physics*, **37**, 957-994, 1019-1050.

[76] Guiasu, S. (1998b). Trial wave functions induced by the minimum mean deviation from statistical equilibrium. *International Journal of Quantum Chemistry*, **68**, 175-190.

[77] Guiasu, S., Leblanc, R., and Reischer, C. (1982). The principle of minimum interdependence. *Journal of Information and Optimization Sciences*, **3**, 149-172.

[78] Guiasu, S. and Shenitzer, A. (1985). The principle of maximum entropy. *Mathematical Intelligencer*, **7**, 42-48.

[79] Haken, H. (1993). Information and physics. In Busch, P., Lahti, P., and Mittelstaedt, P. (eds.) *Symposium on the Foundations of Modern Physics '93: Quantum Measurement, Irreversibility, and the Physics of Information*. World Scientific, Singapore-New Jersey-London-Hong Kong.

[80] Haken, H. and Wolf, H.C. (1987). *Atomic and Quantum Physics*. Springer-Verlag, Berlin-Heidelberg-New York.

[81] Harnwell, G.P. and Livingood, J.J. (1961). *Experimental Atomic Physics*. McGraw-Hill, New York-London.

[82] Hibbert, A. (1996). Energies and oscillator strengths using configuration interaction wave functions. In Bartschat, K. (ed.) *Computational Atomic Physics*. Springer-Verlag, Berlin-Heidelberg-New York, pp.27-64.

[83] Hô, M., Sagar, R.P., Pérez-Jordá, J.M., Smith Jr., V.H., and Esquivel, R.O. (1994). A numerical study of molecular information entropies. *Chemical Physics Letters*, **219**, 15-20.

[84] Hylleraas, E.A. (1929). Neue Berechnung der Energie des Heliums im Grundzustande, sowie des tiefsten Terms von Ortho-Helium, *Zeitschrift für Physik*, **54**, 347-366.

[85] Hylleraas, E.A. (1963). Reminiscences from early Quantum Mechanics of two-electron atoms. *Review of Modern Physics*, **35**, 421-431.

[86] Hylleraas, E.A. and Midtdal, J. (1958). Ground-state energy of two-electron atoms. Corrective results. *Physical Review*, **109**, 1013-1014.

[87] Ingarden, R.S. (1963). Information theory and variational principles in statistical theories. *Bulletin de l'Académie Polonaise des Sciences, Séries des Sciences Mathématiques, Astronomie, et Physique*, **11**, 541-547.

[88] Ingarden, R.S. (1964). Information theory and thermodynamics of light. Part I: Foundation of information theory. *Fortschritte der Physik*, **12**, 567-594.

[89] Ingarden, R.S. (1965). Information theory and thermodynamics of light. Part II: Principles of information thermodynamics. *Fortschritte der Physik*, **13**, 755-805.

[90] Ingarden, R.S. and Urbanik, K. (1961). Information as a fundamental notion of statistical physics. *Bulletin de l'Académie Polonaise des Sciences, Séries des Sciences Mathématiques, Astronomie, et Physique*, **9**, 313-316.

[91] Jaynes, E.T. (1957). Information theory and statistical mechanics. *Physical Review*, **106**, 620-630; **108**, 171-190.

[92] Jaynes, E.T. (1979). Where do we stand on maximum entropy? In R. Levin and M. Tribus (eds.) *The Maximum Entropy Formalism*. MIT Press, Cambridge, Ma., pp.15-118.

[93] Jones, L. and Trutzer, V. (1989). Computationally feasible high-resolution minimum-distance procedures which extend the maximum-entropy method. *Inverse Problems*, **5**, 749-766.

[94] Khinchin, A.I. (1957). *Mathematical Foundations of Information Theory*. Dover Publications, Inc., New York.

[95] Kinoshita, T. (1957). Ground state of the helium atom. *Physical Review*, **105**, 1490-1502.

[96] Kossel, W. (1916). Über Molekülbildung als Folge des Atombaues. *Annalen der Physik (4)*, **49**, 229-362.

[97] Kragh, H. (1982). Erwin Schrödinger and the wave equation: The crucial phase. *Centaurus*, **26**, 154-197.

[98] Kullback, S. (1959). *Information Theory and Statistics*. J. Wiley and Sons, New York.

[99] Kullback, S., and Leibler, R.A. (1951). On information and sufficiency. *Annals of Mathematical Statistics*, **22**, 79-86.

[100] Lancaster, H.O. (1958). The structure of bivariate distributions. *Annals of Mathematical Statistics*, **29**, 719-736.

[101] Lancaster, H.O. (1963). Correlations and canonical forms of bivariate distributions. *Annals of Mathematical Statistics*, **34**, 532-538.

[102] Landé, A. (1919). Das Serienspektrum des Heliums. *Physikalische Zeitschrift*, **20**, 228-234.

[103] Landé, A. (1920). Störungstheorie des Heliums. *Physikalische Zeitschrift*, **21**, 114-122.

[104] Langmuir, I. (1919). The arrangement of electrons in atoms and molecules. *Journal of the American Chemical Society*, **41**, 868-934.

[105] Laplace, P.S. (1812). *Théorie analytique des probabilités*. Courcier, Paris.

[106] Laplace, P.S. (1814). *Essai philosophique sur les probabilités*. Courcier, Paris.

[107] Legendre, A (1805). *Nouvelles Méthodes pour la Détermination des Orbites des Comètes*. Courcier, Paris.

[108] Levine, R.D. and Bernstein, R.B. (1973). Analysis of energy disposal: thermodynamic aspects of the entropy deficiency of a product state distribution. *Chemical Physics Letters*, **22**, 217-221.

[109] Lewis, G.N. (1916). The atom and the molecule. *Journal of the American Chemical Society*, **38**, 762-785.

[110] Lide, D.R. (ed.) (2000). *Handbook of Chemistry and Physics*. (81st ed., 2000-2001). CRC Press, Boca Raton-London-New York-Washington D.C.

[111] Main Smith, J.D. (1924). *Chemistry and Atomic Structure*. D. Van Nostrand Company, New York; Benn, London.

[112] McMullin, E. (1978). *Newton on Matter and Activity*. University of Notre Dame Press, Notre Dame-London.

[113] McQuarrie, D.A. (1983). *Quantum Chemistry*. University Science Books, Mill Valley, California.

[114] Mehra, J. and Rechenberg, H. (1982). *The Historical Development of Quantum Theory.* Volume 1: *The Quantum Theory of Planck, Einstein, Bohr, and Sommerfeld: Its Foundation and the Rise of Its Difficulties, 1900-1925.* Part 1 and Part 2. Springer-Verlag, New York-Heidelberg-Berlin.

[115] Mehra, J. and Rechenberg, H. (1987). *The Historical Development of Quantum Theory.* Volume 5: *Erwin Schrödinger and the Rise of Wave Mechanics.* Springer-Verlag, New York.

[116] Messiah, A. (1964). *Mécanique quantique.* Tome 2. Dunod, Paris.

[117] Moore, W. (1989). *Schrödinger. Life and Thought.* Cambridge University Press, Cambridge, England.

[118] Morrison, R.C. and Parr, R.G. (1991) Approximate density matrices and Husimi functions using the maximum entropy formulation with constraints. *International Journal of Quantum Chemistry,* **39**, 823-837.

[119] Nagaoka, H. (1904). Kinetics of a system of particles illustrating the line and the band spectra and the phenomena of radioactivity. *Philosophical Magazine (6),* **7**, 445-455.

[120] Nagy, Á. and Parr, R.G. (1996). Information entropy as a measure of the quality of an approximate electronic wave function. *International Journal of Quantum Chemistry,* **58**, 323-327.

[121] Nelson, E. (1985). *Quantum Fluctuations.* Princeton University Press, Princeton, New Jersey.

[122] Nicholson, J.W. (1911). A structural theory of chemical elements. *Philosophical Magazine (6),* **22**, 864-889.

[123] Nicholson, J.W. (1914a). Atomic models and X-ray spectra. *Nature,* **92**, 583-584.

[124] Nicholson, J.W. (1914b). The high-frequency spectra of the elements and the structure of the atom. *Philosophical Magazine (6),* **27**, 541-564.

[125] Onicescu, O. (1966). Energia informationala. *Studii si Cercetari Matematice,* **18**, 1419-1420.

[126] Pais, A. (1982). Max Born's statistical interpretation of quantum mechanics. *Science*, **218**, 1193-1198.

[127] Pauli, W. (1925). Über den Zusammenhang des Abschlusses der Elektronengruppen im Atom mit der Komplexstruktur der Spektren. *Zeitschrift für Physik*, **31**, 765-783.

[128] Pearson, K. (1900). On the criterion that a given system of deviations from the probable in the case of a correlated system of variables is such that it can be reasonably supposed to have arisen from random sampling. *Philosophical Magazine* (5th Series), **50**, 157-175.

[129] Pekeris, C.L. (1959). 1^1S and 2^3S states of helium. *Physical Review*, **115**, 1216-1221.

[130] Polkinghorne, J.C. (1986). *The Quantum World*. The Chaucer Press, Bungay, Suffolk.

[131] Preda, V., Bulacu, F., and Bulacu, M. (1999). Pearson global indicator and the quantum mechanics of p-dimensional systems. *International Journal of Theoretical Physics*, **38**, 1469-1480.

[132] Rowlinson, J.S. (1970). Probability, information, and entropy. *Nature*, **225**, 1196.

[133] Rud Nielsen J. (ed.) (1977). *Niels Bohr, Collected Works*. Volume 4: *The Periodic System (1920-1923)*. North-Holland Publishing Company, Amsterdam-New York-Oxford.

[134] Rutherford, E. (1911a). The scattering of α- and β-rays and the structure of atom. *Proceedings of the Literary and Philosophical Society of Manchester (4)*, **55**, 18-20.

[135] Rutherford, E. (1911b). The scattering of α- and β-particles and the structure of atom. *Philosophical Magazine (6)*, **21**, 669-688.

[136] Sanders, F.C. (1999). Recent developments in high-precision computational methods for simple atomic and molecular systems. *Advances in Quantum Chemistry*, **33**, 369-387.

[137] Sarmanov, O.V. and Bratoeva, Z.N. (1967). Probabilistic properties of bilinear expansions of Hermite polynomials. *Theory of Probability and its Applications*, **12**, 470-481.

[138] Scerri, E.C., Kreinovich, V., Wojciechowski, P., and Yager, R.R. (1998). Ordinal explanation of the periodic system of chemical elements. *International Journal of Uncertainty, Fuzziness and Knowledge-Based Systems*, **6**, 387-399.

[139] Schrödinger, E. (1926). Quantisierung als Eigenwertproblem. *Annalen der Physik (4)*, **79**, 361-376, 489-527; **80**, 437-490; **81**, 109-139.

[140] Schrödinger, E. (1928). *Collected Papers on Wave Mechanics*. Blackie and Son, Ltd, London-Glasgow.

[141] Schrödinger, E. (1935). Die gegenwärtige Situation in der Quantenmechanik. *Naturwissenschaften*, **23**, 807-812, 823-828, 844-849.

[142] Schrödinger, E. (1978). *Collected Papers on Wave Mechanics*. (2nd English ed.). Chelsea, New York.

[143] Sears, S.B. and Gadre, S.R. (1981). An information theoretic synthesis and analysis of Compton profiles. *Journal of Chemical Physics*, **75**, 4626-4635.

[144] Sears, S.B., Parr, R.G., and Dinur, U. (1980). On the quantum-mechanical kinetic energy as a measure of the information in a distribution. *Israel Journal of Chemistry*, **19**, 165-173.

[145] Shannon, C.E. (1948). A mathematical theory of communication. *Bell System Technical Journal*, **27**, 379-423; 623-656.

[146] Sheynin, O.B. (1979). C.F. Gauss and the theory of errors. *Archive for History of Exact Sciences*, **9**, 306-324.

[147] Shirai, H. (1998). Reinterpretation of quantum mechanics based on the statistical interpretation. *Foundations of Physics*, **28**, 1633-1662.

[148] Shore, J.E. and Johnson, R.W. (1980). Axiomatic derivation of the principle of maximum entropy and the principle of minimum cross-entropy. *IEEE Transactions of Information Theory*, **26**, 26-37.

[149] Slater, P.B. (1994). Schrödinger-like equations for spin measurements, states, and wave functions, based on the statistical approach of Guiasu. *Canadian Journal of Physics*, **72**, 130-133.

[150] Sommerfeld, A. (1918). Atombau und Röntgenspektren. I. Teil. *Physikalische Zeitschrift*, **19**, 297-307.

[151] Sommerfeld, A. (1919). *Atombau und Spektrallinien*. Fr. Vieweg & Sohn, Braunschweig.

[152] Stigler, S.M. (1981). Gauss and the invention of least squares. *Annals of Statistics*, **9**, 465-474.

[153] Stoner, E.C. (1924). The distribution of electrons among atomic levels. *Philosophical Magazine (6)*, **48**, 719-736.

[154] Striganov, A.R. and Sventitskii, N.S. (1968). *Tables of Spectral Lines of Neutral and Ionized Atoms*. IFI/Plenum, New York-Washington.

[155] Tankard Jr., J.W. (1984). *The Statistical Pioneers*. Schenkman Publishing Company, Cambridge, Massachusetts.

[156] Thomson, J.J. (1904a). On the structure of the atom: an investigation of the stability and periods of oscillations of a number of corpuscles arranged at equal intervals around the circumference of a circle: with application of the results to the theory of atomic structure. *Philosophical Magazine (6)*, **7**, 237-265.

[157] Thomson, J.J. (1904b). On the vibrations of atoms containing 4, 5, 6, 7, and 8 corpuscles and on the effect of a magnetic field on such vibrations. *Proceedings of the Cambridge Philosophical Society*, **13**, 39-48.

[158] Von Neumann, J. (1932). *Mathematische Grundlagen der Quantenmechanik*. Springer-Verlag, Berlin.

[159] Watanabe, S. (1969). *Knowing and Guessing. A Quantitative Study of Inference and Information*. Wiley, New York.

[160] Weber, W. (1871). Über elektrodynamische Massenbestimmungen, insbesondere über das Princip der Erhaltung der Energie. *Abhandlungen der Sächsischen Gesellschaft der Wissneschaften (Leipzig)*, **10**, 1-62.

[161] Weber, W. (1875). Über die Bewegungen der Elektricität in Körpern von molecularer Constitution. *Annalen der Physik (2)*, **156**, 1-61.

[162] Wolfram, S. (1991). *Mathematica. A System of Doing Mathematics by Computer*. (2nd ed.). Addison-Wesley Publishing Co., Redwood City, California.

[163] Yourgrau, W. and Mandelstam, S. (1968). *Variational Principles in Dynamics and Quantum Theory*. Dover Publications, New York.